Jose Pedroza

W9-BOP-839

Rm 247

Jose Pedroza

Step 1. $\dfrac{\text{grams}}{\text{mol wt.}} = \dfrac{1 \text{ mol}}{5 \text{ grams}}$

Step 2 $\dfrac{\text{moles} \zeta i}{\text{moles } O_2} = y$

Step 3 moles ζi @ molecular) = 25

m m egg

$x \cdot y = 2$

$\dfrac{y}{5} = m$

Step 5

Step = 2.5

$\dfrac{100 \text{ ml}}{} \bigg| \dfrac{2 \text{ M HCl}}{1 \text{ m}} \bigg| \dfrac{1 \text{ m}}{.2 \text{ M HCl}}$

$\dfrac{\text{moles}}{\text{liter}} = M$

Step 2 9 moles ϕ)($\dfrac{4 \text{ moles } \phi}{2 \text{ moles } \phi_2 O_3}$)

$\dfrac{\text{moles}}{m} = M$

$1 \times y$

$= 18$

15.9

FUNDAMENTALS OF CHEMISTRY

FUNDAMENTALS
OF CHEMISTRY

ELIZABETH P. ROGERS
UNIVERSITY OF ILLINOIS

BROOKS/COLE PUBLISHING COMPANY
MONTEREY, CALIFORNIA

Brooks/Cole Publishing Company
A Division of Wadsworth, Inc.

Printed in the United States of America

10   9   8   7   6   5   4   3   2   1

Library of Congress Cataloging-in-Publication Data

Rogers, Elizabeth P.
    Fundamentals of chemistry.

    Includes index.
    1. Chemistry.   I. Title.
QD31.2.R64   1986   540   86-12930
ISBN 0-534-06600-3

Sponsoring Editor: Sue Ewing
Editorial Assistant: Lorraine McCloud
Production Coordinator: Fiorella Ljunggren
Manuscript Editor: Patricia Cain
Permissions Editor: Carline Haga
Interior and Cover Design: Victoria A. Van Deventer
Art Coordinator: Judith Macdonald
Interior Illustration: John Foster
Photo Editor: Judy K. Blamer
Typesetting: Progressive Typographers, Inc., Emigsville, Pennsylvania
Cover Printing: The Lehigh Press Company, Pennsauken, New Jersey
Printing and Binding: R. R. Donnelley & Sons Company, Crawfordsville, Indiana

Photographs: **66,** Fundamental Photographs. **215,** Werner H. Müller, Peter Arnold, Inc. **219,** Joel Gordon Photography. **224,** Yoav Levy, Phototake. **231,** Lew Merrim, Monkmeyer Press Photo Service. **287,** Yoav Levy, Phototake.

Preface

To the Instructor

Fundamentals of Chemistry has been written to describe those fundamentals to students with little or no prior formal training in chemistry. It covers the same material that is covered in other preparatory chemistry books but sometimes in a slightly different order or with a slightly different emphasis. The order and the emphasis are those that I have found most useful in teaching beginning chemistry. The material in the book was class-tested at the University of Illinois, where I teach, and modified to reflect the reactions of the some 150 students who have used it.

Pedagogical approach

In writing this text, I have kept firmly in mind a picture of the student who will be using it. I have visualized a capable student who is interested in science but has not yet been successfully introduced to chemistry. This student is a little apprehensive because chemistry has the reputation of being difficult and of requiring a lot of math and a lot of memorization. Word problems are not easy for this student, nor is formal thinking.

Because I believe that chemistry is a reasonable subject and its fundamental concepts accessible to the average person, I have tried to maintain my writing at the same level I would use in describing a concept to an adult of average ability. The accessibility of chemistry is sometimes impaired because chemists observe events a little differently than do lay observers. For this reason, beginners need to be introduced to the types of observations that are made in studying things chemical, to the patterns of thought that are used in building chemical models from these observations, and above all to the vocabulary that a chemist uses. In teaching these patterns of thought, I try to start from where the beginner is, to show where we are going, and to suggest ways of getting there. At the same time, I believe it important that the student realize that chemistry is a developing science, that what we learn today will be subjected to further study and modified as scientists develop new ways of looking at events and new equipment by which to measure properties.

In writing the book, I have been guided by my belief that understanding is more important than the memorization of definitions and the overall picture more meaningful than details and exceptions. Illustrations come from real life and are those with which students are very familiar; for instance, quantization is illustrated with the examples of pay telephones and soft-drink machines, and the energy changes that accompany a chemical reaction are compared with the energy used when bicycling over a hill.

I have tried to begin each new topic by starting at the place where the student is likely to be rather than where an experienced chemist would be. Because I have observed that students have difficulty assimilating several different kinds of mathematical concepts if they are presented in quick succession, I have alternated theoretical concepts with mathematical problems, so that the student's subconscious can assimilate each before trying to build on it. In problem solving, the same steps are used regardless of the type of problem. Once a student arrives at the point where a particular pattern of thinking becomes automatic, he or she has little difficulty in learning stoichiometry.

Organization

The order of topics in the text is fairly standard. It can, of course, be changed to meet the particular needs of a particular course. I should, however, explain my reasons for placing some topics where they are.

- Equations are introduced in Chapter 3. Logically, they should come midbook, but, because any meaningful lab work uses equations, I find that an early discussion of what an equation is and how it is balanced is useful to the student.

- Radioactivity is discussed in Chapter 4 rather than in a separate chapter at the end of the book, where it might be omitted because of time constraints. The simplest concepts of nuclear reactions can be easily grasped after learning the structure of the atom. In this nuclear age every citizen should know something of the characteristics of nuclear reactions, and this exposure may be all the student gets.

- The periodic table is discussed in Chapter 5, the same chapter that describes electron configuration, rather than in a separate chapter. When a student relates electron configuration to the arrangement of elements in the table and to the trends in properties related to position in the table, a whole new level of understanding is achieved. Chemistry starts to hang together. By putting these two topics in the same chapter, this crucial connection is emphasized.

- Acids and bases are introduced in Chapter 5 by relating them to the location of elements in the periodic table. The terms *acid* and *base* are used constantly in chemistry; therefore, students need to understand their meaning early on.

- Bonding is introduced after names and formulas, so that students can use the terminology as bonding is discussed.

- In Chapter 8 — in the second and more thorough discussion of equations — reactions are classified in the traditional way but also in terms of oxidation–reduction. This allows for a brief introduction of oxidation, so that the vocabulary can be learned even if time does not permit the complete coverage of the topic in Chapter 14.

- In Chapter 9 the kinetic molecular theory is presented by arguing from observations, and the gas laws are described later. Although not historically correct, this order seems to work in my classes better than presenting the gas laws first and then deriving the theory from them.

- Acids and bases are discussed in a separate chapter, Chapter 12, to emphasize their importance, to tie together the various properties discussed earlier, and to allow introduction of the Brønsted-Lowry definitions. Lewis acid-base theory is not discussed, because I think that it wouldn't add much at this point to the student's understanding of acids and bases.

- The terms *entropy* and *free energy* are introduced very briefly in Chapter 13. One of the important purposes of a beginning chemistry course is teaching vocabulary. Once a term has been used, however briefly, it becomes less threatening and its meaning is more easily grasped when next encountered.

- At the request of several reviewers, I have included a brief chapter on organic chemistry, mainly as an introduction to vocabulary and broad concepts. This course is not the place for learning IUPAC nomenclature, reaction mechanisms, or organic reactions.

Student aids

To help students organize their study, a number of features have been included:

- Each chapter begins with an overview that lists as informal objectives the principal topics covered in that chapter.

- Each new term is boldfaced when introduced and included in a list of key terms at the end of the chapter.

- Many example questions with solutions are included in the text. Each is followed by a similar problem to be solved by the student. Answers to all these problems are given in the back of the book.

- Each chapter has a summary.

- At the end of each chapter is a set of multiple-choice questions. In this age of machine-graded tests, students need to be exposed to the approach characteristic of multiple-choice questions and experience the difference between the kind of thinking required by multiple-choice questions and that required by essay-type questions.

- Numerous problems, grouped by chapter section, follow the multiple-choice questions. Answers to representative end-of-chapter problems (marked with an asterisk) are provided in the back of the book. Each chapter concludes with review problems that relate to any material covered in that chapter and in earlier ones.

- Several of the numerous tables and algorithms for particular tasks that appear throughout the book have been collected in an appendix that will serve as a reference and quick review of the entire text. This appendix also includes a math review, the basics of graphing, and other techniques that may be required for problem solving.

Supplementary materials

The following supplementary materials are available:

- A student study guide by Mark Bishop of Monterey Peninsula College. It includes suggestions for study, alternative methods for solving problems, and self-tests, as well as solutions to in-chapter problems and selected end-of-chapter problems.

- A lab manual by Kent Backart of Palomar Community College. This is a particularly interesting lab manual because it uses inexpensive and easily obtainable materials.

- An instructor's manual with test items, solutions, and answers to all the other problems in the text. A set of transparency masters is also included in the manual.

Acknowledgments

This preface gives me the opportunity to acknowledge the help I have received in preparing this text. I am particularly grateful to Bruce Thrasher, who first suggested this leisure-time activity and continued to encourage it; to Bill Brown, who gently edited my early efforts; to Sue Ewing, the chemistry editor at Brooks/Cole, who brought the project to completion; and to the staff at Brooks/Cole, who led me through the intricacies of design, copy editing, and production. The many reviewers who made valuable suggestions at each stage include Paul Abajian of Johnson State College, Fidele L. Acorn of Northern Virginia Community College, Philip G. Ansted of Mott Community College, Beatrice Arnowich of Queensborough Community College, R. Owen Asplund of the University of Wyoming, Christina A. Bailey of California Polytechnic State University, Jay Bardole of Vincennes University, John Bauman of the University of Missouri at Columbia, Harold Bender of Clackamas Community

College, Jack Benefield of Valencia Community College, James Birk of Arizona State University, Mark Bishop of Monterey Peninsula College, Carl Bordas of Indiana University of Pennsylvania, Ronald L. Bost of Cooke County College, Kenneth H. Brown of Northwestern Oklahoma State University, Margaret Butler of Catonsville Community College, Robert Byrne of Illinois Valley Community College, Jack Healey of Chabot College, Paul Hunter of Michigan State University, Roy Ketchum of Monroe Community College, Walter Kosiba of San Diego City College, Cortlandt Pierpont of the University of Colorado at Boulder, Carl Prenslow of California State University at Fullerton, Patricia Ann Redden of St. Peter's College, Trudie Slapar of Vincennes University, Michael J. Strauss of the University of Vermont, Vernon Thielmann of Southwest Missouri State University, Harry Unger of Cabrillo College, and Thomas Willard of Florida Southern College.

I am grateful to the students who coped with the first draft of this book and made suggestions to improve its clarity and usefulness. Most of all I acknowledge with profound thanks the patience and understanding shown by my husband, Robert W. Rogers, and my family who were, I fear, sometimes given short shrift as deadlines came and went.

To the Student

This book has been written to help you learn the fundamentals of chemistry. Before you read the chapters, you should read the preface because there I tell about the book and the various features that have been included to help you in your learning—methods for solving problems, key words to focus your thinking, sample problems to solve, multiple-choice questions to help you prepare for tests, as well as what I hope are very clear descriptions of the concepts of chemistry. I hope you will find these features useful, but most of all I hope you will enjoy your introduction to chemistry, for it is a fascinating field that, despite its reputation, is really not difficult to understand. The most important advice I can give you is to keep up with the assignments. What you learn one week lays the foundation for what you will learn the following week. Don't fall behind.

Once again, I hope you will enjoy using this book. If you have comments or suggestions that might help me improve the next edition, please let me know by writing to me in care of the publisher (Brooks/Cole Publishing Company, 555 Abrego, Monterey, California 93940).

Elizabeth P. Rogers

Contents

1

A General Look at Chemistry   1

1.1   What Is Chemistry?   2
1.2   The Kinds of Matter   2
       A. Pure Substances   2
       B. Mixtures   3
1.3   The Properties of Matter   4
1.4   The Law of Conservation of Mass   6
1.5   Energy and the Law of Conservation
       of Energy   7
1.6   A General View of Chemistry   8
       A. Early History and the Scientific
          Approach   8
       B. The Branches of Modern
          Chemistry   10
1.7   Summary   11
       Key Terms   12
       Multiple-Choice Questions   12
       Problems   13

2

Measurement   15

2.1   Systems of Measurement   16
       A. The SI System   16
       B. Mass and Weight   19
2.2   The Recording of Measurements   21
       A. Accuracy and Precision   21
       B. Exponential Notation   22

       C. Uncertainty in Measurement;
          Significant Figures   23
2.3   Conversion Factors and Problem
       Solving by Unit Analysis   30
       A. Conversion Factors   30
       B. Problem Solving by Unit
          Analysis   31
2.4   Density — A Physical Property   34
2.5   Energy Measurements   37
       A. Temperature   37
       B. Specific Heat   42
2.6   Summary   44
       Key Terms   45
       Multiple-Choice Questions   45
       Problems   46
       Review Problems   49

3

Elements   50

3.1   Elements, Compounds, and
       Mixtures   51
3.2   Atoms — The Atomic Theory   51
3.3   The Elements   52
       A. Names and Symbols of the
          Elements   53
       B. Lists of the Elements   54
       C. Distribution of the Elements   57
       D. How Elements Occur in Nature   59

3.4   The Reactions of Elements: Simple
      Equations   63
      A. Writing Chemical Equations   64
3.5   Summary   68
      Key Terms   68
      Multiple-Choice Questions   69
      Problems   70
      Review Problems   72

4

Atomic Structure and the Mole   73

4.1   Subatomic Particles   74
      A. The Electron   74
      B. The Proton   75
      C. The Neutron   75
4.2   Atomic Structure   76
      A. Atomic Number Equals Electrons or
         Protons   76
      B. Mass Number Equals Protons plus
         Neutrons   76
      C. Isotopes   77
      D. The Inner Structure of the Atom   78
4.3   Atomic Weights   82
4.4   The Mole   84
4.5   Radioactivity   86
      A. General Characteristics   86
      B. Radioactive Emissions   87
4.6   Characteristics of Nuclear
      Reactions   89
      A. Equations for Nuclear Reactions   89
      B. Half-Life   91
4.7   Applications of Radioactivity   93
      A. The Use of Tracers   94
      B. Biological Effects of Radiation   95
4.8   The Use of Nuclear Reactions to
      Produce Energy   95
      A. Fission   96
      B. Fusion   98
4.9   Summary   99
      Key Terms   99
      Multiple-Choice Questions   100
      Problems   101
      Review Problems   103

5

Electrons and the Properties of Elements   105

5.1   Radiant Energy   106
5.2   The Energy of an Electron   109
5.3   An Atomic Model   111
      A. Energy Levels   111
      B. Sublevels and Orbitals   112
      C. Our Model and the Spectra of Different
         Elements   117
5.4   The Electron Configuration of
      Atoms   118
      A. Box Diagrams of Electron
         Configuration   120
5.5   The Periodic Table   121
      A. Electron Configuration and the Periodic
         Table   123
      B. Categories of Elements in the Periodic
         Table   124
      C. The Electron Configuration of the
         Noble Gases; Core Notation   126
      D. Valence Electrons   128
5.6   Historic Classification of the
      Elements   129
      A. Families of Elements   130
      B. Historical Development of the Periodic
         Table   131
5.7   Properties That Can Be Predicted
      from the Periodic Table   133
      A. Atomic Radius   133
      B. Ionization Energy   134
      C. The Formation of Ions   136
      D. Metals and Nonmetals; Acids and
         Bases   140
5.8   Summary   142
      Key Terms   143
      Multiple-Choice Questions   144
      Problems   145
      Review Problems   147

6

Compounds I: Names and Formulas   149

6.1   Categories of Compounds   150
6.2   Naming Compounds   151

A. Oxidation Numbers   152
B. Binary Compounds   153
C. Ternary Compounds   157
6.3   Formula Weights   160
A. Calculation of Formula Weights   160
B. Moles of Compounds   162
C. Percent Composition   164
6.4   Empirical Formulas   167
6.5   Empirical versus Molecular
Formulas   169
6.6   Summary   172
Key Terms   172
Multiple-Choice Questions   173
Problems   174
Review Problems   176

A. Interaction between Polar
Molecules   202
B. Water Solutions of Ionic
Compounds   203
C. Electrolytes, Nonelectrolytes, and Weak
Electrolytes   204
D. Other Differences between Ionic and
Covalent Compounds   205
7.6   Summary   206
Key Terms   207
Multiple-Choice Questions   207
Problems   208
Review Problems   210

7

Compounds II: Bonding and Geometry   177

7.1   The Chemical Bond   178
A. Covalent, Polar Covalent, and Ionic
Bonds   178
B. Predicting Bond Type;
Electronegativity   179
C. Single, Double, and Triple
Bonds   182
7.2   Lewis Structures   183
A. The Arrangement of Atoms   183
B. The Number and Placement of
Electrons in Lewis Structures   185
C. Lewis Structures of Ions   189
D. Resonance (Optional)   190
7.3   Bond Angles and the Shapes of
Molecules   192
A. Linear Molecules   192
B. Structures with Three Regions of High
Electron Density around the Central
Atom   193
C. Structures with Four Regions of High
Electron Density around the Central
Atom   194
7.4   The Polarity of Molecules   199
7.5   Interactions of Water with Molecules;
Electrolytes and Nonelectrolytes
202

8

Chemical Reactions and Stoichiometry   211

8.1   Physical versus Chemical
Changes   212
8.2   Kinds of Chemical Changes   213
A. Combination Reactions   214
B. Decomposition Reactions   216
C. Displacement Reactions   218
D. Double-Displacement
Reactions   220
8.3   Oxidation–Reduction: A Second Way
to Classify Reactions   226
A. Identifying Oxidation–Reduction
Reactions   227
B. Combustion Reactions   229
8.4   Mass Relationships in an
Equation   232
A. Simple Problems   232
B. Percent Yield   236
C. Problems Involving a Limiting
Reactant   239
8.5   Energy Changes Accompanying
Chemical Reactions   241
A. Endothermic and Exothermic
Reactions   242
B. The Stoichiometry of Energy
Changes   244
8.6   Summary   246
Key Terms   246
Multiple-Choice Questions   247

Problems   248
Review Problems   251

9

The Properties of Gases   253

9.1   Characteristics of the Solid, Liquid, and Gaseous States   254
      A. Shape and Volume   254
      B. Density   255
      C. Compressibility   255
      D. Inferences about Intermolecular Structure   256
9.2   Kinetic Energy   256
      A. The Distribution of Kinetic Energy   256
      B. Kinetic Energy and Temperature   258
9.3   The Kinetic Molecular Theory   259
9.4   Measuring Gas Samples   260
9.5   The Gas Laws   264
      A. Boyle's Law   264
      B. Charles' Law   266
      C. The Combined Gas Law   268
      D. Avogadro's Hypothesis and Molar Volume   270
      E. The Ideal Gas Equation   272
9.6   Mixtures of Gases; Partial Pressures   275
9.7   Stoichiometry Involving Gases   277
9.8   Real Gases   278
9.9   Summary   280
      Key Terms   281
      Multiple-Choice Questions   281
      Problems   282
      Review Problems   284

10

The Condensed States of Matter   285

10.1   The Kinetic Molecular Theory Applied to the Condensed States   286

10.2   Attractive Forces between Particles   286
       A. Intermolecular Forces in Liquids   287
       B. Interparticle Forces in Solids   289
10.3   Physical Properties of Liquids   292
       A. Vapor Pressure   292
       B. The Specific Heat of Liquids   295
       C. The Molar Heat of Vaporization   296
       D. Surface Tension and Viscosity   298
10.4   Physical Properties of Solids   299
10.5   Transitions from the Solid through the Liquid to the Gaseous State   301
10.6   The Uniqueness of Water   304
10.7   Summary   305
       Key Terms   306
       Multiple-Choice Questions   306
       Problems   308
       Review Problems   310

11

Solutions   311

11.1   The Characteristics of Solutions   312
11.2   Solubility   312
       A. Determining Solubility   313
       B. Saturated Solutions   313
       C. Factors Affecting Solubility   314
11.3   Expressing Concentrations of Solutions   317
       A. Concentration by Mass   317
       B. Concentration by Percent   317
       C. Concentration in Parts per Million (ppm) and Parts per Billion (ppb)   319
       D. Concentration in Terms of Moles   319
11.4   Calculations Involving Concentrations   323
11.5   Titration   326
11.6   Ionic Reactions in Solution   330
11.7   Physical Properties of Solutions   335
       A. Vapor Pressure Lowering   335
       B. Boiling Point Elevation   337

C. Freezing Point Depression   337
D. Osmosis and Osmotic Pressure   337
E. Differences between Colligative
   Properties of Solutions of Ionic and
   Molecular Compounds   339
11.8   Summary   340
        Key Terms   341
        Multiple-Choice Questions   342
        Problems   343
        Review Problems   346

12

Acids and Bases   347

12.1   Definitions of Acids and Bases   348
        A. The Arrhenius Definitions   348
        B. The Brønsted-Lowry
           Definitions   348
12.2   Nomenclature of Acids   351
12.3   Reactions of Acids   353
        A. Reactions with a Base   353
        B. Reactions with a Metal   353
12.4   Ionization of Acids   354
        A. Weak and Strong Acids   354
        B. Polyprotic Acids   355
12.5   Equilibrium in Solutions of Weak
        Acids   357
12.6   Hydrogen Ion Concentration in Acid
        Solutions   360
        A. Hydrogen Ion Concentration in
           Solutions of Strong Acids   360
        B. Hydrogen Ion Concentration in
           Solutions of Weak Acids   361
        C. Changing the Hydrogen Ion
           Concentration in Solutions of Weak
           Acids; The Common-Ion Effect   363
12.7   pH   365
        A. Calculation of pH   365
        B. The Interpretation of pH Values   367
12.8 pK_a 367
12.9   Water as a Weak Acid   368
12.10   Hydrogen Ion Concentration in
        Solutions of the Salts of Weak
        Electrolytes   370
        A. Equilibrium Considerations   370
        B. Brønsted-Lowry Considerations   371

12.11   Summary   372
        Key Terms   373
        Multiple-Choice Questions   373
        Problems   375
        Review Problems   376

13

Reaction Rates and Chemical Equilibrium   377

13.1   Requirements for a Reaction   378
        A. Energy Requirements   378
        B. Requirements at the Molecular
           Level   380
13.2   The Course of a Reaction   381
13.3   The Rate of a Reaction   382
        A. Changing the Rate of a Reaction   383
        B. Other Factors That Affect the Rate of a
           Reaction   385
13.4   Chemical Equilibrium   387
        A. Definition of Chemical
           Equilibrium   387
        B. The Characteristics of Chemical
           Equilibrium   388
        C. The Equilibrium Constant   389
13.5   Shifting Equilibria; Le Chatelier's
        Principle   392
        A. The Effect of Concentration Changes
           on Equilibria   393
        B. The Effect of Pressure Changes on
           Equilibria Involving Gases   394
        C. The Effect of Temperature Changes on
           Equilibria   395
        D. The Effect of Catalysts on
           Equilibria   396
13.6   Equilibria Involving Ions   398
        A. Equilibria of Sparingly Soluble
           Substances   398
        B. Equilibria of Weak Acids; Buffer
           Solutions   401
13.7   Summary   406
        Key Terms   407
        Multiple-Choice Questions   407
        Problems   409
        Review Problems   411

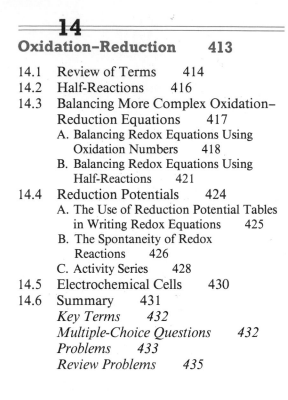

14
Oxidation–Reduction   413

14.1   Review of Terms   414
14.2   Half-Reactions   416
14.3   Balancing More Complex Oxidation–
       Reduction Equations   417
       A. Balancing Redox Equations Using
          Oxidation Numbers   418
       B. Balancing Redox Equations Using
          Half-Reactions   421
14.4   Reduction Potentials   424
       A. The Use of Reduction Potential Tables
          in Writing Redox Equations   425
       B. The Spontaneity of Redox
          Reactions   426
       C. Activity Series   428
14.5   Electrochemical Cells   430
14.6   Summary   431
       Key Terms   *432*
       Multiple-Choice Questions   *432*
       Problems   *433*
       Review Problems   *435*

15
A Brief Look at Organic Chemistry   436

15.1   The General Characteristics of Organic
       Compounds   437

15.2   Hydrocarbons   437
       A. Alkanes   439
       B. Alkenes   441
       C. Alkynes   442
       D. Aromatic Hydrocarbons   442
15.3   Halogens in Organic
       Compounds   443
15.4   Oxygen in Organic
       Compounds   444
       A. Alcohols   444
       B. Ethers   445
       C. Carbonyl Compounds   446
       D. Carboxylic Acids   446
       E. Esters   447
15.5   Nitrogen in Organic
       Compounds   448
15.6   Polyfunctional Compounds   448
15.7   Polymers   450
15.8   Summary   451
       Key Terms   *451*
       Multiple-Choice Questions   *451*
       Problems   *452*

Appendix   455

Answer Section   469

Index   487

FUNDAMENTALS
OF CHEMISTRY

▪ 1 ▪

A General Look at Chemistry

Welcome to the study of chemistry—a vast, complex, and exciting area to enter. By studying chemistry, you will learn a great deal about the natural world, what it is made of, and how it functions at the most fundamental level. This book is an introduction to that study. It introduces you to the vocabulary that chemists use in describing events that are chemical in nature and to characteristics of those events that chemists find significant. We hope that, as you study this material, the vocabulary of chemistry and the chemical way of observing events will become part of your way of describing your world and that you will enjoy the new insights about the chemical nature of that world. One of the ways we have of helping you in this study is to list at the beginning of each chapter the topics covered in the chapter. We hope this list will help to focus your reading and studying of the material covered. In this first chapter, we will discuss:

1. Why chemistry is important to you.
2. Various properties of matter and whether they are physical or chemical in nature.
3. Two important laws of chemistry—the Law of Conservation of Mass and the Law of Conservation of Energy—and their relationship to each other.
4. The development of the scientific approach to the study of matter.
5. The various branches of chemistry.

1.1   What Is Chemistry?

Chemistry is defined as the study of matter and its properties. **Matter** is defined as everything that has mass and occupies space. Although these definitions are acceptable, they do not explain why one needs to know chemistry. The answer to that query is that the world in which we live is a chemical world. Your own body is a complex chemical factory that uses chemical processes to change the food you eat and the air you breathe into bones, muscle, blood, and tissue and even into the energy that you use in your daily living. When illness prevents some part of these processes from functioning correctly, the doctor may prescribe as medicine a chemical compound, either isolated from nature or prepared in a chemical laboratory by a chemist. The world around us is also a vast chemical laboratory. The daily news is filled with reports of acid rain, toxic wastes, the risks associated with nuclear power plants, and the derailment of trains carrying substances such as vinyl chloride, sulfuric acid, and ammonia. Not all chemical news is of disasters. The daily news also carries stories (often in smaller headlines) of new drugs that cure old diseases; of fertilizers, insecticides, and herbicides designed by chemists to allow the farmers to feed our growing populations; and of other new products to make our lives more pleasant. The packages we buy at the grocery store list their contents, including what chemicals the package contains, such as preservatives, and the nutritional content in terms of vitamins, minerals, fats, carbohydrates, and proteins. Everyday life is besieged with chemicals. In beginning the study of chemistry, it is unwise to start with topics as complex as the latest miracle drug. We will begin with the composition of matter and the different kinds of matter. We can then talk about the properties of the different types of matter and the changes that each can undergo. You will learn that each of these changes is accompanied by an energy change and learn the significance of these energy changes.

1.2   The Kinds of Matter

Chemistry is defined as the study of matter. In this introductory text we will not study all types of matter. Rather, we will concentrate on simple substances, the properties that identify them, and the changes they undergo.

A. Pure Substances

A **pure substance** consists of a single kind of matter. It always has the same composition and the same set of properties. For example, baking soda is a single kind of matter, known chemically as sodium hydrogen carbonate. A sample of pure baking soda, regardless of its source or size, will be a white solid containing 57.1% sodium, 1.2% hydrogen, 14.3% carbon, and 27.4% oxygen. The sample will dissolve in water. When heated to 270°C the sample will decompose, giving off carbon dioxide and water vapor and leaving a residue of sodium carbonate. Thus, by definition, baking soda is a pure substance because it has a constant composition and a unique set of properties, some of which we have listed. The properties we have described hold true for any sample of baking soda. These properties are the kinds in which we are interested.

A note about the term *pure;* in this text, the word *pure* means a single substance, not a mixture of substances. As used by the U.S. Food and Drug Administration (USFDA), the term *pure* means "fit for human consumption." Milk, whether whole, 2% fat, or skim, may be "pure" (fit for human consumption) by public health standards, but it is not "pure" in the chemical sense. Milk is a mixture of a great many substances, including water, butterfat, proteins, and sugars. Each of these substances is present in different amounts in each of the different kinds of milk (Figure 1.1).

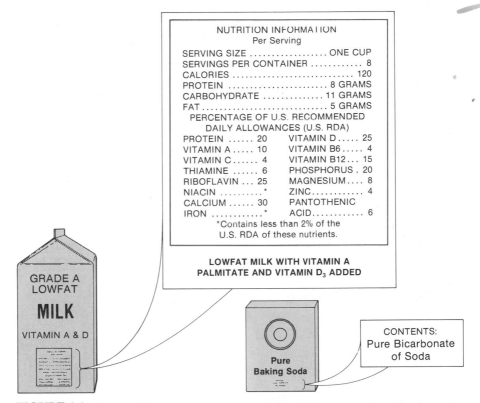

NUTRITION INFORMATION
Per Serving

SERVING SIZE .... ONE CUP
SERVINGS PER CONTAINER 8
CALORIES 120
PROTEIN .... 8 GRAMS
CARBOHYDRATE .... 11 GRAMS
FAT .... 5 GRAMS
PERCENTAGE OF U.S. RECOMMENDED
DAILY ALLOWANCES (U.S. RDA)

PROTEIN 20	VITAMIN D 25
VITAMIN A 10	VITAMIN B6 4
VITAMIN C 4	VITAMIN B12 ... 15
THIAMINE 6	PHOSPHORUS . 20
RIBOFLAVIN ... 25	MAGNESIUM 8
NIACIN ....*	ZINC............ 4
CALCIUM 30	PANTOTHENIC
IRON ....*	ACID........... 6

*Contains less than 2% of the
U.S. RDA of these nutrients.

**LOWFAT MILK WITH VITAMIN A
PALMITATE AND VITAMIN D$_3$ ADDED**

GRADE A
LOWFAT

MILK

VITAMIN A & D

**Pure
Baking Soda**

CONTENTS:
Pure Bicarbonate
of Soda

FIGURE 1.1   Pure substances versus mixtures. The labels on a carton of milk and a box of baking soda show that milk is a mixture and baking soda is a pure substance.

B. Mixtures

A **mixture** consists of two or more pure substances. Most of the matter we see around us is composed of mixtures. Seawater contains dissolved salts; river water contains suspended mud; hard water contains salts of calcium, magnesium, and iron. Both seawater and river water also contain dissolved oxygen, without which fish and other aquatic life could not survive.

Unlike the constant composition of a simple substance, the composition of a mixture can be changed. The properties of the mixture depend on the percentage of each pure substance in it. Steel is an example of a mixture. All steel starts with the pure substance iron. Refiners then add varying percentages of carbon, nickel, chromium, vanadium, or other substances to obtain steels of a desired hardness, tensile strength, corrosion resistance, and so on. The properties of a particular type of steel depend not only on which substances are mixed with the iron but also on the relative percentage of each. One type of chromium–nickel steel contains 0.6% chromium and 1.25% nickel. Its surface is easily hardened, a property that makes it valuable in the manufacture of automobile gears, pistons, and transmissions. The stainless steel used in the manufacture of surgical instruments, food-processing equipment, and kitchenware is also a mixture of iron, chromium, and nickel; it contains 18% chromium and 8% nickel. Steel with this composition can be polished to a very smooth surface and is very resistant to rusting.

You can often tell from the appearance of a sample whether it is a mixture. For example, if river water is clouded with mud or silt particles, you know it is a mixture. If a layer of brown haze lies over a city, you know the atmosphere is mixed with pollutants. However, the appearance of a sample is not always sufficient evidence by which to judge its composition. A sample of matter may look pure without being so. For instance, air looks like a pure substance but is actually a mixture of oxygen, nitrogen, and other gases. Rubbing alcohol is a clear, colorless liquid that looks pure but is actually a mixture of isopropyl alcohol and water, both of which are clear, colorless liquids. As another example, you cannot look at a piece of metal and know whether it is pure iron or a mixture of iron with some other substance such as chromium or nickel.

Figure 1.2 shows the relationships between different kinds of matter.

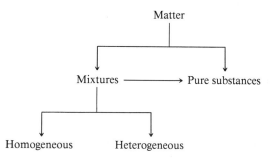

FIGURE 1.2   Classification of matter.

1.3 ## The Properties of Matter

Each kind of matter possesses a number of properties by which it can be identified. In Section 1.2A, we listed some of the properties by which the pure substance baking soda can be identified. These properties fall into two large categories: (1) **physical properties,** those that can be observed without changing the composition of the sample, and (2) **chemical properties,** those whose observation involves a change in composition.

Baking soda dissolves readily in water. If water is evaporated from a solution of baking soda, the baking soda is recovered unchanged; thus, solubility is a physical property. The decomposition of baking soda on heating is a chemical property. You can observe the decomposition of baking soda, but, after you make this observation, you no longer have baking soda. Instead you have carbon dioxide, water, and sodium carbonate. A physical change alters only physical properties, such as size and shape. A chemical change alters chemical properties, such as composition (see Figure 1.3).

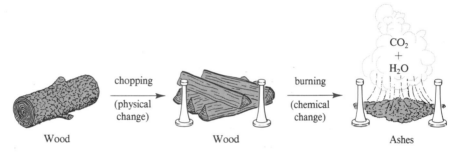

FIGURE 1.3   Physical and chemical properties of matter. Chopping wood physically changes its size but not its composition. Burning wood changes it chemically, turning it into other substances.

This discussion of properties points to another difference between pure substances and mixtures. A mixture can be separated into its components by differences in their physical properties. A mixture of salt and sand can be separated because salt dissolves in water but sand does not. If we add water to a salt–sand mixture, the salt will dissolve, leaving the sand at the bottom of the container. If we pour off the water, the sand will remain. If we boil off the water from the salt solution, we will get the salt by itself. We have separated the two components of the mixture by a difference in their ability to dissolve in water. Solubility is a physical property.

Pure substances, on the other hand, can be separated into their components only by chemical changes. When added to water, the pure substance sodium bicarbonate does not separate into sodium, hydrogen, carbon, and oxygen, although these components of sodium bicarbonate differ greatly in their solubilities in water.

One of the important physical properties of a substance is its **physical state** at room temperature. The three physical states of matter are **solid, liquid,** and **gas.** Most kinds of matter can exist in all three states. You are familiar with water as a solid (ice), a liquid, and a gas (steam) (Figure 1.4). You have seen wax as a solid at room temperature and a liquid when heated. You have probably seen carbon dioxide as a solid (dry ice) and been aware of it as a colorless gas at

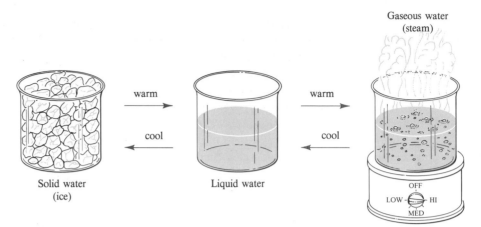

FIGURE 1.4   The three physical states of water: ice (solid), water (liquid), and steam (gas).

higher temperatures. The temperatures at which a given kind of matter changes from a solid to a liquid (its **melting point**) or from a liquid to a gas (its **boiling point**) are physical properties. For example, the melting point of ice (0°C) and the boiling point of water (100°C) are physical properties of the substance water.

Like pure substances, mixtures can exist in the three physical states of solid, liquid, and gas. Air is a gaseous mixture of approximately 78% nitrogen, 21% oxygen, and varying percentages of several other gases. Rubbing alcohol is a liquid mixture of approximately 70% isopropyl alcohol and 30% water. Steel is a solid mixture of iron and other pure substances.

1.4   The Law of Conservation of Mass

The **Law of Conservation of Mass** states that matter can be changed from one form into another, mixtures can be separated or made, and pure substances can be decomposed, but the total amount of mass remains constant. We can state this important law in another way. The total mass of the universe is constant within measurable limits; whenever matter undergoes a change, the total mass of the products of the change is, within measurable limits, the same as the total mass of the reactants.

The formulation of this law near the end of the eighteenth century marked the beginning of modern chemistry. By that time many elements had been isolated and identified, most notably oxygen, nitrogen, and hydrogen. It was also known that, when a pure metal was heated in air, it became what was then called a *calx* (which we now call an oxide) and that this change was accompanied by an increase in mass. The reverse of this reaction was also known: Many calxes on heating lost mass and returned to pure metals. Many imaginative explanations of these mass changes were proposed. Antoine Lavoisier (1743–1794), a French nobleman later guillotined in the revolution, was an amateur chemist with a remarkably analytical mind. He considered the properties of metals and then carried out a series of experiments designed to allow him to measure not just the mass of the metal and the calx but also the mass of the air surrounding the reaction. His results showed that the mass gained by the metal in forming the calx was equal to the mass lost by the surrounding air.

With this simple experiment, in which accurate measurement was critical to the correct interpretation of the results, Lavoisier established the Law of Conservation of Mass, and chemistry became an exact science, one based on careful measurement. For his pioneering work in the establishment of that law and his analytical approach to experimentation, Lavoisier has been called the father of modern chemistry.

Note that this step forward, like so many others in science, depended on technology—in this instance, on the development of an accurate and precise balance (see Figure 1.5).

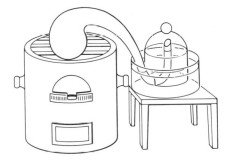

FIGURE 1.5 Lavoisier's apparatus for heating mercury in a confined volume of air (after a drawing by Mme. Lavoisier).

1.5 Energy and the Law of Conservation of Energy

A study of the properties of matter must include a study of energy. **Energy,** defined as the capacity to do work, has many forms. **Potential energy** is stored energy; it may be due to composition (the composition of a battery determines the energy it can release), to position (a rock at the top of a cliff will release energy if it falls to lower ground), or to condition (a hot stone will release heat energy if it is moved to a cooler place). **Kinetic energy** is energy of motion. You are undoubtedly aware that the faster a car is moving, the more damage it does on crashing into an object. Because it is moving faster, it has more kinetic energy and has a greater capacity to do work (in this case, damage).

One of the characteristics of energy is that one form of energy can be converted to another. When wood is burned, some of its potential energy is changed to **radiant energy** (heat and light). Some is changed to kinetic energy as the water and carbon dioxide formed move away from the burning log. Some remains as potential energy in the composition of the water and carbon dioxide produced by the burning. Throughout all these changes, the total amount of energy remains constant. All changes must obey the **Law of Conservation of Energy,** which states that energy can neither be created nor destroyed. An alternative statement is that the total amount of energy in the universe remains constant.

The Law of Conservation of Mass and the Law of Conservation of Energy are interrelated principles. Mass can be changed into energy and energy into mass according to the equation:

$E = mc^2$ where E = energy change

m = mass change (in grams)

c = speed of light (3.00×10^8 m/sec, or 186,000 mi/sec)

This relationship allows us to state the two laws as a single law, called the **Law of Conservation of Mass/Energy:** Energy and mass may be interconverted, but together they are conserved. This law was first stated by Albert Einstein (1879–1955). In most changes, the amount of matter converted to energy is much too small to be detected by even the most sensitive apparatus, and we can say "in this change both mass and energy are separately conserved." It is nevertheless important to be aware of this relationship between mass and energy, because nuclear energy is obtained by just such a conversion of mass to energy.

1.6   A General View of Chemistry
A. Early History and the Scientific Approach

Chemistry as we know it today has its roots in the earliest history of humankind. The ancients were proficient in the arts of metallurgy and dyeing, both of which are chemical in nature. The structure of matter concerned the philosophers of Greece and Rome. The alchemists of the Middle Ages practiced chemistry as they searched for the philosopher's stone that would change "base" matter into gold.

During the eighteenth century, science became a popular hobby of the rich. It was common for men (like Lavoisier) to have laboratories in their homes where they did experiments, considered the implications of the experimental findings, and formulated theories that could be tested by new experiments. These experimentalists met with one another to discuss their work and formulate theories on the nature of matter. This approach to science formed the basis for the pattern of experimentation that was illustrated by Lavoisier's experiment; we call it the **scientific method.**

According to the scientific method, new knowledge and an understanding of the world around us are most reliably gained if the observers organize their work around the following steps:

1. The investigators first define the event or situation they wish to explain. The event may be one of which no studies have been made, or it may be one for which our investigators have hypothesized a new explanation.

2. Careful observations are made about this event. These may be direct observations of nature or observations that others have made.

3. A **hypothesis** or model is constructed that explains or consolidates these observations.

4. New experiments to test the hypothesis are planned and carried out.

5. The original hypothesis is modified to be consistent with both the new and the original observations and capable of predicting the results of further investigations.

A hypothesis that survives extensive testing becomes accepted as a **theory.** Although our present hypotheses and theories are the best we have devised thus far, we have no guarantee that they are final. Regardless of how many experiments have been done to test a given theory and how much data has been

Definition of problem	or	Intuitive hypothesis based on previously collected data

Collection of data specifically aimed at solution of problem or testing intuitive hypothesis
↓
Tentative statement of hypothesis
↓
Collection of more data to test hypothesis
↓
Restatement of hypothesis
↓
Collection of more data to test restatement of hypothesis
↓
Hypothesis becomes theory, which continues to be tested using new data

FIGURE 1.6 One possible series of the steps of the scientific method. Hypothesizing and data collecting continue to alternate for some time before the hypothesis earns the right to be called a theory. Although all scientists collect data and form hypotheses, it is sometimes difficult to describe when each step is taken.

accumulated to support it, a single experiment that can be repeated by other scientists and whose results contradict the theory forces its modification or rejection. Some of our currently accepted theories on the nature of matter may in the future have to be modified or even rejected on the basis of data from new experiments. We must keep an open mind and be ready to accept new data and new theories.

In spite of the vast amounts of new data being collected and the number of new theories being proposed, the understanding of scientific events does not increase at a steady rate. Its forward movement is less like that of a smoothly flowing river than it is like that of a mountain stream, which sometimes rushes ahead and at other times scarcely moves or even wanders off into dead-end swamps. Although there is little doubt that chemical knowledge is expanding, we still have far to go before our understanding of the chemistry of life and of the chemical world in which we live is complete.

An inherent part of the scientific method is the element of creativity. The scientist assembles all of the observations that have been made in a particular area and combines this knowledge in a new way, out of which comes an original and unique hypothesis. For some scientists it is a new concept; for others it is the refinement and clarification of an existing concept. We shall see many examples of creativity in science as we move through this text. Certainly Einstein's equation relating mass and energy is an example of such creativity.

Note the important distinctions between scientific fact and scientific hypothesis and between a scientific law and a scientific theory. A scientific fact is an observed phenomenon, such as the decomposition of sodium hydrogen carbonate on heating (recall Section 1.2A). A scientific hypothesis is an attempt to explain a fact, such as an explanation of why heat causes decomposition. A **scientific law** is the compilation of the observations of many scientific facts, such as the law that all carbonates decompose on heating. A **scientific theory** explains a law. The scientific theory that would accompany the law that all carbonates decompose on heating would explain the relationship between the atoms of a carbonate and how heating changes this relationship so that the carbonate decomposes.

B. The Branches of Modern Chemistry

In recent years chemistry has become a discipline that intrudes on our lives from all directions. There are today hundreds of thousands of practicing chemists. The American Chemical Society in 1984 numbered about 130,000 members. Among the many areas of chemistry are the following.

Analytical chemistry. Analytical chemists devise and carry out tests that determine the amount and identity of the pollutants in our air and water. They also devise the tests by which officials determine the unsanctioned use of drugs and steroids by athletes.

Biological chemistry (biochemistry). Biochemists are concerned with the chemistry of living things. They discovered the composition and function of DNA. They are concerned with the chemical basis of disease and the way our bodies utilize food.

Organic chemistry. Organic chemistry once was defined as the chemistry of substances derived from living matter; that definition is no longer valid. We can say only that the substances organic chemists work with usually contain a great deal of carbon and not many metals. Chemists who work with polymers, petroleum, and rubber are organic chemists.

Inorganic chemistry. Originally inorganic chemists were concerned with minerals and ores — substances not derived from living things — but the exact line separating inorganic chemistry from organic chemistry or from biological chemistry has blurred. For example, some inorganic chemists study the behavior of iron (an inorganic substance) in hemoglobin (an organic substance) in blood (clearly the province of a biochemist).

Chemistry is a broad and exciting field that contains numerous other branches, including nuclear chemistry, physical chemistry, and geochemistry, to name three.

1.7 Summary

Chemistry is the study of matter and its properties. Matter, which includes everything that has mass and occupies space, can be either a pure substance or a mixture. A pure substance has constant composition and constant properties. Mixtures contain two or more pure substances; their properties and composition can vary. Air, steel, and milk are examples of mixtures; water, oxygen, and baking soda are examples of pure substances.

A substance has both physical and chemical properties. Its physical properties can be observed without changing its composition. Observation of the chemical properties of a substance requires a chemical reaction and a change in composition as other pure substances are formed.

The three states of matter are solid, liquid, and gas. Both mixtures and pure substances are found in all of these states. All changes, whether physical or chemical, must conform to the Law of Conservation of Mass and the Law of Conservation of Energy. The formulation of the Law of Conservation of Mass by Lavoisier marked the beginning of modern quantitative chemistry.

All changes, whether chemical or physical, are accompanied by an energy change. The energy involved may be either kinetic (energy of motion) or potential (stored energy). As stated by the Law of Conservation of Mass/Energy, energy may be converted from one type to another, but it may not be destroyed or created except by the interconversion of mass and energy.

In general, the development of chemistry as a science has followed the steps of the scientific method, in which hypotheses are suggested, tested, modified, and retested until they become theories or are discarded. Modern chemistry has many branches, including analytical, biological, organic, and inorganic chemistry.

Key Terms

boiling point (1.3)
chemical properties (1.3)
energy (1.5)
gas (1.3)
hypothesis (1.6A)
kinetic energy (1.5)
Law of Conservation of Energy (1.5)
Law of Conservation of Mass (1.4)
Law of Conservation of Mass/
   Energy (1.5)
liquid (1.3)
matter (1.1)

melting point (1.3)
mixture (1.2B)
physical properties (1.3)
physical state (1.3)
potential energy (1.5)
pure substance (1.2A)
radiant energy (1.5)
scientific law (1.6A)
scientific method (1.6A)
scientific theory (1.6A)
solid (1.3)
theory (1.6A)

Multiple-Choice Questions

MC1. Which of the following does not have constant composition and therefore is not a pure substance?
   a. iron   **b.** distilled water   **c.** gold   **d.** orange juice

MC2. Which of the following is not a physical property of water?
   a. It freezes at $0°C$.
   b. It does not dissolve in gasoline.
   c. It is colorless.
   d. It reacts with iron to form rust.
   e. It dissolves sugar but, when the solution is evaporated, the sugar can be reclaimed.

MC3. Which of the following pure substances is not a solid at room temperature?
   a. platinum   **b.** silver   **c.** oxygen   **d.** gold   **e.** copper

MC4. Which of the following best illustrates the interconvertibility of matter and energy?
   a. a wood fire   **b.** a dynamite explosion
   c. a lead storage battery   **d.** a diesel engine
   e. a nuclear explosion

MC5. Which of the following is not a change from potential energy to kinetic energy?
a. a ball rolling down a hill   b. a car driving on a highway
c. the act of ice skating   d. an elevator going up
e. a bullet flying through the air

MC6. Analytical chemists have shown that pure water, regardless of its source, contains 11% hydrogen. This statement is a:
a. theory.   b. hypothesis.   c. conjecture.   d. fact.
e. guess.

MC7. Which of the following should have the greatest credibility among scientists?
a. hypothesis   b. law   c. theory   d. postulate
e. premonition

MC8. Suppose you analyze a sample of pure water from a deep well in Iowa and conclude that it contains 22% hydrogen (see Question 6). Which would be the *least* sensible thing for you to do?
a. Analyze another sample.   b. Check your apparatus.
c. Ask a friend to analyze the same sample.
d. Call a press conference to announce your finding.
e. Check your calculations.

MC9. Which branch of chemistry would you expect has learned the most about the preparation of substances from coal?
a. inorganic chemistry   b. physical chemistry
c. nuclear chemistry   d. organic chemistry
e. nutritional chemistry

MC10. A chemist who determines the pollutants in the air in a steel mill is probably:
a. an organic chemist.   b. an inorganic chemist.
c. a biochemist.   d. a geochemist.   e. an analytical chemist.

Problems

The answers to problems marked with an asterisk in this and subsequent chapters can be found in the Answer Section at back of the book.

1.1. Which of the following are chemically pure substances? Remember that, regardless of the source, a chemically pure substance always has the same composition.

cherry jello   paint
beer   chocolate ice cream
regular gasoline   cane sugar

1.2. Read the newspaper for a few days. List three stories about chemistry as a pursuit that is beneficial to life and three that portray the harmful effects of chemicals. Try to determine some good aspects of the "harmful" stories; for example, a train derailment might spill vinyl chloride, a toxic gas, but vinyl chloride is used to make plastics that are useful in everyday living. Try also to detect some unfortunate results of the "good" stories; an example would be the unpleasant side effects of a new drug.

* **1.3.** Mix a little vinegar with baking soda. Describe the results. Mix a little water with baking soda. Describe the results. Which is a chemical change, and which is a physical change?

1.4. Watch the world around you for one 8-hour period. Note three chemical and three physical changes that you observed.

1.5. Find three packages whose cover lists the contents. List the ingredients and note which of the products you think are pure substances.

1.6. Light a match. List eight properties of that match, stating whether each is a chemical or physical property. Be sure your list has at least one of each category.

* **1.7.** Obtain a can or bottle of soft drink. Carefully observe what happens when you open it. Are you observing a physical or chemical change? Does the temperature of the drink have any effect on this change?

1.8. Give observations from your own experience to support the statement that air is a mixture.

1.9. Drawing on your own observations, decide whether each of the following events is a physical change (no change in composition or properties) or a chemical change (change in composition and properties). Recall that a physical change is fairly easily reversible, whereas a chemical change is reversible only with difficulty.
   a. milk souring
   b. clothes drying
   c. a tomato ripening
   d. leaves turning color

1.10. You have probably been aware of both carbon dioxide (dry ice) and water in the solid state and in the gaseous state. What happens when each changes from the solid to the gaseous state? Point out any differences between these two transformations.

2

Measurement

Chemists study the properties of matter. They can either estimate properties (hot or cold, large or small) or measure them. Measurement is preferred, particularly in describing physical properties. You know much more about the properties of a steel alloy if you know that a 2-in. cube of this alloy weighs 1.8 lb than if you know only that a small piece of it is quite heavy. In general, scientists use the SI system (Système International) or units derived from the SI system. The properties they measure are those that are useful in identifying the composition of a sample (is it iron or nickel?) rather than those that identify only a particular sample (does it weigh 5 lb?). In addition to properties of mass and volume, scientists measure energy-related properties; for these measurements the Celsius (Centigrade) temperature scale is usually used.

The specific topics we will discuss in this chapter include:

1. Systems of measurement and, in particular, the Système International (SI)—the measurement system of choice for scientists.
2. The difference between mass and weight.
3. The difference between accuracy and precision and how a measurement reflects its precision.
4. The use of unit analysis in solving problems.
5. Several properties calculated from measurements of more than one property—for example, density calculated from measurements of mass and volume or specific heat calculated from measurements of mass, temperature change, and energy change.

2.1   Systems of Measurement

A. The SI System

Measurements in the scientific world and, increasingly, in the nonscientific world are made in **SI (Système International)** units. The system was established in order to allow comparison of measurements made in one country with those made in another. SI units and their relative values were adopted by an international association of scientists meeting in Paris in 1960. Table 2.1 lists the basic SI units and derived units. Notice that **metric units** are part of this system.

The system still in common, nonscientific use in the United States is called the **English system,** even though England, like most other developed countries, now uses metric units. Anyone using units from both the English and SI systems needs to be aware of a few simple relationships between the two systems. These relationships also are given in Table 2.1.

TABLE 2.1   Units of the SI system, units derived from the SI system, and their relationship to several English units

Property being measured	Basic SI unit	Derived units	Relationship to English unit
length	meter (m)	kilometer (km) 1 km = 1000 m · centimeter (cm) 1 cm = 0.01 m	1 m = 39.37 in. 1.61 km = 1 mi 2.54 cm = 1 in.
mass	kilogram (kg)	gram (g) 1 g = 0.001 kg	1 kg = 2.204 lb 453.6 g = 1 lb
volume	cubic meter (m³)	liter (L) 1 L = 0.001 m³ cubic centimeter (cm³, cc) 1 cm³ = 0.001 L milliliter (mL) 1 mL = 1 cm³	1 L = 1.057 qt 946 mL = 1.0 qt
temperature	Kelvin (K)	Celsius (C) K = °C + 273.15	Fahrenheit (F) $°C = \dfrac{°F - 32}{1.8}$
energy	joule (J)	calorie (cal) 1 cal = 4.184 J kilocalorie (kcal) 1 kcal = 1000 cal	$= \frac{5}{9}(°F - 32)$

Two features of the SI system make it easy to use. First, it is a base-10 system; that is, the various units of a particular dimension vary by multiples of

TABLE 2.2 Prefixes used in the SI system

Prefix	Symbol	Base unit multiplied by
mega-	M	1,000,000, or 10^6*
kilo-	k	1,000, or 10^3
deci-	d	0.1, or 10^{-1}
centi-	c	0.01, or 10^{-2}
milli-	m	0.001, or 10^{-3}
micro-	μ	0.000001, or 10^{-6}
nano-	n	0.000000001, or 10^{-9}
pico-	p	0.000000000001, or 10^{-12}

* Exponential notation is discussed in Section 2.2B. Note here that a positive exponent means "raised to a power." Thus 10^2 is 10×10 or "10 squared." A negative exponent means "divided by that power of 10," thus 10^{-2} is $\frac{1}{100}$ or 0.01.

ten. Once a base unit is defined, units larger and smaller than the base unit are indicated by prefixes added to the name of the base unit. Table 2.2 lists some of these **SI prefixes,** along with the abbreviation for each and the numerical factor relating it to the base unit.

Example 2.1 illustrates the use of these prefixes. For each part, the solution is given. Following the example is a similar problem for you to solve, with its answer in the back of the book. By doing this problem you can test whether you understand the method of solution. This text has many examples, all with full solutions and all followed by a similar problem for you to solve. Be sure to do these problems and to check your answers against those given. Only by doing problems on your own will you know whether you can solve them.

Example 2.1   The unit of time in the SI system is the second. How many seconds are in:

    a. one nanosecond   **b.** one kilosecond   **c.** one millisecond

Solution   **a.** From Table 2.2 we learn that the prefix *nano* means "10^{-9}." Therefore one nanosecond is 10^{-9} second.

    b. The prefix *kilo* means "10^3" or "1000." Therefore one kilosecond is 1000 seconds.

    c. The prefix *milli* means "10^{-3}." Therefore one millisecond is 10^{-3} second.

Problem 2.1   The Hertz is an accepted unit of frequency. What is the meaning of:

    a. one picoHertz   **b.** one microHertz   **c.** one megaHertz

The second feature that increases the usefulness of the SI system is the direct relationship between the base units of different dimensions. For example, the unit of volume (cubic meter) is the cube of the unit of length (meter). We shall see later how the unit of mass is related to the unit of volume.

The base unit of length in the SI system is the **meter** (m). The meter, approximately 10% longer than a yard, is equivalent to 39.37 inches, or 1.094 yards. The metric units of length most commonly used in chemistry are listed in Table 2.3 and illustrated in Figure 2.1.

TABLE 2.3   Units of length in the SI system

Unit of length	Relationship to base unit
kilometer (km)	1 km = 1000 m
meter (m)	
decimeter (dm)	10 dm = 1 m
centimeter (cm)	100 cm = 1 m
millimeter (mm)	1000 mm = 1 m
micrometer (μm)	10^6 μm = 1 m
nanometer (nm)	10^9 nm = 1 m

The base unit of volume in the SI system is the cubic meter (m^3). Other commonly used units of volume are the **liter** (L), the cubic centimeter (cm^3 or cc), and the milliliter (mL).

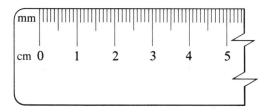

FIGURE 2.1 Each centimeter contains 10 millimeters (shown actual size).

One liter has a volume equal to 0.001 m^3. The nearest unit of comparable volume in the English system is the quart (1.000 L = 1.057 qt). The SI units of volume are summarized in Table 2.4 and illustrated in Figure 2.2. Note particularly that the volume of 1 cm^3 is the same as the volume of 1 mL.

The standard of mass in the SI system is the **kilogram** (kg). A safe in Sèvres, France, holds a metal cylinder with a mass of exactly 1 kg. The mass of that cylinder is the same as the mass of 1000 mL (1 L) of water at 4°C, thereby

TABLE 2.4   Units of volume in the SI system

Unit of volume	Relationship to liter
liter (L)	
milliliter (mL)	1000 milliliters = 1 liter
cubic centimeter (cm³, cc)	1000 cubic centimeters = 1 liter
microliter (μL)	10^6 microliters = 1 liter

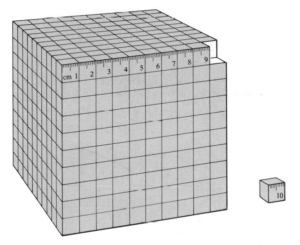

FIGURE 2.2   The large cube measures 10 centimeters on a side and has a volume of 1000 cm³, or 1 L. The small cube next to the large one has a volume of 1 cm³, or 1 mL.

relating mass to volume. The most commonly used SI units of mass are listed in Table 2.5. Notice that the base unit of mass in the SI system is the gram (g), even though the standard of mass in this system is the kilogram.

B. Mass and Weight

In discussing SI units we have used the term *mass* rather than the more familiar term *weight*. **Mass** is a measure of the amount of matter in a particular sample. The mass of a sample does not depend on its location; it is the same whether measured on Earth, on the moon, or anywhere in space. Weight is a measure of the pull of gravity on a sample and depends on where the sample is weighed.

TABLE 2.5   Units of mass in the SI system

Unit of mass	Relationship to base unit
kilogram (kg)	
gram (g)	1000 g = 1 kg
milligram (mg)	1000 mg = 1 g
microgram (μg)	$10^6 \, \mu$g = 1 g

Astronauts traveling in space and landing on the moon have experienced the difference between mass and weight. In Earth's gravitational field at sea level, a particular astronaut may weigh 198 pounds. On the surface of the moon, this astronaut still has the same mass, but his weight (33 pounds) is only one-sixth of what it is on Earth, because the gravitational field of the moon is much weaker than that of the Earth. In outer space the astronaut is weightless, but his mass remains unchanged.

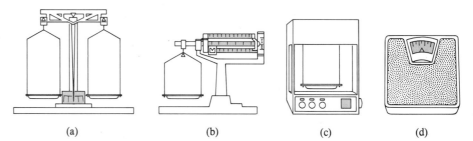

(a)   (b)   (c)   (d)

FIGURE 2.3   Balances of several types and a scale: (a) a classical balance with the weighing pans suspended from a straight beam; (b) a common laboratory balance, which weighs to 0.01 g and hence is called a *centigram balance;* the three beams give rise to the balance's other, less precise name, *triple beam balance;* (c) an electric balance, which weighs rapidly to 0.0001 g; (d) a common bathroom scale, which measures weight by the distortion of a spring.

Weight and mass are measured on different instruments. Mass is measured on a **balance** (Figure 2.3). An object of unknown mass is put at one end of a straight beam and objects of known mass are added to the other end until their mass exactly balances that of the object whose mass is being measured. Because both ends of the beam, at the moment of balancing, are the same distance from the center of the Earth, this measurement is independent of gravity. Weight, on the other hand, is measured on a scale, which determines weight by measuring the distortion of a spring. Such a measurement depends on the pull of gravity. You will weigh less at the top of a mountain than you do in the valley below, because the pull of gravity decreases as you move further from the Earth's center. Nevertheless, your mass is the same in both places.

Despite the clear difference in meaning between the terms *mass* and *weight,* measuring the mass of an object is often called "weighing," and the terms *mass* and *weight* are frequently and incorrectly used interchangeably. Remember that the correct way to describe the amount of matter in a sample is to state its mass.

2.2   **The Recording of Measurements**

A. Accuracy and Precision

Chemistry is an exact science; its development has been based on careful measurements of properties of matter and careful observations of changes in these properties. Measurements in chemistry must be both accurate and precise (Figure 2.4).

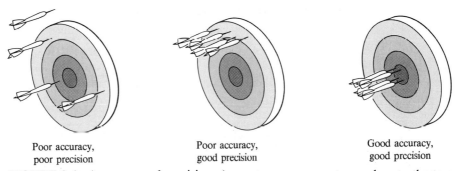

Poor accuracy,   Poor accuracy,   Good accuracy,
poor precision   good precision   good precision

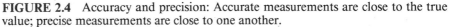

FIGURE 2.4   Accuracy and precision: Accurate measurements are close to the true value; precise measurements are close to one another.

An **accurate** measurement is one that is close to the actual value of the property being measured. The accuracy of a measurement depends on the calibration of the tool used to make the measurement. For example, if you are measuring the distance between two cities by driving between them, an accurate measurement requires that the odometer in your car read 1.00 km for each kilometer driven. If it reads only 0.95 km for each kilometer driven, the accuracy of your measurement will be reduced.

A **precise** measurement is one that can be reproduced. For example, the driving distance between Detroit and Chicago is 493 km. An accurate odometer—one that reads 1.00 km for every kilometer traveled—measures the distance between these two cities as 493 km. However, an odometer that reads 1.00 km for every 0.95 km traveled will measure the Detroit–Chicago distance as 519 km. Each time it is used on the trip, the inaccurate odometer records the same value of 519 km. The odometer reading is precise because it can be reproduced time after time. However, it is not accurate because the odometer itself is not properly calibrated. Note that accuracy requires precision, but precision does not guarantee accuracy.

The importance of obtaining measurements that are both accurate and precise is rarely greater than in a medical laboratory. Patients and doctors alike want to be certain that instruments give readings that are both precise and accurate. For this reason, instruments used in medical laboratories are calibrated each day (and often at the beginning of each shift) against samples of known concentrations. Periodically, accrediting agencies send the laboratories samples to analyze. The results obtained on these samples must be accurate within the range allowed. Precision is not enough; the determinations must also be accurate.

B. Exponential Notation

Measurements in chemistry often involve very large or very small numbers. An example of a very small number is the mass of a hydrogen atom, given by:

Mass of a hydrogen atom = 0.000 000 000 000 000 000 000 001 67 g

At the opposite extreme is the mass of the Earth:

Mass of the Earth = 5,976,000,000,000,000,000,000,000,000 g

Such numbers are hard to deal with, difficult to copy without making mistakes, and almost impossible to name. To simplify the writing and tabulating of such numbers, we use **exponential notation** (also called scientific notation). When exponential notation is used, the measurement is expressed as a number between 1 and 10 multiplied by a power of 10. Table 2.2, which lists the SI prefixes, shows the decimal equivalent of some powers of 10. When exponential notation is used to express a measurement greater than 1, the original number is expressed as a number between 1 and 10 multiplied by 10^n, where n is the number of places the decimal point was moved to the left. We can best understand this notation by studying the following example.

■

Example 2.2	Express the following numbers in exponential notation:
	a. 436,207 **b.** 1,060,435
Solution	**a.** Express 436,207 as a number between 1 and 10 multiplied by 10^n.

$$4.36207 \times 10^n$$

Now determine the value of n. In going from 436,207 to 4.36207×10^n, we moved the decimal point five places to the left; thus, the value of n is 5. Therefore:

$$436,207 = 4.36207 \times 10^5$$

b. Express 1,060,435 as a number between 1 and 10 multiplied by 10^n.

1.060435×10^n

Now determine the value of n. We moved the decimal point six places to the left, so the value of n is 6.

$1{,}060{,}435 = 1.060435 \times 10^6$

Problem 2.2   **a.** Express the mass of the Earth in exponential notation.

b. The planet Pluto is 7,382,000,000 km from Earth at the most distant point in its orbit. Express this distance in exponential notation.

If the number to be expressed in exponential notation is less than 1, the original number is expressed as a number between 1 and 10 multiplied by 10^{-n}, where n equals the number of places the decimal point was moved to the right.

Example 2.3   Express the following numbers in exponential notation:

a. 0.00639   **b.** 0.00001045

Solution   **a.** Express 0.00639 as a number between 1 and 10 multiplied by 10^{-n}.

6.39×10^{-n}

Determine the value of n. In going from 0.00639 to 6.39, we moved the decimal point three places to the right; therefore, the value of n is 3.

6.39×10^{-3}

b. Express 0.00001045 as a number between 1 and 10 multiplied by 10^{-n}.

1.045×10^{-n}

Determine the value of n. In making the change, we moved the decimal point five places to the right, so n equals 5. The number becomes

1.045×10^{-5}

Problem 2.3   **a.** Express the mass of a hydrogen atom in exponential notation.

b. The radius of a proton is 0.0000000154 m. Express this number in exponential notation.

C. Uncertainty in Measurement; Significant Figures

Each time we make a measurement of length, volume, mass, or any other physical quantity, the measurement has some degree of uncertainty. Suppose you have a quantity of liquid whose volume you wish to measure. You are given three different containers in which you might make the measurement—a 50-mL beaker, a 50-mL graduated cylinder, and a 50-mL buret. Figure 2.5 shows these containers, each holding an identical volume of liquid.

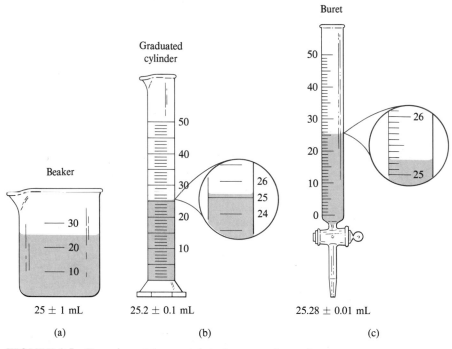

25 ± 1 mL	25.2 ± 0.1 mL	25.28 ± 0.01 mL
(a)	(b)	(c)

FIGURE 2.5   Experimental uncertainty in measuring volume.

Look first at the 50-mL beaker [Figure 2.5(a)]. It has divisions or calibrations every 10 mL. You can see that the level of the liquid in the beaker is between the 20-mL and 30-mL marks. If you look more closely, you can see that the level of liquid is approximately midway between the two marks. You estimate that the volume is 25 mL; however, there is some uncertainty. The volume could be as little as 24 mL or it could be as much as 26 mL. If you record this volume as

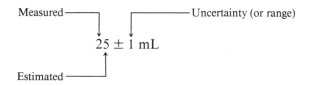

you can show the number you are certain of (20 mL), the number you think is the best estimate (5 mL), and the range within which you are certain the number falls (1 mL), called the uncertainty or range of the reading.

In Figure 2.5(b), the same volume of liquid has been placed in a 50-mL graduated cylinder. Divisions on the cylinder are marked every 1 mL. You can read that the volume is between 25 mL and 26 mL and estimate that it is about 0.2 mL above the 25-mL mark. However, it could be as little as 25.1 mL or as much as 25.3 mL. Therefore, you should record the volume of the liquid as

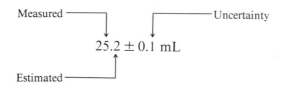

Finally, you measure the liquid in the 50-mL buret [Figure 2.5(c)]. Calibration marks on the buret are 0.1 mL apart. You can read that the volume is between 25.2 mL and 25.3 mL and estimate that it is 0.08 mL above the 25.2-mL mark. Therefore, you should report the volume of the liquid as

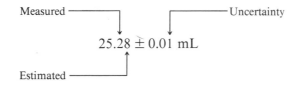

To summarize, the uncertainty of any measurement is assumed to be ± 1 in the last recorded digit. This uncertainty is rarely shown but is understood to be present. For example, if we write a measurement as 372, we understand that the uncertainty is ± 1; if we write 0.017, we understand that the uncertainty is ± 0.001.

Uncertainty in measurements is indicated by the number of significant figures used. **Significant figures** (or significant digits) are all those figures measured plus one that is estimated. Using our volume measurements taken from Figure 2.5, we count the significant figures as follows:

25 mL   contains two significant figures
25.2 mL   contains three significant figures
25.28 mL   contains four significant figures

1. Zero as a significant figure

A zero that serves only to locate the decimal point is not significant; zeros that are not needed to locate the decimal point are significant, for they report a measurement. If the above measurements were given in terms of liters, they would be 0.025 L, 0.0252 L, and 0.02528 L. The number of significant figures in each measurement is the same as before; the zeros have been added only to show the location of the decimal point.

Suppose you had reported the volume of liquid in a buret as 30.50 mL, or 0.03050 L. Are any of these zeros significant? The zeros to the left of the 3 are not significant, for their purpose is to locate the decimal point. The zero between the 3 and the 5 is significant because it shows that the measured volume in that place is 0. The zero after the 5 is also significant. It does not locate the decimal point; rather, it reports a measurement.

The use of exponential notation clarifies the significant figures. Any zero that disappears when a number is expressed exponentially is not significant. For example, the mass of a hydrogen atom has been given as

0.000 000 000 000 000 000 000 001 67 g

In exponential notation this number becomes

1.67×10^{-24} g

Because the zeros in the number have disappeared, we know that they merely showed the location of the decimal point and the magnitude of the number; they were not significant. Similarly, the mass of the Earth expressed exponentially is

5.976×10^{27} g

The zeros shown in the original expression of the measurement (Section 2.2B) have disappeared; they were not significant. Table 2.6 gives further examples.

TABLE 2.6   Significant figures and exponential notation

Number	Exponential expression	Number of significant figures
560,000	5.6×10^5	two (The zeros show only the location of the decimal point.)
560,000.	5.60000×10^5	six (The decimal point in the original number shows that all the zeros are significant.)
30,290	3.029×10^4	four (The first zero is between two digits and is significant. The last shows only the location of the decimal point.)
0.0160	1.60×10^{-2}	three (The first two zeros show the location of the decimal; they are not significant. The last one does not show the location of the decimal point; it reports a measurement and therefore is significant.)

A problem arises when a zero shows both a measurement and the location of the decimal point. The problem is solved by putting a decimal point after such a zero. Thus 250. means that the zero reflects a measurement; 250 means that the zero shows only the magnitude of the number. Similarly, 480,000 means the same as 4.8×10^5, but 480,000. means 4.80000×10^5.

2. Rounding off

When a calculation is performed, the number of significant figures in the numerical answer is determined by the precision of the measurements used in the calculation. It is often necessary to round off the calculated result to the proper number of significant figures. The rules for **rounding off** are:

1. If the digit following the last one to be kept is less than 5, all unwanted digits are discarded. For example, to report the quantity 36.723 mL using four significant figures, write 36.72 mL; to report it using three significant figures, write 36.7 mL.

2. If the digit following the last one to be kept is 5 or greater, the digit to be kept is increased by 1. Thus, to report 38.785 mL using four significant figures, write 38.79 mL; for three significant figures, write 38.8 mL; for two significant figures, write 39 mL.

Example 2.4 Express the following numbers in exponential notation to three significant figures:

    a. 506,251   **b.** 0.005278   **c.** 50,192   **d.** 0.08263

Solution

a. First write 5.06251×10^n. We moved the decimal point five places to the left, so $n = 5$. Three digits must be dropped in order to leave three significant figures. The last digit to be dropped is 2, which is less than 5, so the 6 is not changed. The final number is 5.06×10^5.

b. First write 5.278×10^n. We moved the decimal point three places to the right, so $n = -3$. The zeros are not significant because they only locate the decimal point. The fourth digit, which must be dropped, is 8; therefore the remaining digit must be increased. The final number is 5.28×10^{-3}.

c. First write 5.0192×10^n. We moved the decimal point four places to the left, so $n = 4$. Two digits must be dropped. The fourth digit is 9, so the remaining digit is increased to 2. The final number is 5.02×10^4.

d. First write 8.263×10^n. We moved the decimal point two places to the right, so $n = -2$. The zeros do not show a measurement, so they are dropped. Because 3 is less than 5, no change is made in the last digit kept. The final number is 8.26×10^{-2}.

Problem 2.4 Express the following numbers in exponential notation to three significant figures:

    a. 109,810   **b.** 90,360   **c.** 0.0000006101   **d.** 0.07008

3. The use of significant figures in calculations

Most often the measurements we make are not final answers in themselves. Rather, they are used in further calculations involving addition, subtraction, multiplication, or division. These calculations cannot improve the accuracy of the measurements but must record them.

Multiplication and division. In calculations involving multiplication and division, the answer should contain the same number of significant figures as the measurement in the calculation that contains the fewest significant figures.

Example 2.5   A piece of steel measures 2.6 cm × 5.02 cm × 6.36 cm. What is its volume?

Solution   The volume is the product of the three dimensions

$$2.6 \text{ cm} \times 5.02 \text{ cm} \times 6.36 \text{ cm} = 83.01072 \text{ cm}^3$$

Because the operation is multiplication, the answer can have only as many significant figures as the least accurate measurement going into the calculation. The measurement 2.6 cm has only two significant figures, so the answer must be rounded off to two figures:

$$2.6 \text{ cm} \times 5.02 \text{ cm} \times 6.36 \text{ cm} = 83. \text{ cm}^3$$

Note that we keep the decimal point to clarify that both figures are significant.

Problem 2.5   A car travels 456 miles in 8.5 hours. What is its average speed?

Example 2.6   A sample of straight carbon steel (the kind commonly used to make railroad-track bolts and automobile axles) has a mass of 0.795 g and contains 3.6 × 10^{-3} g carbon. What is the percentage of carbon in this steel alloy?

Solution   We use division for the arithmetic solution to the problem:

$$\frac{3.6 \times 10^{-3} \text{ g}}{0.795 \text{ g}} \times 100\% = 0.45283\%$$

In such an operation, the answer can have only as many significant figures as the least precise measurement used. The number 3.6 × 10^{-3} contains two significant figures, and 0.795 contains three significant figures. The answer, which must contain two significant figures, is 0.45%.

Problem 2.6   Our bodies require tiny amounts of zinc to maintain good health. The body of a person weighing 78 kg should contain 1.61 × 10^{-4} g zinc. What percent of body weight is this requirement?

Addition and subtraction. In calculations involving addition or subtraction, the answer can show only as many decimal places as are common to all the measurements used in the calculation. Note that the location of the decimal point in each of the measurements, rather than the number of significant figures, determines the number of significant figures in the answer.

Example 2.7

A 2.65 mL sample is withdrawn from a bottle containing 375 mL of alcohol. What is the volume of liquid remaining in the bottle?

Solution

We first calculate

$$375 \text{ mL} - 2.65 \text{ mL} = 372.35 \text{ mL}$$

The number 375 in the calculation is reported only to the units place, which is the smallest place common to both numbers. Therefore the correct answer can be reported only to the units place and is given as 372 mL.

Problem 2.7

Three samples of blood are drawn from a patient. One sample has a volume of 0.51 mL, the second a volume of 0.01 mL, and the third a volume of 15.0 mL. What is the total volume drawn from the patient?

Example 2.8

During a trip of 1019 mi, the following amounts of gasoline were purchased: 20.3 gal, 11.6 gal, 15.9 gal, and 5.0 gal. How many miles per gallon were averaged?

Solution

The total amount of gasoline used is 52.8 gal, which has three significant figures. All measurements are given to one decimal place, therefore the sum is significant to the tenths place. To find the average, we use division:

$$\frac{1019 \text{ mi}}{52.8 \text{ gal}} = 19.299242 \text{ mi/gal}$$

In this division operation, the number of significant figures is determined by the number with the fewest significant figures, which is 52.8 gal. The answer must have three significant figures:

$$\text{Answer} = 19.3 \text{ mi/gal}$$

Problem 2.8

In an experiment, a substance was added in small amounts to the starting compound. For a reaction using 16.6 g starting compound, the amounts of added substance were 3.16 mL, 8.90 mL, 7.361 mL, and 5.0 mL. What volume of added substance was used per gram of starting compound?

4. Significant figures and hand calculators

The use of hand-held electronic calculators has greatly increased the importance of understanding significant figures and of observing the rules governing their use. A calculator with an eight-digit display capability may display eight digits in the answer to a calculation, regardless of the number of digits entered. For example, dividing 5.0 by 1.67 on a calculator may give the following answer:

$$\frac{5.0}{1.67} = 2.9940119$$

The correct answer, 3.0, has only two significant figures, as in the least accurate number (5.0) in the problem. All other digits displayed by the calculator are insignificant.

2.3   Conversion Factors and Problem Solving by Unit Analysis

A. Conversion Factors

Measurements made during a chemical experiment are often used to calculate another property. Frequently it is necessary to change measurements from one unit to another—inches to feet, meters to centimeters, or hours to seconds. A relationship between two units that measure the same quantity is a **conversion factor.** For example, the conversion factor between feet and yards is:

3 ft = 1 yd

A conversion factor relates two measurements of the same sample. The measurements may be of the same property (in 3 ft = 1 yd, both measurements are of length) or of different properties of the same sample. In saying that 3 mL alcohol weigh 2.4 g, we are considering two different properties of the same sample—volume and mass. Together these measurements express a conversion factor, for they refer to the same sample and show a relationship between its volume and its mass.

Conversion factors are so named because they offer a way of converting a measurement made in one dimension to another dimension. They do *not* change the original property, only how it is measured. Table 2.1 listed many conversion factors within the metric system and between the metric and English systems.

Conversion factors that define relationships, such as 3 ft = 1 yd or 1 L = 1000 mL, are said to be infinitely significant. This statement means that the number of figures in these factors does not affect the number of significant figures in the answer to the problem.

B. Problem Solving by Unit Analysis

Problem solving by **unit analysis** (or dimensional analysis) is based on the premise that, in an arithmetic operation, units as well as numbers can be cancelled. The idea may be new to you but the method is familiar. For example, if you were asked how many inches are in 6 feet, you would reply without hesitation that 6 feet equals 72 inches. In doing this calculation, you would be using a familiar relationship, or conversion factor, between inches and feet — namely, 12 in. = 1 ft. This relationship can be written as a conversion factor or as a ratio (fraction). Both are equal to 1 because they measure the same distance:

$$\frac{12 \text{ in.}}{1 \text{ ft}} \quad \text{and} \quad \frac{1 \text{ ft}}{12 \text{ in.}}$$

To convert 6 feet to inches, you would use the conversion factor on the left because it allows you to cancel the unit (dimension) you do not want (feet) and arrive at an answer with the unit you do want (inches).

$$6 \text{ ft} \times \frac{12 \text{ in.}}{1 \text{ ft}} = 72 \text{ in.}$$

Note that only the units cancel; the numerical values remain.

Problem solving by unit analysis can be divided into the following steps:

■ **1.** Determine what quantity is wanted and in what units.

■ **2.** Determine what quantity is given and in what units.

■ **3.** Determine what conversion factor (or factors) relates the units given to the units wanted.

■ **4.** Determine how the quantity and units given and the appropriate conversion factors can be combined into a mathematical equation in which the unwanted units cancel and only the wanted units remain.

■ **5.** Perform the mathematical calculations and express the answer using the proper number of significant figures. Next look at the equation again and estimate the answer. If your estimate is close to the calculated answer, all is probably well; if it is quite different, check your calculations. Be sure that you performed all operations correctly and that you properly placed the decimal point.

Example 2.9   How many centimeters are there in 1.63 meters?

Solution   **Wanted**
                  Length in centimeters (? cm)

                  Given
                  1.63 meters. We can write the partial equation:

                  $$? \text{ cm} = 1.63 \text{ m} \times \text{conversion factor}$$

                  Conversion factor
                  We need one that relates meters to centimeters.

                  $$1 \text{ m} = 100 \text{ cm}$$

                  This conversion factor can be written as either

                  $$\frac{1 \text{ m}}{100 \text{ cm}} \quad \text{or} \quad \frac{100 \text{ cm}}{1 \text{ m}}$$

                  Equation
                  By combining these quantities and the appropriate conversion factor, the unit of meters cancels and the desired unit of centimeters remains.

                  $$? \text{ cm} = 1.63 \text{ m} \times \frac{100 \text{ cm}}{1 \text{ m}}$$

                  Arithmetic
                  The calculation gives an answer of 163 cm.

Problem 2.9   A football player runs the opening kickoff back 45 yards. If football were to convert to the metric system, how many meters would this run be?

Many problems require two or more conversion factors to arrive at the units wanted.

Example 2.10   A typical birth weight in the United States is 6.45 pounds. What is this weight in kilograms?

Solution   **Wanted**
                  Weight in kilograms (? kg)

                  Given
                  6.45 pounds. We can start the equation:

                  $$? \text{ kg} = 6.45 \text{ lb} \times \text{conversion factors}$$

Conversion factors

We know from Table 2.1 that 1 lb = 453.6 g. The problem asks for the number of kilograms, not grams; therefore, we also need the conversion factor relating grams and kilograms, 1000 g = 1 kg. Thus, the conversion factors we need in solving this problem can be stated as:

$$\frac{1 \text{ lb}}{453.6 \text{ g}} \quad \text{or} \quad \frac{453.6 \text{ g}}{1 \text{ lb}} \quad \text{and} \quad \frac{1000 \text{ g}}{1 \text{ kg}} \quad \text{or} \quad \frac{1 \text{ kg}}{1000 \text{ g}}$$

Equation

The quantity given and the appropriate conversion factors are combined in the equation so that the unwanted units (lb and g) cancel and only the wanted units (kg) remain:

$$? \text{ kg} = 6.45 \text{ lb} \times \frac{453.6 \text{ g}}{1 \text{ lb}} \times \frac{1 \text{ kg}}{1000 \text{ g}}$$

Arithmetic

The mathematical calculation gives an answer of 2.92572 kg. Expressing this result using only three significant figures (as in 6.45 lb), we arrive at the final answer of 2.93 kg.

Problem 2.10 Aspirin tablets weigh 5.00 grains. One grain equals 2.29×10^{-3} ounce, and 16 ounces equal 1 pound. What is the mass in milligrams of one aspirin tablet?

Example 2.11 Blood donors typically give 1.00 pint of blood during each visit to the blood bank. Calculate this volume in liters.

Solution

Wanted

Volume in liters (? L)

Given

1.00 pint

$$? \text{ L} = 1.00 \text{ pt} \times \text{conversion factors}$$

Conversion factors

1 qt = 2 pt   1 L = 1.057 qt

Equation

$$? \text{ L} = 1.00 \text{ pt} \times \frac{1 \text{ qt}}{2 \text{ pt}} \times \frac{1 \text{ L}}{1.057 \text{ qt}}$$

Answer

0.473 L

Problem 2.11 The gas tank of a car holds 19.5 gallons. How many liters will it hold?

Problem solving by unit analysis is a straightforward method by which you can organize your thinking and approach problems in a systematic way. We will use this method and these same five steps for all the numerical problems that follow.

2.4   Density — A Physical Property

We have said that chemists determine the properties of matter, particularly those properties that help identify the composition of a sample. We can measure the mass and volume of a sample, as was done for several samples of iron with results shown in Table 2.7. However, neither their masses nor their volumes show that all the samples are iron, but all the samples do have the same ratio of mass to volume, as is shown in the far right column. This ratio is called **density.**

$$\text{Density} = \frac{\text{mass}}{\text{volume}}$$

TABLE 2.7   Mass, volume, and density of iron samples

Sample	Volume (cm³)	Mass (g)	Density (g/cm³)
A	1.05	8.25	7.86
B	25.63	201.5	7.862
C	90.7	713	7.86
D	0.02471	0.1942	7.859

All samples of the same kind of matter under the same conditions have the same density. Density is a physical property that characterizes and identifies a particular kind of matter (see Figure 2.6). Table 2.8 lists the densities of some common solids and liquids under normal conditions.

The densities of solids are usually given in grams per cubic centimeter (g/cm³), the densities of gases in grams per liter (g/L), and the densities of liquids in grams per milliliter (g/mL). Recall from Table 2.1 that 1 mL = 1 cm³. Using these units, the density of water is given as 1.000 g/mL at 4°C. Based on the information in Table 2.8, we can make some basic observations. The densities of most metals are greater than that of water. The densities of liquids vary; some are less dense than water, whereas others are more dense. For example, the density of gasoline is about 30% less than that of water, and the density of chloroform is about 50% greater.

Densities vary with temperature. For example, the density of water at 4°C is 1.000 g/mL and at 80°C is 0.9718 g/mL; the density of oxygen is 1.43 g/L at 0°C and 1.10 g/L at 80°C. Except for water, the densities in Table 2.8 are given at 0°C.

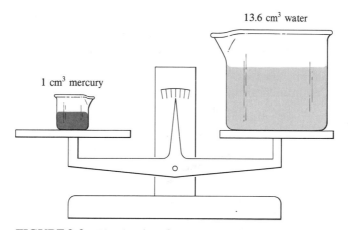

FIGURE 2.6 The density of mercury (13.6 g/mL) compared with the density of water (1.000 g/mL). 1 mL mercury balances 13.6 mL water. Mercury is one of the heaviest liquids known.

TABLE 2.8 Densities of some common solids and liquids under normal conditions

Metals (g/cm³)		Other solids (g/cm³)		Liquids (g/mL)	
aluminum	2.70	bone	1.85	chloroform	1.49
gold	19.32	butter	0.86	ethyl alcohol	0.791
magnesium	1.74	cork	0.24	gasoline	0.67
mercury	13.59	diamond	3.51	water (4°C)	1.000
sodium	0.97	sugar	1.59		

Density is a conversion factor that relates mass to volume. If you know two of the three quantities (mass, volume, and density), you can calculate the third.

Example 2.12

Uranium is a heavy metal. A 13.65-g sample of uranium has a volume of 0.72 cm³. What is the density of uranium?

Solution

Wanted
Density in grams per cubic centimeter (? g/cm³)

Given
13.65 g uranium has a volume of 0.72 cm³.

Conversion factor

Density = mass/volume

Equation

$$? \text{ g/cm}^3 = \frac{13.65 \text{ g}}{0.72 \text{ cm}^3}$$

$$= 18.958333 \text{ g/cm}^3$$

Rounding off to two significant figures, we get:

Density of uranium $= 19 \text{ g/cm}^3$

Problem 2.12 A piece of granite with a volume of 59.3 cm³ weighs 156.6 g. What is the density of granite?

Example 2.13 A sample of ethyl alcohol has a mass of 2.02 g. What is the volume of this sample in milliliters?

Solution **Wanted**
Volume in milliliters (? mL)

Given
2.02 g. This gives the equation

$$? \text{ mL} = 2.02 \text{ g} \times \text{conversion factors}$$

Conversion factors
From Table 2.8 we know that the density of ethyl alcohol is 0.791 g/mL, which can be restated as: 1.00 mL ethyl alcohol has a mass of 0.791 g.

Equation

$$? \text{ mL} = 2.02 \cancel{\text{ g}} \times \frac{1.00 \text{ mL}}{0.791 \cancel{\text{ g}}}$$

Answer
2.55 mL

Problem 2.13 A sample of lead has a mass of 16.5 g. What is the volume of this sample? (The density of lead is 11.3 g/cm³.)

Often, particularly in discussing fluids, specific gravity is reported rather than density. The **specific gravity** (sp gr) of a substance is the ratio of its density to that of a reference substance:

$$\text{Specific gravity} = \frac{\text{density of substance}}{\text{density of reference substance}}$$

Generally, water is the reference substance for comparing solids and liquids, and air is the reference substance for comparing gases.

A value of specific gravity must state the temperature at which the densities were measured. Specific gravity has no units, because the density units cancel in its calculation. For example, we calculate the specific gravity of benzene at 20°C as follows:

$$\text{sp gr}_4^{20} = \frac{\text{density of benzene at } 20°C}{\text{density of water at } 4°C}$$

$$= \frac{0.8784 \text{ g/mL}}{1.0000 \text{ g/mL}}$$

$$= 0.8784$$

2.5 Energy Measurements

We have learned that chemistry is concerned with the properties of matter and with the energy changes that matter undergoes. We have discussed properties related to the mass and volume of a sample of matter. In this section we examine properties related to energy. Energy is measured either in **joules** (J) or in **calories** (cal), where the conversion factor relating the two units is:

4.184 J = 1 cal

The terms *kilojoule* (kJ), 1000 J, and *kilocalorie* (kcal), 1000 cal, are also commonly used. The large calorie (Calorie) used in nutrition is equal to one kilocalorie.

The amount of heat energy associated with a particular sample is dependent on its temperature, its mass, and its composition. Let us consider temperature before discussing its relationship to the energy of a sample.

A. Temperature

Temperature measures how hot or cold a sample is relative to something else, usually an arbitrary standard.

1. Temperature scales

Temperature is measured with a thermometer and is most commonly reported using one of three different scales: **Fahrenheit** (F), **Celsius** (C) (sometimes called centigrade), and **Kelvin** (sometimes called absolute).

The relationship between temperatures on these three scales is straightforward if you understand how a thermometer is constructed and calibrated. Two essential features of a thermometer are: (1) it contains a substance that expands when heated and contracts when cooled, and (2) it has some means to measure the expansion and contraction. In the thermometer with which you may be most familiar, the substance that expands and contracts is mercury. In order to measure its expansion or contraction, the mercury is confined within a small, thin-walled glass bulb connected to a very narrow or capillary tube. When the temperature increases, the mercury expands and its level in the capillary tube rises. This increase in height is proportional to the increase in temperature.

A thermometer is calibrated in the following manner. First, the mercury bulb of a new thermometer is immersed in a mixture of ice and water. When the height of the mercury in the column remains constant, a mark is made. This mark is one reference point. The ice–water mixture is then heated to boiling and kept at that temperature while the height of the mercury in the column rises to a new constant level. Another mark is made on the column at this level; this mark is a second reference point.

Further steps depend on whether this thermometer will measure temperature on the Celsius, Fahrenheit, or Kelvin scale. If the Celsius scale is to be used, the reference point for the ice–water mixture is labeled 0°C and that for boiling water is labeled 100°C. The distance between these two reference points is divided into 100 equal segments. If the thermometer is to measure temperature on the Fahrenheit scale, the reference point for the ice–water mixture is labeled 32°F and that for boiling water is labeled 212°F. The distance between these two points is divided into 180 equal segments. If the thermometer is to measure temperature on the Kelvin scale, the ice–water reference point is labeled 273.15 K, the boiling-water reference point is labeled 373.15 K, and the distance between these two marks is divided equally into 100 segments. Notice that K does not use a degree symbol. The symbol K means "degrees Kelvin." As you can see, the temperatures measured by any of these thermometers

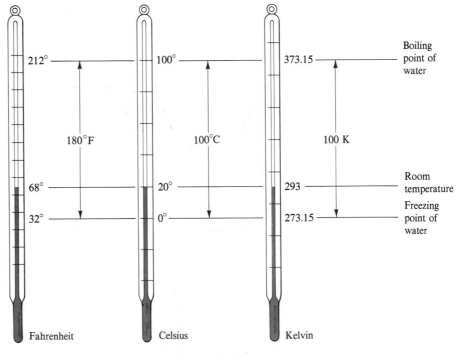

FIGURE 2.7 Fahrenheit, Celsius, and Kelvin thermometers.

do not differ; the difference is in the units with which each temperature is reported. The relationships between the three temperature scales are illustrated in Figure 2.7.

2. Conversions between the temperature scales

A temperature reading on any one of the three scales can be converted to a reading on any other. First, consider a conversion from degrees Celsius to degrees Fahrenheit. Figure 2.7 shows that, between the temperature readings of the ice–water and boiling-water marks, there are 180 Fahrenheit degrees but only 100 Celsius degrees. This relationship can be written as a conversion factor:

$$180°F = 100°C \quad\text{or}\quad \frac{180°F}{100°C} = \frac{9°F}{5°C} = \frac{1.8°F}{1°C}$$

In other words, a temperature increase of 9 Fahrenheit degrees is equivalent to an increase of 5 Celsius degrees. Figure 2.7 also shows that the numerical values assigned to the two ice–water reference points differ by 32 degrees; a reading of 0° on the Celsius scale corresponds to a reading of 32° on the Fahrenheit scale. Combining these facts in an equation, we get:

$$°F = \tfrac{9}{5}(°C) + 32 \quad\text{or}\quad °F = 1.8(°C) + 32$$

This equation can be rearranged to give the Fahrenheit to Celsius conversion equation:

$$°C = \tfrac{5}{9}(°F) - 32 \quad\text{or}\quad °C = \frac{(°F - 32)}{1.8}$$

Example 2.14
a. A recommended temperature setting for household hot-water heaters is 140°F. What is this temperature on the Celsius scale?

b. The boiling point of pure ethyl alcohol is 78.5°C. What is its boiling point on the Fahrenheit scale?

Solution
a. $°C = \dfrac{(140 - 32)}{1.8} = 60°C$

b. $°F = 1.8(78.5) + 32 = 173°F$

Note that in part b we give the answer to three significant figures, the same number of figures as in the original temperature. Recall that conversion factors are considered to be exact, therefore they do not affect the significant figures of the final answer.

Problem 2.14
Convert the following temperatures:

a. 68°F to Celsius   b. 45°C to Fahrenheit

What is the relationship between the Celsius and Kelvin scales? Because each scale has exactly 100 divisions, or degrees, between the ice–water temperature and the boiling-water temperature, the temperature change represented by a Celsius degree is the same as that represented by a Kelvin. The scales differ in the readings at the ice–water reference point; the reading is 0° on the Celsius scale and 273.15 on the Kelvin scale. Therefore, to convert a Celsius temperature to a Kelvin temperature, simply add 273.15.

$$K = {}^\circ C + 273.15 \qquad \text{or} \qquad {}^\circ C = K - 273.15$$

Remember that K is not preceded by the degree symbol (°). The symbols for Fahrenheit and Celsius do require the degree symbol; for example, we write 212°F and 100°C, but 373.15 K.

Example 2.15   Convert the following temperatures:

    a. 28°C to Kelvin   **b.** 310 K to Celsius   **c.** 45°F to Kelvin

Solution   **a.** The conversion factor is

$$K = {}^\circ C + 273.15$$

Therefore the Kelvin temperature is:

$$K = 28^\circ C + 273.15 = 301 \text{ K}$$

(Because the numbers were added, the last significant figure is in the units place.)

b. Use the same equation as in part a, but rearrange to solve for °C.

$${}^\circ C = K - 273.15$$

So the Celsius temperature is

$${}^\circ C = 310 \text{ K} - 273.15 = 37^\circ C$$

c. A conversion from Fahrenheit to Kelvin must be done in two steps — first to Celsius and then to Kelvin; there is no direct route from Fahrenheit to Kelvin.

$${}^\circ C = \frac{(45^\circ F - 32)}{1.8} = 7.2^\circ C$$

$$K = 7.2^\circ C + 273.15 = 280.4 \text{ K}$$

Problem 2.15   Convert the following temperatures:

    a. 105°C to Kelvin   **b.** 230 K to Celsius   **c.** 98°F to Kelvin

3. Melting points and boiling points

Among the data used to identify a substance are the temperatures at which it changes state. The **melting point** (mp) of a substance is the temperature at which it changes from a solid to a liquid (or from a liquid to a solid, in which case it may be called the freezing point). The **boiling point** (bp) of a substance is the temperature at which under normal conditions the substance changes from a liquid to a gas. The melting points and boiling points of several substances are shown in Table 2.9.

TABLE 2.9 The melting point, boiling point, and physical state at 20°C of several substances

Substance	Melting point, °C	Boiling point, °C	Physical state at 20°C
propane	−190	−42	gas
chloroform	−64	62	liquid
sodium chloride (table salt)	801	1413	solid
quartz	1610	2230	solid

The far right column in Table 2.9 shows the physical state of several substances under normal conditions. The **physical state** is predictable from the melting and boiling points of a substance. A substance that boils below room temperature, 20°C, will be a gas under normal conditions, one that melts below room temperature and boils above it will be a liquid, and one that melts above room temperature will be a solid.

Example 2.16   The boiling point of sulfur dioxide is given as − 10°C and its melting point as −72.7°C under normal conditions. What is its physical state under normal conditions?

Solution   Sulfur dioxide both melts and boils below 20°C. It must be a gas under normal conditions.

Problem 2.16   Under normal conditions, naphthalene melts at 80.5°C and boils at 218°C. What is the physical state of naphthalene at room temperature under normal conditions?

B. Specific Heat

When energy in the form of heat is added to a sample, the resulting temperature change depends on the sample's mass and composition. We are aware of this dependence on composition when we notice that a piece of iron left in bright sunshine quickly becomes too hot to touch, whereas a sample of water with the same mass left in the same location for the same length of time becomes only pleasantly warm. The difference is due to the difference in composition and is expressed quantitatively in the specific heats of the two materials. The **specific heat** (sp ht) of a substance is the amount of energy required to raise the temperature of a 1-g sample by 1 degree Celsius. Typically specific heat has units of joules per gram°C (J/g°C).

The specific heat of iron is 0.4525 J/g°C; that is, 0.4525 J is required to raise the temperature of 1 g iron by 1°C. The specific heat of water is 4.184 J/g°C, so 4.184 J are required to raise the temperature of 1 g water by 1°C. Each kind of matter has a unique specific heat. Several are listed in Table 2.10.

Specific heat is a conversion factor that relates energy input to sample mass, composition, and temperature change.

TABLE 2.10   Specific heat of some common substances

Metal	J/g°C	Liquid	J/g°C
silver	0.238	benzene	1.70
copper	0.385	ether	2.21
gold	0.130	ethyl alcohol	2.43
iron	0.453	ice (solid)	2.06
lead	0.159	water	4.18

Example 2.17   How much energy measured in joules is required to raise the temperature of 56 g water from 20°C to 81°C?

Solution   **Wanted**
The amount of energy measured in joules (? J)

Given
A 56-g sample and a temperature change of 61°C. We know that the number of joules depends on both the mass of the sample and the size of the temperature change. Therefore, we start the equation:

$$? \, J = 56 \text{ g} \times 61°C \times \text{conversion factor}$$

Conversion factor
The specific heat of water (from Table 2.10) is:

$$\frac{4.184 \text{ J}}{1.0 \text{ g} \times 1.0°\text{C}}$$

Equation

$$? \text{ J} = 56 \text{ g} \times 61°\text{C} \times \frac{4.184 \text{ J}}{1.0 \text{ g} \times 1.0°\text{C}}$$

Answer

$$\text{J} = 14293 \text{ J} = 1.4 \times 10^4 \text{ J}$$

Problem 2.17   How many joules are required to change the temperature of 235 g iron from 19°C to 85°C?

The specific heat of a substance can be calculated if we know the amount of energy required to cause a measured temperature change in a sample of known mass.

Example 2.18   A sample of aluminum weighing 56 g requires 2.98 kJ to raise the temperature from 22°C to 82°C. Calculate the specific heat of aluminum.

Solution   Specific heat has units of joules per (grams × temperature change in degrees C). We are given a value for each of these quantities in the problem. Substituting them into the definition for specific heat, we get:

$$\text{Sp ht of aluminum} = \frac{2.98 \text{ kJ}}{56 \text{ g} \times 60°\text{C}} \times \frac{1000 \text{ J}}{1 \text{ kJ}} = 0.887 \text{ J/g}°\text{C}$$

Problem 2.18   It requires 4.05 kJ to raise the temperature of 29.5 g magnesium from 22°C to 157°C. Calculate the specific heat of magnesium.

With other sets of data we can calculate the final temperature of a sample.

Example 2.19   Calculate the final temperature of a 56-g sample of water, originally at 25°C, to which is added 4.05 kJ of heat energy.

Solution   **Wanted**
The final temperature in °C. Use two steps to solve this problem. First, calculate the temperature change; second, add this value to the initial temperature of 25°C.

Given

A 56-g sample of water at 25°C and addition of 4.05 kJ of heat energy. Remember that the number of joules required depends on the mass of the sample, the temperature change, and the specific heat.

Conversion factor

$$\text{Specific heat} = \frac{\text{joules}}{\text{grams} \times \text{temperature change}}$$

Rearranging this relationship, we get:

$$\text{Temperature change} = \frac{\text{joules}}{\text{grams} \times \text{specific heat}}$$

We also need the specific heat of water:

$$\frac{4.184 \text{ J}}{1.0 \text{ g} \times 1.0°C}$$

Equation

$$? \text{ C} = \frac{4.05 \text{ kJ}}{56 \times \dfrac{4.184 \text{ J}}{1.0 \text{ g} \times 1.0°C}} = \frac{4.05 \text{ kJ}}{56 \text{ g}} \times \frac{1.0 \text{ g} \times 1.0°C}{4.184 \text{ J}} \times \frac{1000 \text{ J}}{1 \text{ kJ}}$$

$$= 17°C \quad \text{(the temperature change)}$$

The final temperature is

$$17°C + 25°C = 52°C$$

Problem 2.19   Calculate the final temperature if 1.83 kJ are added to 54.3 g iron at 26°C.

2.6   Summary

Progress in chemistry and other sciences depends in large part on precise and accurate measurements. Scientific measurements are made in the internationally used SI system. Related units commonly used are the meter, the liter, and the gram. Prefixes are used to identify smaller and larger units.

The mass of an object is preferred to its weight, for mass is independent of gravity. Measurements must be precise (reproducible) and accurate. Exponential notation is used to indicate the degree of accuracy and the number of significant digits in the measurement, particularly in reporting very large or very small numbers. The results of calculations based on measurements must reflect the number of significant digits in the measurements.

Calculations often involve the use of conversion factors that relate one property to another—for example, density—or one unit to another—for example, feet to meters. Density, which relates the mass of a sample to its volume, is a property useful in identifying the composition of the sample.

Energy, the ability to do work, is measured in joules or calories. Temperature, which is energy-related, can be measured on three different scales — Kelvin, Celsius, and Fahrenheit. The temperature at which under normal conditions a substance melts or boils is a physical property useful in identifying substances. Specific heat reports the amount of energy required to change the temperature of 1 gram of a substance by 1 degree Celsius.

Key Terms

accurate (2.2A)

balance (2.1B)

boiling point (2.5A3)

calorie (2.5)

Celsius (2.5A)

conversion factor (2.3A)

density (2.4)

English system (2.1A)

exponential notation (2.2B)

Fahrenheit (2.5A)

joule (2.5)

Kelvin (2.5A)

kilogram (2.1A)

liter (2.1A)

mass (2.1B)

melting point (2.5A3)

meter (2.1A)

metric units (2.1A)

physical state (2.5A3)

precise (2.2A)

rounding off (2.2C2)

significant figures (2.2C)

SI prefixes (2.1A)

SI (Système International) (2.1A)

specific gravity (2.4)

specific heat (2.5B)

unit analysis (2.3B)

Multiple-Choice Questions

MC1. Which of the following is the smallest mass?

    a. 12.19×10^{-3} g   **b.** 0.0165 mg   **c.** 923.1 μg

    d. 2.3440×10^{-6} kg   **e.** 39,000 cg

MC2. Evaluate $\dfrac{2.892 \times 10^8 \times 1.5 \times 10^{-3}}{6.02 \times 10^{-2}}$

    a. 2.7×10^6 **b.** 7.2×10^6 **c.** 7.2×10^8 **d.** 7.2×10^3

    e. 7.2×10^{-6}

MC3. Which of the following operations will give an answer containing three significant figures?

    1. $1.06 \text{ g} \times \dfrac{1.4 \text{ cm}^3}{0.11 \text{ g}} =$ **2.** $6.022 \times 10^{23} \times 1.19 \text{ mol} =$

    3. $\dfrac{5.2 \times 6.3 \times 2.9}{1.43} =$ **4.** $2.54 - 1.95 =$

    a. all   **b.** 2 and 3   **c.** 2 only   **d.** 2 and 3   **e.** 3 and 4

MC4. What is the volume of a box with dimensions of 15.6 cm $\times$ 0.312 m $\times$ 59.1 mm?
  a. 2.88×10^3 cm^3 **b.** 288 cm^3 **c.** .0288 m^3
  d. 2.88×10^5 cm^3 **e.** none

MC5. The density of ethanol is 0.789 g/cm^3. What is its density in lb/in.^{3} (1 lb = 454 g; 1 in. = 2.54 cm)?
  a. 4.41×10^{-3} lb/in.^{3} **b.** 1.12×10^{-2} lb/in.^{3}
  c. 2.85×10^{-2} lb/in.^{3} **d.** 9.10×10^{-2} lb/in.^{3}
  e. 4.57×10^{-2} lb/in.^{3}

MC6. A metal has a density three times that of water. When a piece of this metal is submerged in a graduated cylinder containing 22 mL water, the water level rises to 71 mL. What is the mass of this metal? (Remember the rules for significant figures.)
  a. 16 g   **b.** 49 g   **c.** 300 g   **d.** 147 g   **e.** none

MC7. The melting point of gallium is 29.8°C; its boiling point is 2403°C. At what temperature on the Fahrenheit scale does gallium melt?
  a. 61.8°F   **b.** 48.5°F   **c.** 16.5°F   **d.** −3.96°F   **e.** 85.6°F

MC8. At what Kelvin temperature does gallium become a gas?
  a. 1335 K   **b.** 2676 K   **c.** 2130 K   **d.** 2793 K
  e. 2776 K

MC9. On a hot day (31°C), a sample of gallium would be:
  a. solid.   **b.** liquid.   **c.** gaseous.   **d.** crystalline.
  e. Not enough information is given.

MC10. To raise the temperature of 216 g indium from 12°C to 55°C, 2.16 kJ are required. What is the specific heat of indium?
  a. 0.234 J/g°C   **b.** 23.4 J/g°C   **c.** 0.113 J/g°C
  d. Not enough data are given.   **e.** 1.12 J/g°C

Problems

2.1 Systems of Measurement

2.20. Convert the following to metric units.
  a. 1.0 cup (4 cups = 1.0 qt) to milliliters
  b. 1.5 acres (1.0 a = 4840 sq yd) to square meters
  c. $\frac{1}{4}$ lb to grams
  d. 100 yd to meters
  e. 2.5 gal to liters
  f. the speed of sound (1.085×10^3 ft/sec) to meters per second

*2.21. Carry out the following conversions. Be certain your answers have the correct number of significant figures.

  a. 39 cm to inches
  b. 145 g to pounds
  c. 13.0 yd to meters
  d. 1.6 qt to liters
  e. 166 km to miles
  −f. 565 lb to kilograms
  g. 12.6 in. to centimeters
  h. 250 mL to quarts
  −i. 3.65 sq ft to sq meters
  j. 2.63 m to feet

2.22. Carry out the following conversions. Be certain your answers have the correct number of significant figures.

a. 163 g to kilograms
b. 28.4 cm^3 to milliliters
c. 3.0 L to milliliters
d. 14 mg to grams
e. 0.034 mg to nanograms
f. 0.95 mL to microliters

2.23. Carry out the following conversions. Be certain your answers have the correct number of significant figures.
a. 7.6 g to milligrams
b. 72 kg to grams
c. 17.8 cm^3 to liters
d. 16.8 cm to meters
e. 20.6 m to millimeters
f. 34.3 L to milliliters

2.24. Carry out the following conversions. Be certain your answers have the correct number of significant figures.
a. 6,246 m to kilometers
b. 1,963 g to milligrams
c. 15,960 g to kilograms
d. 235,616 cm to meters
e. 22,412 mL to liters
f. 1,963,529 mg to kilograms

*2.25. Convert the following to metric units.
a. $\frac{1}{2}$ gal to liters
b. 4.0 min/mi to minutes per kilometer
c. 2.3 lb to kilograms
d. 7 ft, 6 in. to meters
e. 1 qt to liters
f. 6.5 million tons to metric tons
(1 ton = 2000 lb, 1 metric ton = 10^3 kg)

2.2 Recording Measurements

2.26. Report the following numbers in exponential notation using three significant figures.
a. 160,502   b. 0.006059
c. 132.419   d. 1.00605
e. 0.0006004   f. 800.923

2.27. Report the following numbers in exponential notation using three significant figures.

a. 0.000650   b. 678,000
c. 4,263,529   d. 0.43689
e. 0.1000234   f. 1,007,855

*2.28. Carry out the following calculations, and report your answers using the correct number of significant figures.
a. $396 + 1.05 + 16,203.526 + 2,900. =$
b. $14.70 \times 0.0025 \times 9.2 =$
c. $29.62 - 1.009 =$
d. $0.00159 + 0.01956 =$
e. $35.78 + 32.0 + 5.765 =$
f. $45,000 + 987 + 54.8 + 0.786 =$

*2.29. Carry out the following calculations, and report your answers using the correct number of significant figures.
a. $14,705 + 0.0001 + 20.35 =$
b. $65 \div 3,256 =$
c. $0.002395 \times 12.625 \times 6.02 =$
d. $13.49(1.23 + 44.6) =$
e. $8.89 \div 0.68 =$

2.3 Conversion Factors

*2.30. Are you exceeding the speed limit if you are driving 97 km/hr in a 55-mi/hr zone?

2.31. A person weighs 96.4 kg and is 1.90 m tall. What are this person's weight and height as measured in the English system (ft, in., lb)?

2.32. Calculate your weight and height as measured in the metric system (kg, cm).

2.33. According to the Food and Nutrition Board of the National Research Council, the recommended weight for a woman 5 ft, 4 in. tall is 122 ± 10 lb. Calculate the woman's height in centimeters and her recommended weight in kilograms.

2.34. Driving in Europe, you plan to buy 6.3 gal gasoline. The pump registers in the metric system. How many liters will you buy?

2.4 Density

2.35. Calculate the density of the following samples:

	Mass	Volume	Density
a.	13.6 g	21.9 mL	——
b.	4.6 g	1.2 L	——
c.	155.1 g	13.2 mL	——
d.	5.23 g	6.9 mL	——

*2.36. Calculate the missing quantity.

	Mass	Volume	Density
a.	——	23 mL	1.45 g/mL
b.	5.6 g	——	0.831 g/mL
c.	——	11.4 mL	5.4 g/mL
d.	——	0.54 L	1.3 g/mL

2.37. A sample of liquid weighing 19.8 g has a volume of 25 mL. Identify the liquid using the data in Table 2.8.

2.38. A sample of magnesium has a volume of 7.43 mL and a density of 1.74 g/cm³. What is the mass of the sample?

2.39. At 20°C the volume of a colorless liquid is 9.43 mL and its density is 0.789 g/mL. What is the mass of the sample? Could this liquid be water?

*2.40. A piece of iron weighs 6.53 g and has a density of 7.86 g/cm³. What is the volume of the piece of iron?

$1.8(13) + 32$

2.5A Temperature

2.41. Complete the following temperature chart.

	°F	°C	K
a.	55.	13	286
b.	——	165	——
c.	——	——	450
d.	——	−40	——
e.	−25	——	——

*2.42. Normal body temperature is 98.6°F. What is normal body temperature on the Celsius scale? on the Kelvin scale?

2.43. If body temperature is elevated on the Fahrenheit scale from 98.6°F to 104.6°F, by how many degrees is it elevated on the Celsius scale?

2.44. The boiling point of benzene is 80.1°C. At what temperature on the Fahrenheit scale does benzene boil?

2.5B Specific Heat

2.45. The specific heat of ethyl alcohol is 2.43 J/g°C. How much energy is needed to raise the temperature of 75 g ethyl alcohol from 21°C to 37°C (body temperature)?

2.46. The specific heat of magnesium is 0.983 J/g°C. How much energy must be added to 7.15 g magnesium to raise its temperature by 15°C?

2.47. The average specific heat of dry air is 1.00 J/g°C.
  a. If 523 J heat are added to 75 g dry air at 20°C, what is the final temperature of the air?
  b. The average specific heat of water is 4.18 J/g°C. If 523 J are added to 75 g water at 20°C, what is the final temperature of the water sample?
  c. Why are temperature variations between day and night more extreme in a desert than near an ocean?

*2.48. a. How many joules are required to raise the temperature of 15.6 g water from 5°C to 15°C?
  b. How many calories are required to raise the temperature of 15.6 g ice from −15°C to −5°C? (Use 2.05 J/g°C for the specific heat of ice.)

2.49. How many joules must be removed from 555 g water to cool the sample from 25°C to 0°C?

Review Problems

*2.50. A person has a mass of 113 lb. What volume (in liters) of mercury has the same mass?

2.51. The Earth is 9.29×10^7 mi from the sun.
a. What is this distance in meters?
b. Light travels 3.0×10^8 m/sec. How long does it take light to travel from the sun to the Earth?

2.52. When "burned" as a metabolic fuel, 1 g body fat is equivalent to 32.2 kJ energy. How many kilojoules of energy must an adult use to lose the equivalent of 1 lb body fat?

2.53. Lime has a density of 3.25 g/cm^3. What is the volume of 10.0 lb lime? The melting point of lime is 2614°C. What is its physical state at room temperature under normal conditions?

2.54. Pure ammonia melts at -78°C and boils at -33°C. Under normal conditions 10.0 L pure ammonia has a mass of 7.710 g. What is the density of pure ammonia and its physical state under normal conditions? Is liquid household ammonia pure ammonia?

*2.55. The rocket fuel hydrazine melts at 1.4°C and boils at 113.5°C. A 25.00-mL sample of hydrazine has a mass of 25.3 g. What is the density of hydrazine and its physical state at room temperature under normal conditions?

2.56. The city of Chicago declares an ozone alert when the concentration of ozone in the air reaches 137 μg/m^3.
a. Express this ozone concentration in micrograms per liter and in nanograms per liter.
b. The lung capacity per breath of an adult is about 2 qt. What is the mass of ozone taken into the lungs per breath by an adult breathing air that contains 137 μg ozone per cubic meter of air?

* 2.57. For an adult in good health and with an adequate diet, the concentration of ascorbic acid (vitamin C) in the blood is about 0.2 mg per 100. mL blood.
a. What is this concentration in grams per liter? in milligrams per milliliter?
b. The average person has about 5 L blood. Calculate the total number of milligrams of ascorbic acid in the blood of an adult in good health eating an adequate diet.
c. According to the National Research Council, the Recommended Daily Allowance for ascorbic acid is 45 mg. How does this amount compare with the number of milligrams of ascorbic acid in the blood of a healthy adult with an adequate diet?

2.58. The following table lists the energy expended by various activities, given in terms of kilojoules per kilogram of body weight per hour. The rates are applicable to either sex.

Activity	*kJ*/*kg/hr*
bicycling (moderate speed)	31.8
carpentry	54.9
lying still (awake)	0.5
playing Ping-Pong	21.2
running	33.7
sawing wood	27.4
sitting quietly	1.9
skating	16.8

a. Calculate the number of kilojoules expended per hour by a 160-lb male in each of these activities.
b. Calculate the number of kilojoules expended per hour by a 130-lb female in each of these activities.
c. How many hours must each run to burn off the equivalent of one pound of body fat? (See problem 2.52.)

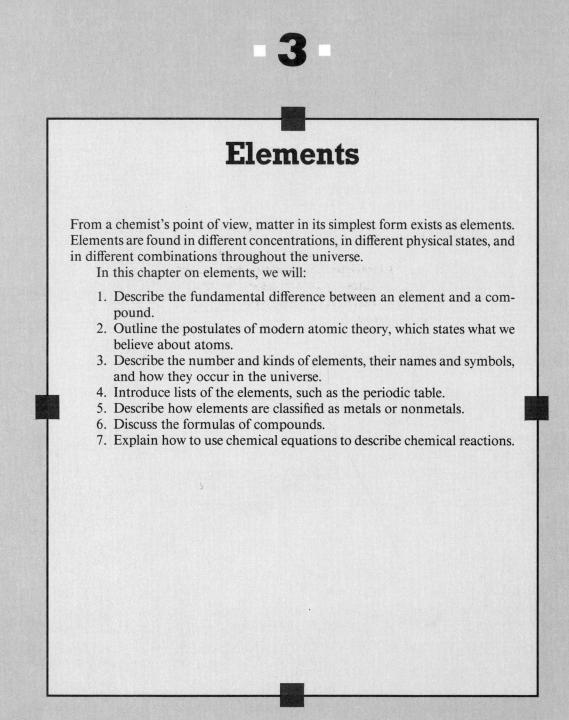

· 3 ·

Elements

From a chemist's point of view, matter in its simplest form exists as elements. Elements are found in different concentrations, in different physical states, and in different combinations throughout the universe.

In this chapter on elements, we will:

1. Describe the fundamental difference between an element and a compound.
2. Outline the postulates of modern atomic theory, which states what we believe about atoms.
3. Describe the number and kinds of elements, their names and symbols, and how they occur in the universe.
4. Introduce lists of the elements, such as the periodic table.
5. Describe how elements are classified as metals or nonmetals.
6. Discuss the formulas of compounds.
7. Explain how to use chemical equations to describe chemical reactions.

3.1 Elements, Compounds, and Mixtures

A pure substance can be either an element or a compound. **Elements** are those pure substances that cannot be decomposed by ordinary chemical means such as heating, electrolysis, or reaction. Gold, silver, and oxygen are examples of elements. **Compounds** are pure substances formed by the combination of elements; they can be decomposed by ordinary chemical means. Baking soda (discussed in Section 1.2A) is a compound; it contains the elements sodium, hydrogen, carbon, and oxygen, and it decomposes on heating. Mercuric oxide is another compound; it contains the elements mercury and oxygen, and on heating it decomposes to those elements.

Compounds differ from mixtures in that the elements in a compound are held together by chemical bonds and cannot be separated by differences in their physical properties. The components of a mixture are not joined together by any chemical bonds, and, as was shown in Section 1.3, they can be separated from one another by differences in their physical properties.

Figure 3.1 reviews the relations between the different kinds of matter. Notice that mixtures can be separated into their components by differences in physical properties. Compounds can be separated into their components only by chemical change.

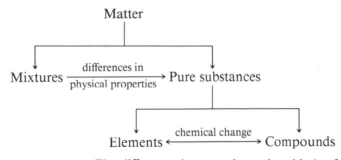

FIGURE 3.1   The differences between the various kinds of matter.

3.2 Atoms—The Atomic Theory

By the end of the eighteenth century, experimenters had well established that each pure substance had its own characteristic set of properties such as density, specific heat, melting point, and boiling point. Also established was the fact that certain quantitative relationships, such as the Law of Conservation of Mass, governed all chemical changes. But there was still no understanding of the nature of matter itself. Was matter continuous, like a ribbon from which varying amounts could be snipped, or was it granular, like a string of beads from which only whole units or groups of units could be removed? Some scientists believed strongly in the continuity of matter, whereas others believed equally strongly in granular matter; but both reasonings were based solely on speculation.

In 1803, an English schoolmaster named John Dalton (1766–1844) summarized and extended the then-current theory of matter. The postulates of his theory, changed only slightly from their original statement, form the basis of modern **atomic theory.** Today, we express these four postulates as:

1. Matter is made up of tiny particles called **atoms.** (A typical atom has a mass of approximately 10^{-23} g and a radius of approximately 10^{-10} m.)

2. Over 100 different kinds of atoms are known; each kind is an element. (A list of the elements is on the inside back cover of this text.) All the atoms of a particular element are alike chemically but can vary slightly in mass and other physical properties. Atoms of different elements have different masses.

3. Atoms of different elements combine in small, whole-number ratios to form compounds. For example, hydrogen and oxygen atoms combine in a ratio of $2:1$ to form the compound water, H_2O. Carbon and oxygen atoms combine in a ratio of $1:2$ to form the compound carbon dioxide, CO_2. Iron and oxygen atoms combine in a ratio of $2:3$ to form the familiar substance rust, Fe_2O_3.

4. The same atoms can combine in different whole-number ratios to form different compounds. As just noted, hydrogen and oxygen atoms combined in a $2:1$ ratio form water; combined $1:1$, they form hydrogen peroxide, H_2O_2 (Figure 3.2). Carbon and oxygen atoms combined in a $1:2$ ratio form carbon dioxide; combined in a $1:1$ ratio, they form carbon monoxide, CO.

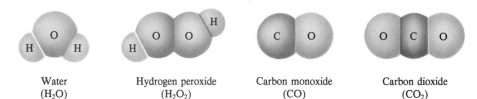

| Water | Hydrogen peroxide | Carbon monoxide | Carbon dioxide |
| (H_2O) | (H_2O_2) | (CO) | (CO_2) |

FIGURE 3.2   Atoms of the same elements combine in different ratios to form different compounds.

3.3   The Elements

Elements are pure substances. The atoms of each element are chemically distinct and different from those of any other element. Approximately 110 elements are now known. By 1980, 106 of these had been unequivocally characterized and accepted by the **International Union of Pure and Applied Chemistry (IUPAC).** Since that time, elements 107 and 109 have been identified among the products of a nuclear reaction. The search for new elements continues in many laboratories around the world; new elements may be announced at any time.

A. Names and Symbols of the Elements

Each element has a name. Many of these names are already familiar to you — gold, silver, copper, chlorine, platinum, carbon, oxygen, and nitrogen. The names themselves are interesting. Many refer to a property of the element. The Latin name for gold is *aurum,* meaning "shining dawn." The Latin name for mercury, *hydrargyrum,* means "liquid silver."

The practice of naming an element after one of its properties continues. Cesium was discovered in 1860 by the German chemist Bunsen (the inventor of the Bunsen burner). Because this element imparts a blue color to a flame, Bunsen named it cesium from the Latin word *caesius,* meaning "sky blue."

Other elements are named for people. Curium is named for Marie Curie (1867–1934), a pioneer in the study of radioactivity. Marie Curie, a French scientist of Polish birth, was awarded the Nobel Prize in Physics in 1903 for her studies of radioactivity. She was also awarded the Nobel Price in Chemistry in 1911 for her discovery of the elements polonium (named after Poland) and radium (Latin, *radius,* "ray").

Some elements are named for places. The small town of Ytterby in Sweden has four elements named for it: terbium, yttrium, erbium, and ytterbium. Californium is another example of an element named for the place where it was first observed. This element does not occur in nature. It was first produced in 1950 in the Radiation Laboratory at the University of California, Berkeley, by a team of scientists headed by Glenn Seaborg. Seaborg was also the first to identify curium at the metallurgical laboratory of the University of Chicago (now Argonne National Laboratory) in 1944. Seaborg himself was named a Nobel laureate in 1951 in honor of his pioneering work in the preparation of other unknown elements.

Each element has a **symbol,** one or two letters that represent the element much as your initials represent you. The symbol of an element represents one atom of that element. For 14 of the elements, the symbol consists of one letter. With the possible exceptions of yttrium (Y) and vanadium (V), you are probably familiar with the names of all elements having one-letter symbols. These elements are listed in Table 3.1. For 12 of these elements, the symbol is the first letter of the name.

Potassium was discovered in 1807 and named for potash, the substance from which potassium was first isolated. Potassium's symbol, K, comes from *kalium,* the Latin word for potash. Tungsten, discovered in 1783, has the symbol W, for wolframite, the mineral from which tungsten was first isolated.

Most other elements have two-letter symbols. In these two-letter symbols, the first letter is always capitalized and the second is always lowercased. Eleven

TABLE 3.1   Elements with one-letter symbols

Symbol	Element	Symbol	Element
B	boron	P	phosphorus
C	carbon	K	potassium
F	fluorine	S	sulfur
H	hydrogen	W	tungsten
I	iodine	U	uranium
N	nitrogen	V	vanadium
O	oxygen	Y	yttrium

elements have names (and symbols) beginning with the letter C. One of these, carbon, has a one-letter symbol, C. The other ten have two-letter symbols (see Table 3.2).

TABLE 3.2   Elements whose name begins with the letter C

Symbol	Element	Symbol	Element
Cd	cadmium	Cl	chlorine
Ca	calcium	Cr	chromium
Cf	californium	Co	cobalt
C	carbon	Cu	copper
Ce	cerium	Cm	curium
Cs	cesium		

B. Lists of the Elements

While you study chemistry, you will often need a list of the elements. This and most other chemistry books provide two such lists. On the inside of the back cover of this text the elements are listed alphabetically by name. The list also includes the symbol, the atomic number, and the atomic weight of the element. The significance of atomic numbers and weights will be discussed in Chapter 4. For now it is sufficient to know that each element has a number between 1 and 110 called its *atomic number.* This number is as unique to the element as its name or symbol.

The second list, called the **periodic table,** arranges the elements in order of increasing atomic number in rows of varying length. The significance of the length of the row and the relation among elements in the same row or column will be discussed in Chapter 5. The periodic table appears on the inside of the front cover of this text. Throughout the book we will refer to the periodic table, because it contains an amazing amount of information. For now you need only be aware that elements in the same column have similar properties and that the heavy stair-step line that crosses the table diagonally from boron (B) to astatine (At) separates the metallic elements from the nonmetallic elements. The periodic table is also shown in Figure 3.3. The screened areas mark the elements you will encounter most often in this text.

1. Metals and nonmetals

Metals appear below and to the left of the heavy diagonal line in the periodic table. The characteristic properties of a metal are:

1. It is shiny and lustrous.
2. It conducts heat and electricity.
3. It is ductile and malleable; that is, it can be drawn into a wire and can be hammered into a thin sheet.
4. It is a solid at 20°C. Mercury is the only exception to this rule; it is a liquid at room temperature. Two other metals, gallium and cesium, have melting points close to room temperature (19.8°C and 28.4°C).

Nonmetals vary more in their properties than do metals; some may even have one or more of the metallic properties listed. Some nonmetals are gaseous; chlorine and nitrogen are gaseous nonmetals. At 20°C one nonmetal, bromine, is a liquid, and others are solids — for example, carbon, sulfur, and phosphorus.

Example 3.1

Give the symbol of the following elements. Tell whether each is a metal or nonmetal. If it is a metal, tell its physical state at 20°C.

　　a. fluorine　　**b.** bismuth　　**c.** potassium

Solution

The symbol of the element is found in the alphabetical list of elements. Notice that the symbol for potassium is K, not P. Now find the elements in the periodic table. Both bismuth and potassium are below the diagonal line, so they are metals. Fluorine is a nonmetal, as it appears above the line. Because mercury is the only metal that is liquid at 20°C, potassium and bismuth must be solids at 20°C. We can draw up a chart:

Element	Symbol	Metal or nonmetal	Physical state
fluorine	F*	nonmetal	no way of knowing
bismuth	Bi	metal	solid
potassium	K	metal	solid

* Notice that fluorine is the only element in its column of the periodic table to have a one-letter symbol.

Atomic number → 6
Symbol ← C
Atomic weight → 12.01

I	II											III	IV	V	VI	VII	VIII
1 H 1.008																	2 He 4.00
3 Li 6.94	4 Be 9.01											5 B 10.81	6 C 12.01	7 N 14.01	8 O 16.00	9 F 19.00	10 Ne 20.18
11 Na 22.99	12 Mg 24.31											13 Al 26.98	14 Si 28.09	15 P 30.97	16 S 32.06	17 Cl 35.45	18 Ar 39.95
19 K 39.10	20 Ca 40.08	21 Sc 44.96	22 Ti 47.88	23 V 50.94	24 Cr 52.00	25 Mn 54.94	26 Fe 55.85	27 Co 58.93	28 Ni 58.69	29 Cu 63.55	30 Zn 65.38	31 Ga 69.72	32 Ge 72.59	33 As 74.92	34 Se 78.96	35 Br 79.90	36 Kr 83.80
37 Rb 85.47	38 Sr 87.62	39 Y 88.91	40 Zr 91.22	41 Nb 92.91	42 Mo 95.94	43 Tc (98)	44 Ru 101.07	45 Rh 102.91	46 Pd 106.42	47 Ag 107.87	48 Cd 112.41	49 In 114.82	50 Sn 118.69	51 Sb 121.75	52 Te 127.60	53 I 126.90	54 Xe 131.29
55 Cs 132.91	56 Ba 137.33	57 La 138.91	72 Hf 178.49	73 Ta 180.95	74 W 183.85	75 Re 186.21	76 Os 190.2	77 Ir 192.22	78 Pt 195.08	79 Au 196.97	80 Hg 200.59	81 Tl 204.38	82 Pb 207.2	83 Bi 208.98	84 Po (209)	85 At (210)	86 Rn (222)
87 Fr (223)	88 Ra 226.03	89 Ac 227.03	104 Unq (261)	105 Unp (262)	106 Unh (263)	107 Uns (262)	108	109 Une									

Lanthanides	58 Ce 140.12	59 Pr 140.91	60 Nd 144.24	61 Pm (145)	62 Sm 150.36	63 Eu 151.96	64 Gd 157.25	65 Tb 158.92	66 Dy 162.50	67 Ho 164.93	68 Er 167.26	69 Tm 168.93	70 Yb 173.04	71 Lu 174.97
Actinides	90 Th 232.04	91 Pa 231.04	92 U 238.03	93 Np 237.05	94 Pu (244)	95 Am (243)	96 Cm (247)	97 Bk (247)	98 Cf (251)	99 Es (252)	100 Fm (257)	101 Md (258)	102 No (259)	103 Lr (260)

FIGURE 3.3 The periodic table of the elements. The symbols of the most common elements appear in the screened areas.

Problem 3.1   List the symbols of the following elements. Tell whether each is a metal or a nonmetal. If it is a metal, tell its physical state at 20°C.

      a. strontium   **b.** phosphorus   **c.** chromium

C. Distribution of the Elements

The known elements are not equally distributed throughout the world. Only 91 are found in either the Earth's crust, oceans, or atmosphere; the others have been produced in laboratories. Traces of some but not all of these elements have been found on Earth or in the stars. The search for the others continues. You might read of its success or of the isolation of new elements as you are taking this course.

TABLE 3.3   Distribution of elements in the Earth's crust, oceans, and atmosphere

Element	Percent of total mass	Element	Percent of total mass
oxygen	49.2	chlorine	0.19
silicon	25.7	phosphorus	0.11
aluminum	7.50	manganese	0.09
iron	4.71	carbon	0.08
calcium	3.39	sulfur	0.06
sodium	2.63	barium	0.04
potassium	2.40	nitrogen	0.04
magnesium	1.93	fluorine	0.03
hydrogen	0.87	all others	0.49
titanium	0.58		

Table 3.3 lists the 18 elements that are most abundant in the Earth's crust, oceans, and atmosphere, along with their relative percentages of the Earth's total mass. One of the most striking points about this list is the remarkably uneven distribution of the elements (see Figure 3.4). Oxygen is by far the most abundant element. It makes up 21% of the volume of the atmosphere and 89% of the mass of water. Oxygen in air, water, and elsewhere constitutes 49.2% of the mass of the Earth's crust, oceans, and atmosphere. Silicon is the Earth's second most abundant element (25.7% by mass). Silicon is not found free in nature but occurs in combination with oxygen, mostly as silicon dioxide (SiO_2), in sand, quartz, rock crystal, amethyst, agate, flint, jasper, and opal, as well as in various silicate minerals such as granite, asbestos, clay, and mica. Aluminum is the most abundant metal in the Earth's crust (7.5%). It is always found combined in nature. Most of the aluminum used today is obtained by processing

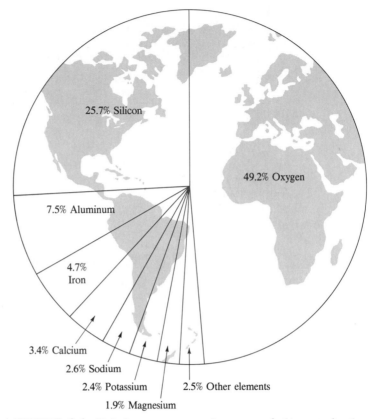

FIGURE 3.4   Relative percentages by mass of elements in the
Earth's crust, oceans, and atmosphere.

bauxite, an ore that is rich in aluminum oxide. These three elements (oxygen,
silicon, and aluminum) plus iron, calcium, sodium, potassium, and magnesium make up more than 97% of the mass of the Earth's crust, oceans, and
atmosphere. Another surprising feature of the distribution of elements is that
several of the metals that are most important to our civilization are among the
rarest; these metals include lead, tin, copper, gold, mercury, silver, and zinc.

The distribution of elements in the cosmos is quite different from that on
Earth. According to present knowledge, hydrogen is by far the most abundant
element in the universe, accounting for as much as 75% of its mass. Helium and
hydrogen together make up almost 100% of the mass of the universe.

Table 3.4 lists the biologically important elements—those found in a
normal, healthy body. The first four of these elements—oxygen, carbon, hydrogen, and nitrogen—make up about 96% of total body weight (see Figure
3.5). The other elements listed, although present in much smaller amounts, are
nonetheless necessary for good health.

TABLE 3.4 Biologically important elements (amounts given per 70-kg body weight) *154.12 lbs.*

Major elements	Approximate amount (kg)	Elements present in less than 1-mg amounts (listed alphabetically)
oxygen	45.5	arsenic
carbon	12.6	chromium
hydrogen	7.0	cobalt
nitrogen	2.1	copper
calcium	1.0	fluorine
phosphorus	0.70	iodine
magnesium	0.35	manganese
potassium	0.24	molybdenum
sulfur	0.18	nickel
sodium	0.10	selenium
chlorine	0.10	silicon
iron	0.003	vanadium
zinc	0.002	

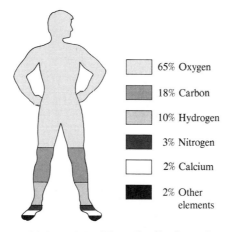

65% Oxygen

18% Carbon

10% Hydrogen

3% Nitrogen

2% Calcium

2% Other elements

FIGURE 3.5 The distribution (by mass) of elements in the human body.

D. How Elements Occur in Nature

Elements occur as single atoms or as groups of atoms chemically bonded together. The nature of these chemical bonds will be discussed in Chapter 7. Groups of atoms bonded together chemically are called **molecules** or **formula units.**

Molecules may contain atoms of a single element, or they may contain atoms of different elements (in which case the molecule is of a compound). Just as an atom is the smallest unit of an element, a molecule is the smallest unit of a compound—that is, the smallest unit having the chemical identity of that compound.

Let us consider how the elements might be categorized by the way they are found in the universe.

1. The noble gases

Only a few elements are found as single, uncombined atoms; Table 3.5 lists these elements. Under normal conditions, all of these elements are gases; collectively, they are known as the **noble gases.** They are also called **monatomic gases,** meaning that they exist, uncombined, as single atoms (*mono* means "one"). The formula for each of the noble gases is simply its symbol. When the formula of helium is required, the symbol He is used. The subscript 1 is understood.

TABLE 3.5   The noble gases

Symbol	Element
He	helium
Ne	neon
Ar	argon
Kr	krypton
Xe	xenon
Rn	radon

2. Metals

Pure metals are treated as though they existed as single, uncombined atoms even though a sample of pure metal is an aggregate of billions of atoms. Thus, when the formula of copper is required, its symbol, Cu, is used to mean one atom of copper.

3. Nonmetals

Some nonmetals exist, under normal conditions of temperature and pressure, as molecules containing two, four, or eight atoms. Those nonmetals that occur as **diatomic** (two-atom) **molecules** are listed in Table 3.6. Thus, we use O_2 as the formula for oxygen, N_2 for nitrogen, and so on. Among the nonmetals, sulfur exists as S_8 and phosphorus is found as P_4. For other nonmetals (those not listed in Table 3.5 or 3.6) a monatomic formula is used—for example, As for arsenic and Se for selenium.

TABLE 3.6 Diatomic elements

Formula	Name	Normal state
H_2	hydrogen	colorless gas
N_2	nitrogen	colorless gas
O_2	oxygen	colorless gas
F_2	fluorine	pale yellow gas
Cl_2	chlorine	greenish yellow gas
Br_2	bromine	dark red liquid
I_2	iodine	violet black solid

Example 3.2

For each of the following elements, tell whether it is a metal or a nonmetal and give the formula it has in the uncombined state.

    a. iodine   **b.** calcium   **c.** carbon   **d.** xenon

Solution

To carry out this exercise, use the following steps:

  1. Look up the symbol of each element.

  2. Find its location in the periodic table and decide whether it is a metal or a nonmetal.

  3. Look through the preceding section and determine into which category it falls. Prepare a table of answers.

Element	Symbol	Metal or nonmetal	Formula	Reference section
iodine	I	nonmetal	I_2	3.3D3
calcium	Ca	metal	Ca	3.3D2
carbon	C	nonmetal	C	3.3D3
xenon	Xe	nonmetal	Xe	3.3D1

Problem 3.2

For each of the following elements, tell whether it is a metal or a nonmetal and give the formula it has in the uncombined state.

    a. nitrogen   **b.** strontium   **c.** chromium   **d.** helium

4. Compounds

Although many elements can occur in the uncombined state, all elements except some of the noble gases are also found combined with other elements in compounds. In Section 3.1 we defined a compound as a substance that can be decomposed by ordinary chemical means. A compound can also be defined as a pure substance that contains two or more elements. The composition of a compound is expressed by a formula that uses the symbols of all the elements in the compound. Each symbol is followed by a **subscript,** a number that shows how many atoms of the element occur in one molecule (the simplest unit) of the compound; the subscript 1 is not shown. Water is a compound with the formula H_2O, meaning that one molecule (or formula unit) of water contains two hydrogen atoms and one oxygen atom. The compound sodium hydrogen carbonate has the formula $NaHCO_3$, meaning that a single formula unit of this compound contains one atom of sodium, one atom of hydrogen, one atom of carbon, and three atoms of oxygen. Notice that the symbols of the metals in sodium hydrogen carbonate are written first, followed by the nonmetals, and that, of the nonmetals, oxygen is written last. This order is customary.

Sometimes a formula will contain a group of symbols enclosed in parentheses as, for example, $Cu(NO_3)_2$. The parentheses imply that the group of atoms they enclose act as a single unit. The subscript following the parenthesis means that the group is taken two times for each copper atom.

Example 3.3	A molecule of hydrogen sulfide contains two atoms of hydrogen and one atom of sulfur. Write the formula for hydrogen sulfide.
Solution	**1.** Write the symbols of each element: H S
	2. Follow each symbol with a subscript indicating the number of atoms per molecule: H_2S
Problem 3.3	Ethanol is composed of two atoms of carbon, six atoms of hydrogen, and one atom of oxygen. Write the formula for ethanol.

Example 3.4	The formula of magnesium hydroxide is $Mg(OH)_2$. What is the composition of a unit of magnesium hydroxide?
Solution	The symbols tell us that magnesium hydroxide contains magnesium, oxygen, and hydrogen. The parentheses show that a combination of one oxygen atom with one hydrogen atom acts as a unit and that two of these units are present in a formula unit of magnesium hydroxide. The formula unit of magnesium hydroxide contains one magnesium atom, two oxygen atoms, and two hydrogen atoms.
Problem 3.4	What is the composition of a formula unit of calcium nitrate, $Ca(NO_3)_2$?

The properties of a compound are quite unlike those of the elements from which it is formed. This fact is apparent if we compare the properties of carbon dioxide, CO_2 (a colorless gas used in fire extinguishers), with those of carbon (a black, combustible solid) and oxygen (a colorless gas necessary for combustion). The properties of compounds are discussed in greater detail in Chapter 6.

3.4   The Reactions of Elements: Simple Equations

A study of chemistry involves the study of chemical changes or, as they are more commonly called, chemical reactions. Examples of **chemical reactions** are: the combination of elements to form compounds, the decomposition of compounds (such as sodium hydrogen carbonate or mercuric oxide), and reactions between compounds, such as the reaction of vinegar (a solution of acetic acid) with baking soda (sodium hydrogen carbonate). Reactions are usually described using **chemical equations.** Equations may be expressed in words: Mercuric oxide decomposes to mercury and oxygen. Using formulas, we state this reaction as:

$$2\ HgO \longrightarrow 2\ Hg + O_2$$

A chemical equation has several parts: The **reactants** are those substances with which we start (here mercuric oxide, HgO, is the reactant). The arrow ($\rightarrow$) means "reacts to form" or "yields." The **products** are those substances formed by the reaction (here mercury and oxygen are the products). The numbers preceding the formulas are called **coefficients.** Sometimes the physical state of the reaction components are shown; we use a lowercase, italic letter in parentheses following the substance to show its state. For example, if the equation for the decomposition of mercuric oxide were written as:

$$2\ HgO(s) \longrightarrow 2\ Hg(l) + O_2(g)$$

we would know that the mercuric oxide was a solid, the mercury was a liquid, and the oxygen was a gas when the equation was carried out. The same equation is repeated below with all the parts labeled:

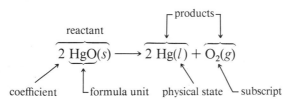

Table 3.7 lists the parts of an equation and the notations commonly used.

TABLE 3.7   Parts of an equation

Reactants	The starting substances, which combine in the reaction. (Formulas must be correct.)
Products	The substances that are formed by the reaction. (Formulas must be correct.)
Arrows	
$\rightarrow$	Found between reactants and products, means "reacts to form."
$\nrightarrow$	Used between reactants and products to show that the equation is not yet balanced.
$(\uparrow)$	Placed after the formula of a product that is a gas.
$(\downarrow)$	Placed after the formula of a product that is an insoluble solid, also called a **precipitate.**
Physical state	Indicates the physical state of the substance whose formula it follows.
	(g) Indicates that the substance is a gas
	(l) Indicates that the substance is a liquid
	(s) Indicates that the substance is a solid
	(aq) Means that the substance is in aqueous (water) solution
Coefficients	The numbers placed in front of the formulas to balance the equation.
Conditions	Words or symbols placed over or under the horizontal arrow to indicate conditions used to cause the reaction.
	Δ Heat is added
	hv Light is added
	elec Electrical energy is added

A. Writing Chemical Equations

A correctly written equation obeys certain rules.

1. The formulas of all reactants and products must be correct.

Correct formulas must be used. An incorrect formula would represent a different substance and therefore completely change the meaning of the equation. For example, the equation

$$2\,H_2O_2 \longrightarrow 2\,H_2O + O_2$$

describes the decomposition of hydrogen peroxide. This reaction is quite different from the decomposition of water, which is described by the equation

$$2\,H_2O \longrightarrow 2\,H_2 + O_2$$

When an uncombined element occurs in an equation, the guidelines in Section 3.3D (parts 1, 2, and 3) should be used to determine its formula.

2. An equation must be balanced by mass.

An equation is balanced by mass when the number of atoms of each element in the reactants equals the number of atoms of that element in the products. For example, the equation shown for the decomposition of water has four atoms of hydrogen in the two molecules of water on the reactant side and four atoms of hydrogen in the two molecules of hydrogen gas on the product side; therefore, hydrogen is balanced. It has two atoms of oxygen in the two reacting molecules of water and two atoms of oxygen in the single molecule of oxygen produced; therefore, oxygen is also balanced.

$$2 \, H_2O \longrightarrow 2 \, H_2 + O_2$$

four (2×2) H atoms on the left $=$ four (2×2) H atoms on the right

two (2×1) O atoms on the left $=$ two (1×2) O atoms on the right

When the atoms are balanced, the mass is balanced and the equation obeys the Law of Conservation of Mass.

You can write and balance equations in three steps:

■ **1.** Write the correct formulas of all the reactants. Use a plus sign (+) between the reactants and follow the final reactant with an arrow. After the arrow, write the correct formulas of the products, separating them with plus signs.

■ **2.** Count the number of atoms of each element on each side of the equation. Remember that all elements present must appear on both sides of the equation.

■ **3.** Change the coefficients as necessary so that the number of atoms of each element on the left side of the equation is the same as that on the right side. Only the coefficients may be changed to balance an equation; the subscripts in a formula must never be changed.

Example 3.5   The brilliant white light in some fireworks displays is produced by burning magnesium in air. The magnesium reacts with oxygen in the air to form magnesium oxide, MgO. Write the balanced chemical equation for this reaction.

Solution   **1.** Write the correct formulas for all reactants, followed by an arrow; then write the correct formulas of the products. Magnesium is a metal, so the symbol Mg is used. Oxygen is a diatomic nonmetal, so we use the formula O_2. The formula of the product is given as MgO. Together, these formulas give:

$$Mg + O_2 \overset{\,/}{\longrightarrow} MgO$$

The line through the arrow indicates that the equation is not yet balanced.

FIGURE 3.6   A flashbulb contains pure magnesium sealed in an atmosphere of pure oxygen. When the flashbulb is set off, the electric current heats the magnesium causing it to react vigorously with the oxygen to produce magnesium oxide. The accompanying blinding light is energy released by the reaction.

2. Count the atoms on each side:

	Reactants	*Products*
magnesium	1	1
oxygen	2	1

3. Change the coefficients in the equation so that the atoms of each element are balanced. We can balance the atoms by placing a coefficient of 2 in front of Mg and in front of MgO. Therefore, the balanced equation for the combustion of magnesium is:

$$2 \, Mg + O_2 \longrightarrow 2 \, MgO$$

Problem 3.5   Sodium reacts with chlorine to form table salt, NaCl (sodium chloride). Write a balanced chemical equation for this reaction.

$2NaCl_2 \rightarrow 2Na + Cl$

Example 3.6   Write a balanced equation for the combustion of ethanol, C_2H_6O, to carbon dioxide and water. **Combustion** means combination with oxygen. The formula of carbon dioxide, CO_2, is implied in its name (*di* means "two"). The formula of water is written as H_2O.

Solution   **1.** Write the correct formulas for reactants and products:

$$C_2H_6O + O_2 \overset{\longrightarrow}{\not\;} CO_2 + H_2O$$

2. Count the atoms:

	Reactants	*Products*
carbon	2	1
hydrogen	6	2
oxygen	3	3

3. Balance the equation by mass. We will start with carbon because it is often easiest to begin with the element present in the fewest compounds. Two carbon atoms appear on the left, so we need two carbon atoms on the right; we write 2 CO_2. Next we balance hydrogen: Six hydrogens are on the left, so we write 3 H_2O on the right.

$$C_2H_6O + O_2 \xrightarrow{\quad/\quad} 2\, CO_2 + 3\, H_2O$$

Finally, we balance the oxygen. We now have seven oxygens on the right and only three on the left. Changing the coefficient in front of O_2 from 1 to 3 will give four more oxygens on the left. The equation is then balanced:

$$C_2H_6O + 3\, O_2 \longrightarrow 2\, CO_2 + 3\, H_2O$$

Problem 3.6   Propane, C_3H_8, burns in air to form carbon dioxide and water. Write the balanced chemical equation for this reaction.

■

Example 3.7   The rusting of iron is actually the chemical reaction of iron with oxygen to form the oxide of iron, Fe_2O_3. Write a balanced chemical equation for this reaction.

Solution   **1.** Write the formulas:

$$Fe + O_2 \xrightarrow{\quad/\quad} Fe_2O_3$$

2. Count the atoms:

	Reactants	*Products*
iron	1	2
oxygen	2	3

3. Balance the equation:

Oxygen is a troublesome element in this equation because the number of oxygen atoms must be divisible by both 2 and 3. The lowest possible number is 6, so we write:

$$Fe + 3\ O_2 \overset{/}{\longrightarrow} 2\ Fe_2O_3$$

The equation is balanced when we write four iron atoms on the left.

$$4\ Fe + 3\ O_2 \longrightarrow 2\ Fe_2O_3$$

Problem 3.7   Dinitrogen pentoxide, N_2O_5, reacts with water, H_2O, to form nitric acid, HNO_3. Write the balanced chemical equation for this reaction.

3.5   Summary

Elements and compounds make up the universe. The smallest unit of an element is an atom. The properties of atoms are described by the atomic theory, first proposed in 1803. Elements occur as single atoms, as groups or molecules of atoms of a single element, or in compounds. Compounds are pure substances formed by the chemical combination of atoms of more than one element. Compounds have fixed properties and composition.

We now know of slightly more than 100 elements. Each has a name, a symbol, and a number. The elements are very unevenly distributed throughout the universe.

A chemical reaction occurs when atoms of different elements combine to form compounds or when a compound decomposes to elements or to simpler compounds. Chemical equations can be written to describe these events. Equations must contain correct formulas and must be balanced by mass.

Key Terms

atomic theory (3.2)
atoms (3.2)
chemical equations (3.4)
chemical reactions (3.4)
coefficients (3.4)
combustion (3.4A2)
compounds (3.1)
diatomic molecules (3.3D3)
elements (3.1)
formula units (3.3D)
International Union of Pure and
    Applied Chemistry (IUPAC) (3.3)

metals (3.3B1)
molecules (3.3D)
monatomic gases (3.3D1)
noble gases (3.3D1)
nonmetals (3.3B1)
periodic table (3.3B)
precipitate (3.4)
products (3.4)
reactants (3.4)
subscript (3.3D4)
symbol (3.3A)

Multiple-Choice Questions

Use the following groups of elements to answer Questions 1–4.

    a. none of these groups
    b. chromium, nitrogen, oxygen, sulfur
    c. sodium, iron, potassium, silver
    d. nitrogen, iodine, phosphorus, fluorine
    e. carbon, tin, lead, silicon

MC1. Which of these groups of elements contains only elements whose symbol does not begin with the same letter as its name?

MC2. Which of these groups contains only metals?

MC3. Which of these groups has only one-letter symbols?

MC4. Which group contains only diatomic elements?

MC5. The atomic theory states that atoms of two different elements can combine in more than one ratio to form several compounds. Such a series would be:
    a. NaCl, KCl, LiCl   **b.** C_2H_2, C_2H_4, C_2H_6
    c. NaCl, $MgCl_2$, $AlCl_3$ **d.** NaBr, NaI, NaCl   **e.** none of these

Balance the equations in Questions 6–9 using whole-number coefficients.

MC6. Potassium forms an oxide containing one oxygen atom for every two atoms of potassium. What is the coefficient of oxygen in the balanced equation for the reaction of potassium with oxygen to form this oxide?
    a. 0   **b.** 1   **c.** 2   **d.** 3   **e.** 4

MC7. In the balanced equation for the decomposition of water into its elements, what is the sum of the coefficients?
    a. 2   **b.** 3   **c.** 4   **d.** 5   **e.** 6

MC8. In the balanced equation for the reaction of boron with fluorine to form boron trifluoride (BF_3), what is the coefficient of boron?
    a. 0   **b.** 1   **c.** 2   **d.** 3   **e.** 4

MC9. In the balanced equation for the reaction of oxygen with nitrogen to form dinitrogen oxide (N_2O), what is the coefficient of oxygen?
    a. 1   **b.** 2   **c.** 3   **d.** 4   **e.** 5

MC10. How many of the following element–symbol pairs are *incorrect?*

sodium, Na	fluorine, Fl	boron, Bo
calcium, Ca	neon, Ne	manganese, Mg
silicon, Si	potassium, P	silver, Ag

    a. 0   **b.** 1   **c.** 2   **d.** 3   **e.** 4

Problems

3.1 Elements, Compounds, and Mixtures

*3.8. Classify the members of the following list of pure substances as elements or compounds.

mercury	hydrogen
lime (calcium oxide)	neon
table salt	water

3.2 Atomic Theory

3.9. State the postulates of the atomic theory.

3.10. Nitrogen and oxygen combine to form the following compounds: nitrogen oxide, NO; nitrogen dioxide, NO_2; dinitrogen trioxide, N_2O_3; and dinitrogen tetroxide, N_2O_4; as well as others. Tell how this series of compounds is predicted by the atomic theory.

3.3 The Elements

3.11. Give the symbol of each of the following elements. Try not to look in your text. Classify each as a metal or nonmetal.
a. phosphorus   **b.** oxygen
c. cobalt   **d.** calcium
e. chlorine   **f.** bromine
g. potassium   **h.** copper
i. magnesium

3.12. Give the name of the element for each of the following symbols. Try to answer without looking them up. Classify each as a metal or nonmetal.
a. Na   **b.** S   **c.** I   **d.** Ce
e. Al   **f.** N   **g.** Cl   **h.** Mn
i. Zn   **j.** F

3.13. Make a list of the names and symbols of those elements found in Table 3.3 but not in Table 3.4. Do the same for those in Table 3.4 that are not in Table 3.3. Classify each as a metal or nonmetal.

*3.14. From a list of the elements, write down each element whose name does not begin with the same letter as does its symbol.

Match these elements with the appropriate Latin or German name from the following list:

natrium	hydrargyrum
ferrum	stibium
stannum	wolfram
aurum	kalium
argentum	plumbum

3.15. Look up each of these elements in a dictionary and determine the origin of its name. What characterizes the atomic numbers of the group?

americium	californium
berkelium	nobelium
neptunium	lawrencium
einsteinium	

3.16. The elements in column VII of the periodic table are known collectively as the halogens. List the names and symbols of the halogens.

3.17. The elements in column VI of the periodic table are occasionally known as the chalcogens. List these elements by name and symbol.

3.18. List the elements with one-letter symbols. Tell whether each is a metal or nonmetal and whether it is diatomic.

3.19. Write the formulas and names of all compounds mentioned in the discussion of atomic theory (Section 3.2).

*3.20. The atomic composition of the following compounds is given. Write the formula of the compound.

sodium bromide:	1 atom of sodium, 1 atom of bromine
methane:	1 atom of carbon, 4 atoms of hydrogen
aspirin:	9 atoms of carbon, 8 atoms of hydrogen, 4 atoms of oxygen

ammonia: 1 atom of nitrogen, 3 atoms of hydrogen

urea: 2 atoms of nitrogen, 1 atom of carbon, 4 atoms of hydrogen, 1 atom of oxygen

3.21. Formulas of the following compounds are given. State their atomic composition.

sodium hydroxide, NaOH
potassium chloride, KCl
sulfuric acid, H_2SO_4
silver nitrate, $AgNO_3$
calcium carbonate, $CaCO_3$
zinc sulfide, ZnS

3.22. The following compounds may be familiar to you. For each, both its common name and its correct chemical name are given. From the formula, tell the composition of each.

Common name	Chemical name	Formula
dry ice	carbon dioxide	CO_2
blue vitriol	copper sulfate	$CuSO_4$
muriatic acid	hydrochloric acid	HCl

3.4 The Reactions of Elements: Simple Equations

All the equations in this section should be balanced using whole-number coefficients.

3.23. Balance the following equations. (All formulas given are correct.)
a. $H_2 + O_2 \longrightarrow H_2O$
b. $Mg + HCl \longrightarrow H_2 + MgCl_2$
c. $CuSO_4 + Fe \longrightarrow FeSO_4 + Cu$
d. $NaOH + H_2SO_4 \longrightarrow$
$Na_2SO_4 + H_2O$
e. $HBr \longrightarrow H_2 + Br_2$

3.24. The following equations describe reactions that form oxygen. All the formulas given are correct. Balance each equation.

a. $HgO \longrightarrow Hg + O_2$
b. $KClO_3 \longrightarrow KCl + O_2$
c. $Na_2O_2 + H_2O \longrightarrow NaOH + O_2$
d. $H_2O \longrightarrow H_2 + O_2$

*3.25. The following compounds can be prepared by the reaction of oxygen with the appropriate element. Write a balanced equation for the preparation of each of these compounds.
a. calcium oxide, CaO
b. silicon dioxide, SiO_2
c. nitric oxide, NO
d. magnesium oxide, MgO
e. the oxide of phosphorus, P_4O_{10}

*3.26. On heating, the following compounds decompose to elements. Write a balanced equation for each of these reactions.
a. hydrogen iodide, HI
b. silver oxide, Ag_2O
c. phosphorus trichloride, PCl_3

3.27. Using whole-number coefficients, balance the following equations. All formulas are correct:
a. $Ca(OH)_2 + HCl \longrightarrow CaCl_2 + H_2O$
b. $KOH + H_2S \longrightarrow K_2S + H_2O$
c. $NaOH + HNO_3 \longrightarrow$
$NaNO_3 + H_2O$
d. $Mg(OH)_2 + HCl \longrightarrow MgCl_2 + H_2O$
e. $NaOH + CO_2 \longrightarrow Na_2CO_3 + H_2O$

3.28. Using whole-number coefficients, balance the following equations. All formulas are correct.
a. $N_2O_5 + H_2O \longrightarrow HNO_3$
b. $P_4O_{10} + H_2O \longrightarrow H_3PO_4$
c. $CaO + H_2O \longrightarrow Ca(OH)_2$
d. $Li_2O + H_2O \longrightarrow LiOH$
e. $Al_2O_3 + H_2O \longrightarrow Al(OH)_3$

Review Problems

3.29. What volume of bromine ($d = 3.10$ g/mL) will weigh 2.95 kg?

3.30. A cube of osmium measuring 1.5 cm on a side weighs 75.9 g. What is the density of osmium?

*__3.31.__ You have a piece of iron measuring $2.0 \times 4.6 \times 3.2$ cm and a piece of nickel measuring $1.8 \times 2.6 \times 2.1$ cm. The density of iron is 7.9 g/cm^3, and the density of nickel is 8.9 g/cm^3. Which metal piece is heavier and by how much?

3.32. The mileage rating for a car is 35 mi/gal. What is this rating in kilometers per liter?

*__3.33.__ At what temperature are the Fahrenheit and Celsius temperature readings numerically the same but opposite in sign?

▪ 4 ▪

Atomic Structure and the Mole

Modern atomic theory emerged in 1803 with the publication of Dalton's postulates concerning the existence and behavior of atoms (Section 3.2). Since then many scientists using the results of ingenious experiments have refined and elaborated those postulates. We now have a model structure for atoms that describes their composition and explains the relationship between atoms of different elements. This chapter discusses atoms and, in so doing, will:

1. Describe the properties of electrons, protons, and neutrons.
2. Describe how the composition of an atom can be determined from its atomic and mass numbers.
3. Describe the experiment that provided definitive proof of the nuclear structure of an atom.
4. Define atomic weight and show how it is related to the isotopic distribution of an element.
5. Define the mole and show how the concept of moles can be used to relate the mass of a sample to the number of atoms the sample contains.
6. Describe the properties of radioactive nuclei: how they disintegrate, the products of their disintegration, and applications of these processes.

4.1   Subatomic Particles

An atom is very small. Its mass is between 10^{-21} and 10^{-23} g. A row of 10^7 atoms (10,000,000 atoms) extends only 1.0 mm. We know that atoms contain many different **subatomic particles** such as electrons, protons, and neutrons, as well as mesons, neutrinos, and quarks. The atomic model used by chemists requires knowledge of only electrons, protons, and neutrons, so our discussion is limited to them.

A. The Electron

An electron is a tiny particle with a mass of 9.108×10^{-28} g and a negative charge. All neutral atoms contain electrons. The electron was discovered and its properties defined during the last quarter of the nineteenth century. The experiments that proved its existence were studies of the properties of matter in gas-discharge or cathode-ray tubes.

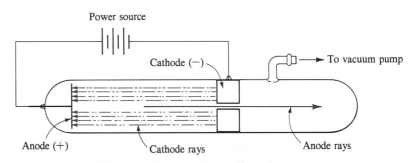

FIGURE 4.1   Diagram of a cathode-ray tube.

Figure 4.1 is a diagram of a cathode-ray tube. This apparatus consists of a glass tube sealed at both ends. Within the tube are two metal plates called electrodes, which are connected to an outside power supply. If the tube is full of air or some other gas, no current flows between the electrodes, regardless of how large a voltage is applied from the power source. If the tube has been partially evacuated before sealing (that is, almost all the gas has been pumped out of it), the application of a high voltage from the power source across the two electrodes gives rise to a glow inside the tube, and, simultaneously, a current begins to flow between the electrodes. We need not discuss in detail the various experiments performed with this apparatus; we will only state the conclusions drawn from them. The current is carried by streams of tiny particles given off by the negative electrode, called the cathode. The positive electrode is called the anode. The tiny particles are called **electrons.**

In these experiments, the presence of these electrons and their properties did not change if the metal of the electrode was changed, nor were any changes observed in their properties when different gases were used in the tube. Eventually, the experimenters became convinced that all matter contains electrons.

Each electron carries a single, negative electric charge and has a mass of 9.108×10^{-28} g. Because the mass of an atom is approximately 10^{-23} g, the mass of an electron is negligible compared to that of an atom.

B. The Proton

Gas-discharge tubes of slightly different design were used to identify small, positively charged particles that moved from the positive electrode (anode) to the negative electrode (cathode). The mass and charge of these particles varied but were always a simple multiple of the mass and charge of the positive particle observed when the gas-discharge tube contained hydrogen. The particle formed from hydrogen is called the **proton.**

The mass of a proton is 1.6726×10^{-24} g, or about 1836 times the mass of an electron. The proton carries a positive electrical charge that is equal in magnitude to the charge of the electron but opposite in sign. All atoms contain one or more protons.

C. The Neutron

The third subatomic particle of interest to us is the **neutron.** Its mass of 1.675×10^{-24} g is very close to that of the proton. A neutron carries no charge. With the exception of the lightest atoms of hydrogen, all atoms contain one or more neutrons.

The properties of these three subatomic particles are summarized in Table 4.1. The third column of the table lists the relative masses of these particles.

TABLE 4.1   Properties of the proton, the neutron, and the electron

Particle	Actual mass (g)	Relative mass (amu)	Relative charge
proton	1.6726×10^{-24}	1.007	$+1$
neutron	1.6749×10^{-24}	1.008	0
electron	$9.108 \ \times 10^{-28}$	5.45×10^{-4}	-1

Because the actual masses of atoms and subatomic particles are so very small, we often describe their masses by comparison rather than in SI units, hence the term *relative mass.* If a proton is assigned a mass of 1.007, then a neutron will have a relative mass of 1.008 and an electron a mass of 5.45×10^{-4}. When talking about relative masses, we use the term **atomic mass unit** (amu). Using this unit, a proton has a mass of 1.007 amu, a neutron a mass of 1.008 amu, and an electron a mass of 5.45×10^{-4} amu. Charges, too, are given relative to one another. If a proton has a charge of $+1$, then an electron has a charge of -1.

4.2   Atomic Structure

A. Atomic Number Equals Electrons or Protons

Each element has an **atomic number.** The atomic numbers are listed along with the names and symbols of the elements on the inside cover of the text. The atomic number equals the charge on the nucleus. It therefore also equals the number of protons in the nucleus and also equals numerically the number of electrons in the neutral atom. The atomic number has the symbol Z.

Different elements have different atomic numbers; therefore, atoms of different elements contain different numbers of protons (and electrons). Oxygen has the atomic number 8; its atoms contain 8 protons and 8 electrons. Uranium has the atomic number 92; its atoms contain 92 protons and 92 electrons.

The relationship between atomic number and the number of protons or electrons can be stated as follows:

$$\text{Atomic number} = \text{number of protons per atom}$$
$$= \text{number of electrons per neutral atom}$$

B. Mass Number Equals Protons plus Neutrons

Each atom also has a mass number, denoted by the symbol A. The **mass number** of an atom is equal to the number of protons plus the number of neutrons that it contains. In other words, the number of neutrons in any atom is its mass number minus its atomic number:

$$\text{Number of neutrons} = \text{mass number} - \text{atomic number}$$

or

$$\text{Mass number} = \text{number of protons} + \text{number of neutrons}$$

The atomic number and the mass number of an atom of an element can be shown by writing, in front of the symbol of the element, the mass number as a superscript and the atomic number as a subscript:

$$^{\text{mass number}}_{\text{atomic number}}\text{Symbol of element} \quad \text{or} \quad ^{A}_{Z}\text{X}$$

For example, an atom of gold (symbol Au), with atomic number 79 and mass number of 196 is denoted as:

$$^{196}_{79}\text{Au}$$

Example 4.1   What is the composition of a silver atom, in terms of protons, neutrons, and electrons?

$$^{107}_{47}\text{Ag}$$

Solution   The atomic number of this atom is given by the subscript 47; it therefore contains 47 protons and 47 electrons. The mass number of this atom is given by the superscript 107; it therefore contains $(107 - 47)$, or 60, neutrons.

Problem 4.1   What is the composition of an atom of phosphorus?

$$^{31}_{15}\text{P}$$

C. Isotopes

Although all atoms of a given element must have the same atomic number, they need not all have the same mass number. For example, some atoms of carbon (atomic number 6) have a mass number of 12, others have a mass number of 13, and still others have a mass number of 14. These different kinds of atoms of the same element are called isotopes. **Isotopes** are atoms that have the same atomic number (and are therefore of the same element) but different mass numbers. The composition of atoms of the naturally occurring isotopes of carbon are shown in Table 4.2.

TABLE 4.2 The naturally occurring isotopes of carbon

Isotope	Protons	Electrons	Neutrons
$^{12}_{6}\text{C}$	6	6	6
$^{13}_{6}\text{C}$	6	6	7
$^{14}_{6}\text{C}$	6	6	8

The various isotopes of an element can be designated by using superscripts and subscripts to show the mass number and the atomic number. They can also be identified by the name of the element with the mass number of the particular isotope. For example, as an alternative to

$$^{12}_{6}\text{C}, \quad ^{13}_{6}\text{C}, \quad \text{and} \quad ^{14}_{6}\text{C}$$

we can write carbon-12, carbon-13, and carbon-14.

About 350 isotopes occur naturally on Earth, and another 1500 have been produced artificially. The isotopes of a given element are by no means equally abundant. For example, 98.89% of all carbon occurring in nature is carbon-12,

1.11% is carbon-13, and only a trace is carbon-14. Some elements have only one naturally occurring isotope. Table 4.3 lists the naturally occurring isotopes of several common elements, along with their relative abundance.

TABLE 4.3   Relative abundance of naturally occurring isotopes of several elements

Isotope	Abundance (%)	Isotope	Abundance (%)
hydrogen-1	99.985	silicon-28	92.21
hydrogen-2	0.015	silicon-29	4.70
hydrogen-3	trace	silicon-30	3.09
carbon-12	98.89	chlorine-35	75.53
carbon-13	1.11	chlorine-37	24.47
carbon-14	trace	phosphorus-31	100
nitrogen-14	99.63	iron-54	5.82
nitrogen-15	0.37	iron-56	91.66
oxygen-16	99.76	iron-57	2.19
oxygen-17	0.037	iron-58	0.33
oxygen-18	0.204	aluminum-27	100

Example 4.2   The three isotopes of hydrogen are:

    a. hydrogen-1 (protium)   **b.** hydrogen-2 (deuterium)

    c. hydrogen-3 (tritium)

Give the symbol and atomic composition of each of these isotopes.

Solution   **a.** $_1^1H$: 1 proton, 0 neutrons, 1 electron

    b. $_1^2H$: 1 proton, 1 neutron, 1 electron

    c. $_1^3H$: 1 proton, 2 neutrons, 1 electron

Problem 4.2   Naturally occurring uranium is 99.3% uranium-238 and 0.7% uranium-235. Give the composition of an atom of each isotope.

D. The Inner Structure of the Atom

So far, we have discussed electrons, protons, and neutrons and ways to determine how many of each a particular atom contains. The question remains: Are these particles randomly distributed inside the atom like blueberries in a muffin, or does an atom have some organized inner structure? At the beginning of the twentieth century, scientists were trying to answer this question. Various theories had been proposed, but none had been verified by experiment. In our discussion of the history of science, we suggested that, at various points in its development, science has marked time until someone performed a key experiment that provided new insights. In the history of the study of atoms, a key experiment was performed in 1911 by Ernest Rutherford and his colleagues.

1. Forces between bodies

Our understanding of the conclusions drawn from Rutherford's experiment depends on a knowledge of the forces acting between bodies. Therefore, before discussing his experiment, a brief review of these forces is in order. First is the force of **gravity** that exists between all bodies. Its magnitude depends on the respective masses and on the distance between the centers of gravity of the two interacting bodies. You are familiar with gravity; it acts to keep your feet on the ground and the moon in orbit. **Electrical forces** also exist between charged particles. The magnitude of the electrical force between two charged bodies depends on the charge on each body and on the distance between their centers. If the charges are of the same sign (either positive or negative), the bodies repel each other; if the charges are of opposite sign, the bodies attract each other. **Magnetic forces,** a third type, are similar to electrical forces. Each magnet has two poles—a north pole and a south pole. When two magnets are brought together, a repulsive force exists between the like poles and an attractive force between the unlike poles. The magnetic and electrical forces can interact in the charged body. These three forces were known at the end of the nineteenth century when the structure of the atom came under intensive study.

2. Rutherford's experiment

Let us describe **Rutherford's experiment.** In 1911, it was generally accepted that the atom contained electrons and protons but that they were probably not arranged in any set pattern. Rutherford wished to establish whether a pattern existed. He hoped to gain this information by studying how the protons in the atom deflected the path of another charged particle shot through the atom. For his second particle, he chose alpha (α) particles. An **alpha particle** contains two protons and two neutrons, giving it a relative mass of 4 amu and a charge of $+2$. An alpha particle is sufficiently close in mass and charge to a proton that its path would be changed if it passed close to the proton.

In the experiment, a beam of alpha particles was directed at a piece of gold foil, so thin as to be translucent and, more importantly for Rutherford, only a few atoms thick. The foil was surrounded by a zinc sulfide screen that flashed each time it was struck by an alpha particle. By plotting the location of the flashes, it would be possible to determine how the path of the alpha particles through the atom was changed by the protons in the atom.

The three paths shown in Figure 4.2 (paths A, B, and C) are representative of those observed. Most of the alpha particles followed path A; they passed directly through the foil as though it were not there. Some were deflected slightly from their original path, as in path B; and an even smaller number bounced back from the foil as though they had hit a solid wall (path C).

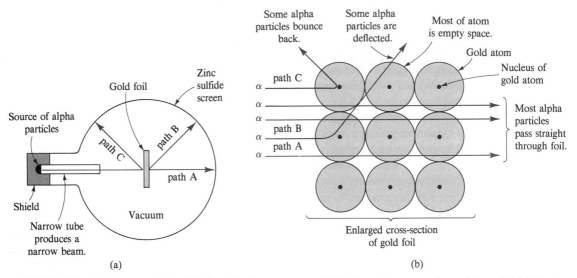

FIGURE 4.2   (a) Cross-section of Rutherford's apparatus. (b) Enlarged cross-section of the gold foil in the apparatus, showing the deflection of alpha particles by the nuclei of the gold atoms.

Although you may be surprised that any alpha particles passed through the gold foil, Rutherford was not. He had expected that many would pass straight through (path A). He had also expected that, due to the presence in the atom of positively charged protons, some alpha particles would follow a slightly deflected path (path B). The fact that some alpha particles bounced back (path C) is what astounded Rutherford and his co-workers. Path C suggested that the particles had smashed into a region of dense mass and had bounced back. To use Rutherford's analogy, the possibility of such a bounce was as unlikely as a cannonball bouncing off a piece of tissue paper.

3. Results of the experiment

Careful consideration of the results and particularly of path C convinced Rutherford (and the scientific community) that an atom contains a very small, dense nucleus and a large amount of extranuclear space. According to Rutherford's theory, the **nucleus of an atom** contains all the mass of the atom and therefore all the protons. The protons give the nucleus a positive charge. Because like charges repel each other, positively charged alpha particles passing close to the nucleus are deflected (path B). The nucleus, containing all the protons and neutrons, is more massive than an alpha particle; therefore, an alpha particle striking the nucleus of a gold atom bounces back from the collision, as did those following path C.

Outside the nucleus, in the relatively enormous extranuclear space of the atom, are the tiny electrons. Because electrons are so small relative to the space they occupy, the extranuclear space of the atom is essentially empty. In Rutherford's experiment, alpha particles encountering this part of the atoms in the gold foil passed through the foil undeflected (path A).

If the nucleus contains virtually all the mass of the atom, it must be extremely dense. Its diameter is about 10^{-12} cm, about 1/10,000 that of the whole atom. Given this model, if the nucleus were the size of a marble, the atom with its extranuclear electrons would be 300 m in diameter. If a marble had the same density as the nucleus of an atom, it would weigh 3.3×10^{10} kg.

This model of the nucleus requires the introduction of a force other than those discussed earlier, one that will allow the protons, with their mutually repelling positive charges, to be packed close together in the nucleus, separated only by the uncharged neutrons. These **nuclear forces** seem to depend on interactions between protons and neutrons. Some are weak and some are very strong. Together they hold the nucleus together, but they are not yet understood.

The model of the atom based on Rutherford's work is, of course, no more than a model; we cannot see these subatomic particles or their arrangement within the atom. However, this model does give us a way of thinking about the atom that coincides with observations made about its properties. We can now determine not only what subatomic particles a particular atom contains but also whether or not they are in its nucleus. For example, an atom of carbon-12

$$^{12}_{6}C$$

contains 6 protons and 6 neutrons in its nucleus and 6 electrons outside the nucleus.

Example 4.3 Iodine-131 is used in thyroid therapy. What is the composition of an atom of this isotope? In which part of the atom are these particles located?

Solution The atomic number of iodine is 53 (from the table of the elements). Therefore, an atom of iodine-131 contains 53 protons in the nucleus and 53 electrons outside the nucleus. The mass number of this isotope is 131 (from its name). Hence, in addition to 53 protons, the nucleus of an atom of iodine-131 contains $(131 - 53)$, or 78, neutrons.

Problem 4.3 Cobalt-60 is used in cancer therapy. What is the composition of an atom of this isotope? In which part of the atom are these particles located?

We have two distinct parts of an atom — the nucleus and the extranuclear space. The nucleus of an atom does not play any role in chemical reactions, but it does participate in radioactive reactions. (Such reactions are discussed later in this chapter.) The chemistry of an atom depends on its electrons — how many there are and how they are arranged in the extranuclear space.

| 4.3 | **Atomic Weights** |

The **atomic weight** (or atomic mass) of an element is the average relative mass of the naturally occurring atoms of that element. Both the periodic table and the alphabetical list of the elements show the atomic weights of the elements. The atomic weight of an element is based on the variety of naturally occurring isotopes of that element and the relative abundance of each.

A collection of naturally occurring carbon atoms contains 98.89% carbon-12 atoms and 1.11% carbon-13 atoms, along with a trace percentage of carbon-14 atoms. The atomic weight of carbon (12.01) reflects the relative abundance of these three isotopes. The atomic weight of chlorine (35.45) reflects the fact that 75.53% of naturally occurring chlorine is chlorine-35 and 24.47% is chlorine-37.

The atomic weight of some elements is given as a whole number enclosed in parentheses. These elements are unstable; that is, their nuclei decompose radioactively. The number in parentheses is the mass number of the most stable or best-known isotope of that element.

Atomic weights are measured in atomic mass units. One atomic mass unit is defined as $\frac{1}{12}$ the mass of an atom of carbon-12. With this reference standard, no element has an atomic weight less than unity. The approximate atomic weight of an element can be calculated if the relative abundance of its isotopes is known.

Example 4.4

Naturally occurring rubidium is 72.15% rubidium-85, with an atomic mass of 84.91 amu, and 27.85% rubidium-87, with an atomic mass of 86.79 amu. What is the approximate atomic weight of rubidium?

Solution

Wanted
The approximate atomic weight, or the average mass of a naturally occurring atom of rubidium

Given
Rubidium occurs as 72.15% rubidium-85 and 27.85% rubidium-87. Suppose you have a sample of naturally occurring rubidium that contains 10,000 atoms. (By using 10,000 atoms, our calculations will deal only in whole numbers of atoms.) You can calculate how many of the atoms are rubidium-85.

$$\frac{72.15}{100} \times 10,000 \text{ atoms} = 7,215 \text{ atoms of rubidium-85}$$

The total mass of these atoms is:

$$7,215 \text{ atoms} \times \frac{84.91 \text{ amu}}{1 \text{ atom}} = 6.126 \times 10^5 \text{ amu}$$

You can also calculate the number of atoms of rubidium-87 and the total mass of these atoms.

$$\frac{27.85}{100} \times 10{,}000 \text{ atoms} = 2{,}785 \text{ atoms of rubidium-87}$$

$$2{,}785 \text{ atoms} \times \frac{86.79 \text{ amu}}{1 \text{ atom}} = 2.417 \times 10^5 \text{ amu}$$

The total mass of the sample is:

$$6.126 \times 10^5 \text{ amu} + 2.417 \times 10^5 \text{ amu} = 8.543 \times 10^5 \text{ amu}$$

The average mass of a rubidium atom in the sample is:

$$\frac{8.543 \times 10^5 \text{ amu}}{10{,}000 \text{ atoms}} = 85.43 \text{ amu} \quad \text{(per atom)}$$

Problem 4.4 Naturally occurring copper is 69.09% copper-63, with an atomic mass of 62.93 amu, and 30.91% copper-65, with an atomic mass of 64.93 amu. Calculate the approximate atomic weight of copper.

The next example illustrates a second way of calculating atomic weight from isotopic distribution.

Example 4.5 Naturally occurring bromine is 50.54% bromine-79, with an atomic mass of 78.92 amu, and 49.46% bromine-81, with an atomic mass of 80.92. Calculate the approximate atomic weight of bromine.

Solution

Wanted
The approximate average mass of naturally occurring bromine atoms

Given
Of the total collection of bromine atoms, 50.54% will have a mass of 78.92 amu. Their contribution to the total mass of the bromine will be:

$$\frac{50.54}{100} \times 78.92 \text{ amu} = 39.89 \text{ amu}$$

The other 49.46% will have a mass of 80.92 amu. Their contribution to the mass will be:

$$\frac{49.46}{100} \times 80.92 \text{ amu} = 40.02 \text{ amu}$$

The total mass will be $39.89 + 40.02 = 79.91$ amu, the approximate atomic weight of bromine or the average mass of a naturally occurring bromine atom.

Problem 4.5   Element 114 is a heavy element not yet found to be naturally occurring. Suppose that it is found and that 64% of it has a mass number of 288 and 36% has a mass number of 280. What would be the appropriate atomic weight of element 114?

■

If the identity of the naturally occurring isotopes and the atomic weight of an element are known, it is possible to estimate which of the isotopes is most abundant.

■

Example 4.6   The atomic weight of magnesium is 24.31. Magnesium occurs naturally as magnesium-24, magnesium-25, and magnesium-26. Which of these isotopes is most plentiful in nature?

Solution   The atomic weight of an element is the average mass of its naturally occurring atoms. The atomic weight of magnesium is slightly above 24 and much less than 25 or 26. The isotope magnesium-24 must be the most abundant in nature.

Problem 4.6   The atomic weight of neon is 20.182. Neon occurs in nature as neon-20, neon-21, and neon-22. Which of these isotopes is most plentiful in nature?

■

4.4   The Mole

As we have observed, atoms are very small—too small to be weighed or counted individually. Nevertheless, we often need to know how many atoms (or molecules, or electrons, and so on) a sample contains. To solve this dilemma, we use a counting unit called **Avogadro's number,** named after the Italian scientist Amedeo Avogadro (1776–1856):

Avogadro's number $= 6.02 \times 10^{23}$

Just as an amount of 12 is described by the term *dozen,* Avogadro's number is described by the term **mole.** A dozen eggs is 12 eggs; a mole of atoms is 6.02×10^{23} atoms. Avogadro's number can be used to count anything. You could have a mole of apples or a mole of Ping-Pong balls. You can get some idea of the magnitude of Avogadro's number by considering that a mole of Ping-Pong balls would cover the surface of the Earth with a layer approximately 60 miles thick. Avogadro's number is shown here to three significant figures, which is the degree of accuracy usually required in calculations. Actually, the number of items in a mole has been determined to six or more significant figures, the exact number depending on the method by which the number was determined.

One mole of any substance contains 6.02×10^{23} units of that substance. Equally important is the fact that one mole of a substance has a mass in grams numerically equal to the formula weight of that substance. Thus, one mole of an element has a mass in grams equal to the atomic weight of that element and contains 6.02×10^{23} atoms of the element. For those elements that do not occur as single atoms — that is, the diatomic gases, sulfur, and phosphorus — it is important to be certain that you specify what you are talking about. One mole of atoms of oxygen has a mass of 16 g, as 16 is the atomic weight of oxygen, and contains 6.02×10^{23} atoms of oxygen. One mole of oxygen gas, which has the formula O_2, has a mass of 32 g and contains 6.02×10^{23} molecules of oxygen but 12.04×10^{23} ($2 \times 6.02 \times 10^{23}$) atoms, because each molecule of oxygen contains two oxygen atoms.

These definitions allow a new definition of atomic weight: The atomic weight of an element is the mass in grams of one mole of naturally occurring atoms of that element.

$$\text{Formula weight} = \frac{\text{grams}}{\text{mole}}$$

Using these relationships, we can calculate the number of atoms in a given mass of an element or the mass of a given number of atoms.

Example 4.7

The atomic weight of sulfur is 32.06. How many moles of sulfur are contained in 5.05 g sulfur?

Solution

Wanted
Moles of sulfur (? mol)

Given
5.05 g S

Conversion factor
1 mol S = 32.06 g S

Equation
$$? \text{ mol S} = 5.05 \text{ g S} \times \frac{1 \text{ mol S}}{32.06 \text{ g S}}$$

Answer
0.158 mol S

handwritten:
$1.62 \text{ C} \times \frac{12.01 \text{ g C}}{1 \text{ m C}} =$

$12.01 \text{ g C} = 1 \text{ mole C} = 6.02 \times 10^{23} \text{ C atom}$

$\text{Given} \times \text{C.F.} = \text{Ans}$ How many atoms

$\dfrac{7.83 \text{ g} \times 6.02 \times 10^{23}}{12.01 \text{ g C}} = \text{Ans}$

Problem 4.7

The atomic weight of carbon is 12.01. What is the mass of 1.62 mol carbon?

Example 4.8

What mass of copper contains 5.14×10^{22} atoms of copper?

Solution

Wanted
Mass of copper (? g Cu)

Given

5.14×10^{22} atoms Cu

Conversion factors

  1 mole of copper weighs 63.54 g

  1 mole of anything contains 6.02×10^{23} items

Equation

$$? \text{ g Cu} = 5.14 \times 10^{22} \text{ atoms Cu} \times \frac{1 \text{ mol atoms}}{6.02 \times 10^{23} \text{ atoms}} \times \frac{63.54 \text{ g Cu}}{1 \text{ mol Cu}}$$

Answer

5.43 g Cu

Problem 4.8   A sample of uranium contains 5.15×10^{20} atoms. What is the mass of the sample?

Example 4.9   Calculate the number of atoms of lead in 4.26 lb lead.

Solution

Wanted

? atoms Pb

Given

4.26 lb Pb

Conversion factors

  1.00 lb = 453.6 g (Table 2.1)

  1.00 mol Pb = 207.19 g Pb (atomic weight)

  1.00 mole contains 6.02×10^{23} atoms (Avogadro's number)

Equation

$$? \text{ atoms Pb} = 4.26 \text{ lb Pb} \times \frac{453.6 \text{ g}}{1.00 \text{ lb}} \times \frac{1.00 \text{ mol Pb}}{207.19 \text{ g Pb}} \times \frac{6.02 \times 10^{23} \text{ atoms}}{1.00 \text{ mol}}$$

$$= 5.61 \times 10^{24} \text{ atoms Pb}$$

Problem 4.9   Calculate the number of atoms in 9.86 lb copper.

4.5   Radioactivity

A. General Characteristics

From the discussions in the previous section, we know that the atoms of any element have two distinct parts: the nucleus, which contains the protons and neutrons, and the extranuclear space, which contains the electrons. The electrons in the atom, particularly those farthest from the nucleus, determine the chemical properties of the element. We will discuss electrons and the chemical properties of elements in detail in Chapter 5. In the remainder of this chapter, we will describe properties of the nucleus and, in particular, the characteristics of nuclear decay, which is also called **radioactivity** or radioactive decay of the nucleus.

In nuclear decay, the nuclei of radioactive atoms decay spontaneously to form other nuclei, a process that always results in a loss of energy and often involves the release of one or more small particles. Some atoms are naturally radioactive. Others that are normally stable can be made radioactive by bombarding them with subatomic particles. Often, one isotope of an element is radioactive and others of the same element are stable. A radioactive isotope is called a **radioisotope.**

Radioactivity is a common phenomenon. Of the 350 isotopes known to occur in nature, 67 are radioactive. Over a thousand radioactive isotopes have been produced in the laboratory. Every element, from atomic number 1 through number 109, has at least 1 natural or artificially produced radioactive isotope. Of the 3 known isotopes of hydrogen, one is radioactive — hydrogen-3, more commonly known as tritium. Oxygen, the Earth's most abundant element, has 8 known isotopes, 5 of which are radioactive (oxygen-13, -14, -15, -19, and -20). Iodine, an element widely used in nuclear medicine, has 24 known isotopes ranging in mass from 117 to 139 amu. Of these, only iodine-127 is stable; this isotope is the only naturally occurring one. Uranium has 14 known isotopes, all of which are radioactive.

B. Radioactive Emissions

Nuclei undergoing nuclear decay release various kinds of emissions. We will discuss three of these emissions: alpha particles, beta particles, and gamma rays. All three are forms of **ionizing radiation,** so called because their passage through matter leaves a trail of ions and molecular debris.

1. Alpha (α) particles

An alpha particle is identical to a helium atom that has been stripped of its two electrons; thus, an alpha particle contains two protons and two neutrons. Because an alpha particle has no electrons to balance the positive charge of the two protons, it has a charge of $+2$ and can be represented as He^{2+}. If a particle has a charge, whether negative or positive, it can be shown as a superscript. Thus He^{2+} means a helium atom that has lost two electrons and has a $+2$ charge. The symbol O^{2-} means an oxygen atom that has added two electrons and thus has a charge of -2. Atoms that have acquired a charge by losing or gaining electrons are called **ions.**

Besides He^{2+}, other symbols for this particle are

$$^4_2He \quad \text{and} \quad ^4_2\alpha$$

When ejected from a decaying nucleus, alpha particles interact with all matter in their path, whether it be photographic film, lead shielding, or body tissue, stripping electrons from other atoms as they go. In their wake, they leave a trail of positive ions (atoms from which electrons have been removed) and free electrons. A single alpha particle, ejected at high speed from a nucleus, can create up to 100,000 ions along its path before it gains two electrons to become a neutral helium atom.

In air, an alpha particle travels about 4 cm before gaining the two electrons. Within body tissue, its average path is only a few thousandths of a centimeter. An alpha particle is unable to penetrate the outer layer of human skin. Because of this limited penetrating power, external exposure to alpha particles is not nearly as serious as internal exposure. If a source of alpha emissions is taken internally, the alpha radiation can do massive damage to the surrounding tissue; therefore alpha emitters are never used in nuclear medicine.

2. Beta (β) particles

A **beta particle** is a high-speed electron ejected from a decaying nucleus; it carries a charge of -1. (Section 4.6A discusses how a nucleus can eject an electron even though it does not contain electrons.) A beta particle is represented as

$$_{-1}^{0}e \quad \text{or} \quad _{-1}^{0}\beta$$

Like alpha particles, beta particles cause the formation of ions by interacting with whatever matter is in their path. Beta particles are far less massive than alpha particles and carry a charge with only half the magnitude of that of the alpha particle. (This property depends only on the size of the charge, not its sign.) Thus beta particles produce less ionization and travel farther through matter before combining with a positive ion to become a neutral particle. The path of a beta particle in air can be 100 times that of an alpha particle. About 25 cm of wood, 1 cm of aluminum, or 0.5 cm of body tissue will stop a beta particle.

Because beta particles cause less ionization than alpha particles, beta particles are more suitable for use in radiation therapy, since the likelihood of damage to healthy tissue is greatly reduced. Beta emitters such as calcium-46, iron-59, cobalt-60, and iodine-131 are widely used in nuclear medicine.

3. Gamma (γ) rays

The release of either alpha or beta particles from a decaying nucleus is generally accompanied by the release of nuclear energy in the form of gamma rays, represented as

$$_{0}^{0}\gamma$$

Gamma rays have no charge or mass and are similar to X rays. Even though they bear no charge, gamma rays are able to produce ionization as they pass through matter. The degree of penetration of gamma rays through matter is much greater than that of either alpha or beta particles. The path length of a gamma ray can be as much as 400 m in air and 50 cm through tissue. Because of their penetrating power, gamma rays are especially easy to detect. Virtually all radioactive isotopes used in diagnostic nuclear medicine are gamma emitters. Each of the beta emitters listed in the previous paragraph is also a gamma emitter. Additional gamma emitters commonly used in nuclear medicine include chromium-51, arsenic-74, technetium-99, and gold-198.

The characteristics of alpha particles, beta particles, and gamma rays are summarized in Table 4.4.

TABLE 4.4 Characteristics of radioactive emissions

Name	Symbol	Charge	Mass (amu)	Penetration through matter
alpha particle	$^4_2\alpha$	$+2$	4	4.0 cm air 0.005 cm tissue no penetration through lead
beta particle	$^0_{-1}e$ or $^0_{-1}\beta$	-1	5.5×10^{-4}	6–300 cm air 0.006–0.5 cm tissue 0.0005–0.03 cm lead
gamma ray	$^0_0\gamma$	0	0	400 m air 50 cm tissue 3 cm lead

4.6 Characteristics of Nuclear Reactions

A. Equations for Nuclear Reactions

Radioactivity is the decay or disintegration of the nucleus of an atom. During the process, either alpha or beta particles may be emitted. Energy, in the form of gamma rays, may also be released by this process, and a different atom is formed. This new atom may be of a different element, or a different isotope of the same element. All of these characteristics and more can be shown by using an equation to describe the radioactive process.

Like a chemical equation, a nuclear equation must be balanced. First, the total mass of the products must equal the total mass of the reactants. Second, the total charge of the reactants (the sum of their atomic numbers) must equal the total charge of the products.

Consider the equation for the decay of radium-226 to radon-222, with the simultaneous loss of an alpha particle and energy in the form of a gamma ray. Radium-226 is the reactant; radon, an alpha particle, and a gamma ray are the products. The equation is:

$$\underset{\text{radium-226}}{^{226}_{88}\text{Ra}} \longrightarrow \underset{\text{radon-222}}{^{222}_{86}\text{Rn}} + \underset{\substack{\text{alpha}\\\text{particle}}}{^{4}_{2}\text{He}} + \underset{\text{energy}}{^{0}_{0}\gamma}$$

In the notation for particles, the superscript shows the mass of the particle, and the subscript shows the charge. The charge on each of these particles is its atomic number. The equation is balanced with respect to mass because the sum of the masses of the reactants (226) equals the sum of the masses of the products (222 + 4 + 0). The equation is balanced with respect to charge because the sum of the atomic numbers of the reactants (88) equals the sum of the atomic numbers of the products (86 + 2 + 0). The energy change accompanying the reaction is shown by the release of gamma rays.

A similar equation can be written for nuclear decay by beta emission. Iodine-131 is a beta emitter commonly used in nuclear medicine. The equation for its decay is:

$$^{131}_{53}I \longrightarrow \, ^{131}_{54}Xe + \, ^{0}_{-1}\beta + \, ^{0}_{0}\gamma$$

Note that both the charge and the mass are balanced and that iodine-131 emits both a gamma ray and a beta particle. For this reason, iodine-131 is known as a beta-gamma emitter. Carbon-14, the isotope widely used in radiodating of archaeological artifacts containing carbon, is also a beta emitter:

$$^{14}_{6}C \longrightarrow \, ^{14}_{7}N + \, ^{0}_{-1}\beta$$

How can nuclei give off beta particles (high-energy electrons) if the nucleus has no electrons? The process is not yet clearly understood, but it may occur through the disintegration of a neutron to form a proton and the emitted electron:

$$\text{neutron} \longrightarrow \text{proton} + \text{electron}$$
$$^{1}_{0}n \longrightarrow \, ^{1}_{1}H + \, ^{0}_{-1}e$$

The electron is ejected and the proton remains in the nucleus. In beta emission, the atomic number of the product nucleus is one greater than that of the reactant nucleus because the nucleus now contains one more proton. The mass of the product nucleus is approximately the same as that of the reactant nucleus because an electron's mass is negligible with respect to that of a proton.

Emission of a gamma ray changes neither the mass nor the charge of the nucleus. It accompanies the rearrangement of a nucleus from a less stable, more energetic nuclear configuration to a more stable, less energetic form. The identity and mass of the nucleus stay the same. The changes caused by the emission of the three types of radiation are summarized in Table 4.5.

Given the atomic number and mass number of a radioactive isotope and the type of radiation emitted during its decay, we can easily predict the mass number, atomic number, and identity of the new element formed.

Example 4.10   Cobalt-60 decays by emission of a beta particle. Predict the atomic number and mass number of the isotope formed. Which element has this atomic number?

Solution First, write an equation showing the radioactive decay.

$$^{60}_{27}\text{Co} \longrightarrow \; ^{\text{mass number}}_{\text{atomic number}}\text{element} + ^{0}_{-1}\beta$$

Second, determine the mass number and atomic number of the new isotope. Because mass number is unchanged in beta emission (see Table 4.5), the isotope formed must also have a mass number of 60. The atomic number increases by 1 in beta emission (Table 4.5), so the isotope formed must have an atomic number of 28. Therefore, write:

$$^{60}_{27}\text{Co} \longrightarrow \; ^{60}_{28}\text{element} + ^{0}_{-1}\beta$$

Third, consult a table of the elements and find the element whose atomic number is 28; this element is nickel. Thus, the complete equation for the radioactive decay of cobalt-60 is:

$$^{60}_{27}\text{Co} \longrightarrow \; ^{60}_{28}\text{Ni} + ^{0}_{-1}\beta$$

Problem 4.10 Radium-226 decomposes with the loss of an alpha particle. Write the equation for this reaction and identify the atom produced.

TABLE 4.5 **Changes in atomic number and mass number resulting from the emission of an alpha particle, beta particle, or gamma ray**

	Change in	
Radiation emitted	*Atomic number*	*Mass number*
alpha particle	-2	-4
beta particle	$+1$	0
gamma ray	0	0

B. Half-Life

The rate of decay of a radioactive isotope (also called a radioisotope) is measured in terms of its half-life. **Half-life** is defined as the length of time required for half the sample to decay. The length of half-lives varies from fractions of a second for some isotopes to billions of years for others. Table 4.6 lists half-lives and modes of decay for several isotopes.

Iodine-131 has a half-life of 8.1 days. If you start today with a 25-mg sample of iodine-131, after 8.1 days that sample will contain only 12.5 mg iodine-131. At the end of 16.2 (2×8.1) days, the sample will contain only 6.25 mg iodine-131. Of course, the matter in the sample does not disappear; it

Isotope	Emissions	Half-life
hydrogen-3	$_{-1}^{0}\beta$	12.3 years
carbon-14	$_{-1}^{0}\beta$	5730 years
calcium-47	$_{-1}^{0}\beta, _{0}^{0}\gamma$	4.5 days
cobalt-60	$_{-1}^{0}\beta, _{0}^{0}\gamma$	5.26 years
gold-198	$_{-1}^{0}\beta, _{0}^{0}\gamma$	2.7 days
iodine-131	$_{-1}^{0}\beta, _{0}^{0}\gamma$	8.1 days
iron-59	$_{-1}^{0}\beta, _{0}^{0}\gamma$	45.1 days
molybdenum-99	$_{-1}^{0}\beta, _{0}^{0}\gamma$	67.0 hours
phosphorus-32	$_{-1}^{0}\beta$	14.3 days
sodium-24	$_{-1}^{0}\beta, _{0}^{0}\gamma$	15.0 hours

TABLE 4.6   Half-lives of some radioisotopes

changes to another element, the product of the radioactive decay of iodine-131. Figure 4.3 shows the amounts of iodine-131 remaining after the passage of several half-lives, given an initial sample containing 25 mg of the isotope.

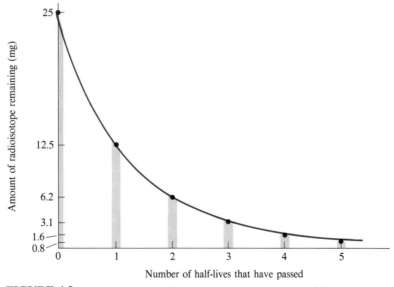

FIGURE 4.3   Rate of decay of iodine-131 as a function of time.

Knowing the identity of a radioisotope, its half-life, and the type of radiation it emits, you can determine the identity of the product and calculate the amount formed in a given period of time.

Example 4.11   Phosphorus-32 is a beta emitter with a half-life of 14.3 days.

a. Write the equation for the nuclear decay of phosphorus-32.

b. A package containing 5 mg phosphorus-32 was in transit for 28 days and 14.5 hours. How much phosphorus-32 did the package contain on arrival?

Solution   a. In beta emission, the mass number remains the same and the atomic number increases by 1. The element with atomic number 16 is sulfur. Thus, the equation is:

$$^{32}_{15}P \longrightarrow {}^{32}_{16}S + {}^{0}_{-1}\beta$$

b. How many half-lives have elapsed?

$$28 \text{ days } | \text{ } 14.5 \text{ hours} \left(\frac{1 \text{ day}}{24 \text{ hours}} \right) = 28.6 \text{ days}$$

$$28.6 \text{ days} \times \frac{1 \text{ half-life}}{14.3 \text{ days}} = 2 \text{ half-lives}$$

Because 5 mg phosphorus-32 would decrease by half during the first half-life (to 2.50 mg) and the remaining amount would decrease by half in the second half-life, only one-fourth (1.25 mg) of the original sample would be phosphorus-32. The rest of the sample would be sulfur-32.

Problem 4.11   Strontium-90, a major product of the disintegration of uranium, decays by beta emission and has a half-life of 28 years. The equation for its decay is:

$$^{90}_{38}Sr \longrightarrow {}^{90}_{39}Y + {}^{0}_{-1}\beta$$

Once absorbed into the bones, strontium-90 emits beta rays, damaging the bone marrow. The maximum permissible dose of strontium-90 for an adult is 6.9×10^{-9} g. If a man received this dose at age 20, and it was all incorporated into his bone tissue, what mass of strontium-90 would be found in his body after death at the age of 104?

4.7   Applications of Radioactivity

The use of radioisotopes is widespread in chemistry, biology, medicine, and other areas of science and industry. In all applications of radioactivity, the following characteristics of nuclear decay must be considered:

1. The chemical properties and reactions of a radioisotope are exactly the same as those of a nonradioactive isotope of the same element.

2. Radiation can be detected some distance from its source.

3. Each radioisotope has a characteristic half-life.

4. Radioactive emissions interfere with normal cell growth.

Several uses of radioisotopes in the health and biological sciences illustrate ways in which they can provide information that would be difficult or impossible to obtain by any other means.

A. The Use of Tracers

Because the chemical properties and reactions of a radioisotope are exactly the same as those of a nonradioactive isotope of the same element, a radioisotope can be substituted for a stable isotope of the same element in a molecule or compound, without changing the chemical properties of the compound. Such a compound is said to be "tagged" or "labeled." Because the radiation emitted by the radioisotope can be detected some distance from the radiating atom, the progress of those atoms through the body can be followed, or traced. Hence, such labeled compounds are called **tracers.**

Iodine-131 is the most commonly used radioisotope for the study of iodine metabolism in humans. The thyroid gland has a remarkable ability to extract iodine from the bloodstream and use it to produce the thyroid hormones, thyroxine and triiodothyronine. These two hormones have a direct effect on the body's metabolism. Very small amounts of iodine-131, in the form of sodium iodide, can be injected into the bloodstream and, within minutes, will begin to concentrate in the thyroid gland. By monitoring the accumulation of radioactivity, it is possible to estimate the size and shape of the thyroid gland and to determine whether any part of it is functioning abnormally (Figure 4.4).

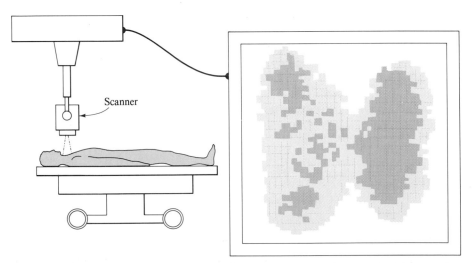

Scanner

FIGURE 4.4   The uptake of iodine by the thyroid gland can be measured by tracing atoms of the radioisotope iodine-131. The plot at the right of the figure shows the location in the thyroid of the source of each radioactive emission detected by the counter. Note that the radioactivity counts show more iodine in one lobe of the thyroid gland than in the other.

Several other radioisotopes commonly used in nuclear medicine are listed in Table 4.7. Notice that all have comparatively short half-lives and that all are pure gamma or beta-gamma emitters (see Section 4.5B).

TABLE 4.7   Radioisotopes used in nuclear medicine for clinical diagnosis

Radioisotope	Radiation Particle emitted	Half-life	Used to study
chromium-51	$_0^0\gamma$	27.8 days	red blood cells
cobalt-57	$_{-1}^0\beta,\ _0^0\gamma$	270 days	absorption, storage, and metabolism of vitamin B-12
iron-59	$_{-1}^0\beta,\ _0^0\gamma$	45.1 days	red blood cells
strontium-87	$_0^0\gamma$	2.8 hours	bone metabolism

B. Biological Effects of Radiation

Although the exact manner in which radiation damages tissues and cells is not fully understood, cellular damage clearly can occur any time alpha or beta particles or other ionizing radiation passes through the cell. The effects of ionizing radiation range from minor damage of specific molecules to the death of cells. Ionizing radiation is especially damaging to the cell nucleus— particularly nuclei undergoing rapid division and nuclei of younger, less mature cells. Many types of cancer cells are especially sensitive to gamma radiation because they are growing rapidly and are therefore less mature than cells of surrounding noncancerous tissue. This sensitivity is the reason behind the use of radiation to destroy cancer cells. The apparatus used for such treatments is designed to focus the radiation sharply on the cancer cells, thus minimizing the damage to nearby healthy cells (Figure 4.5).

4.8   The Use of Nuclear Reactions to Produce Energy

In Section 4.7 we listed four characteristics of radioactivity and nuclear decay that form the basis for the use of radioisotopes in the health and biological sciences. A fifth characteristic of nuclear reactions is that they release enormous amounts of energy. The first nuclear reactor to achieve controlled nuclear disintegration was built in the early 1940s by Enrico Fermi and his colleagues at the University of Chicago. Since that time, a great deal of effort and expense has gone into developing nuclear reactors as a source of energy. The nuclear reactions presently used or studied by the nuclear power industry fall into two categories: fission reactions and fusion reactions.

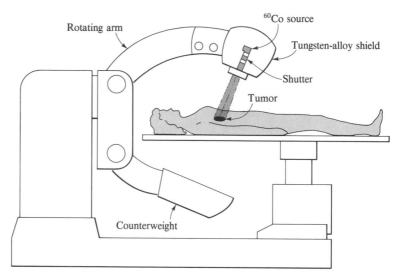

FIGURE 4.5   Cancer treatment with cobalt-60. The source moves along a circular track, rotating the radioactive beam around the patient, so that only the tumor receives continuous radiation.

A. Fission

In **nuclear fission,** a large nucleus is split into two medium-sized nuclei. Only a few nuclei are known to undergo fission. Nuclear power plants currently in use depend primarily on the fission of uranium-235 and plutonium-239.

When a nucleus of uranium-235 undergoes fission, it splits into two smaller atoms and, at the same time, releases neutrons ($_0^1$n) and energy. Some of these neutrons are absorbed by other atoms of uranium-235. In turn, these atoms split apart, releasing more energy and more neutrons. A typical reaction is:

$$_{92}^{235}U \; + \; _0^1n \longrightarrow [_{92}^{236}U] \longrightarrow \; _{56}^{139}Ba + \; _{36}^{94}Kr + 3\; _0^1n + energy$$

The brackets around $_{92}^{236}U$ indicate that it has a highly unstable nucleus. Under proper conditions, the fission of a few nuclei of uranium-235 sets in motion a chain reaction (Figure 4.6) that can proceed with explosive violence if not controlled. In fact, this reaction is the source of energy in the atomic bomb.

In nuclear power plants, the energy released by the controlled fission of uranium-235 is collected in the reactor and used to produce steam in a heat exchanger. The steam then drives a turbine to produce electricity. Energy generation can be regulated by inserting control rods between the fuel rods in the reactor to absorb excess neutrons, thereby controlling the rate of the chain reaction. A typical nuclear power plant in operation today uses about 2 kg

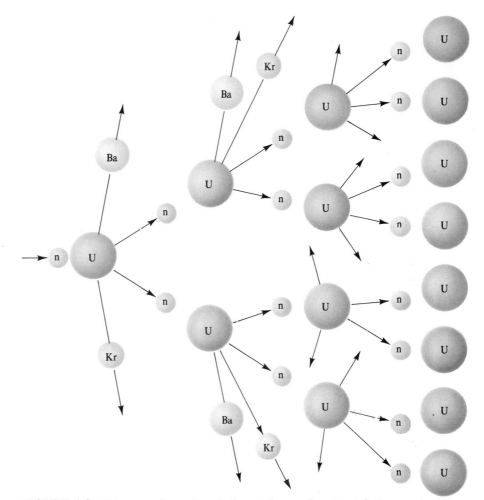

FIGURE 4.6   Diagram of a nuclear fission chain reaction. Each fission results in two (or more) neutrons that can react with other uranium atoms so that the number of nuclear fissions occurring soon reaches an enormous number.

uranium-235 to generate 1000 megawatts of electricity. About 5600 tons $(5.1 \times 10^6$ kg) of coal are required to produce the same amount of electricity in a conventional power plant.

Uranium-235 (natural abundance 0.71%) is very scarce and difficult to separate from uranium-238 (natural abundance 99.28%). The much more abundant uranium-238 does not undergo fission and therefore cannot be used as a fuel for nuclear reactors. However, if uranium-238 is bombarded with neutrons (from uranium-235, for example), it absorbs a neutron and is transformed into uranium-239. This isotope undergoes beta emission to generate neptunium-239, which, in turn, undergoes another beta emission to produce plutonium-239:

$$^{238}_{92}U + ^{1}_{0}n \longrightarrow ^{239}_{92}U$$
$$^{239}_{92}U \longrightarrow ^{239}_{93}Np + ^{0}_{-1}\beta$$
$$^{239}_{93}Np \longrightarrow ^{239}_{94}Pu + ^{0}_{-1}\beta$$

Plutonium-239 also undergoes fission, with the production of more energy and more neutrons. These neutrons can then be used to breed more plutonium-239 from uranium-238. Thus, a so-called breeder reactor can produce its own supply of fissionable material. Several breeder reactors are now functioning in Europe.

Nuclear reactors using fissionable materials pose several serious risks to the environment. First is the everpresent danger that leaks, accidents, or acts of sabotage will release radioactive materials from the reactor into the environment. This problem has been a continuing concern since the accident at the Three-Mile Island nuclear plant in March 1979 — a concern that has increased in the wake of the disaster at the Chernobyl reactor in the Soviet Union in April 1986, where radioactive fallout spread within days across the globe. Second, many of the products of nuclear fission are themselves radioactive. The radioactivity from spent nuclear fuel and from the products of nuclear fission will remain lethal for thousands of years; the safe disposal of these materials is a problem that has not yet been solved. Third, obsolete generating plants also present a problem to future generations, for they contain much radioactive material. One suggestion has been to encase such plants in concrete for 100 years or more. Although it may become necessary, this solution is hardly simple or permanent.

B. Fusion

Nuclear fusion, the other process currently under study for the generation of atomic energy, depends on the putting together or fusing of two nuclei to form a single nucleus. One of the most promising fusion reactions generates energy by the fusion of two deuterium (hydrogen-2) atoms to form an atom of helium-3:

$$^{2}_{1}H + ^{2}_{1}H \xrightarrow{\text{energy}} ^{3}_{2}He + ^{1}_{0}n + \text{energy}$$

Such reactions require enormously high energy to force the two positively charged nuclei close together enough to fuse. Once the nuclei fuse, however, much more energy is released than is required for the reaction. Nuclear fusion occurs in the core of the sun, where the temperature is approximately 40 million degrees Celsius. Unfortunately, scientists have not yet found a way to produce and control nuclear fusion on Earth. Controlled nuclear fusion produces almost no radioactive wastes and would therefore be a nonpolluting source of energy. A massive effort is being made in this country and abroad to find ways to harness this energy source.

4.9 Summary

The subatomic particles of most interest to a chemist are electrons, protons, and neutrons. Electrons have a relative charge of -1 and a relative mass of 5.45×10^{-4} amu. Protons have a relative charge of $+1$ and a relative mass of 1.007 amu. Neutrons have no charge and a relative mass of 1.008 amu. Each atom is a mix of these three particles. The atomic number of an atom is equal to the number of protons in the nucleus of that atom and also to the number of electrons outside the nucleus. Most of the mass of an atom is concentrated in the nucleus, where neutrons and protons are found. The total number of particles in the nucleus (neutrons plus protons) is equal to the mass number of the atom. Atoms of the same element always contain the same number of protons but can differ slightly in the number of neutrons. This variation explains the existence of isotopes, which are atoms of the same element that vary slightly in mass. Rutherford's gold-foil experiment, carried out in 1911, confirmed the existence of the nucleus.

The atomic weight of an element is the average of the relative masses of the naturally occurring atoms of that element. To three significant figures, one mole of an element contains 6.02×10^{23} atoms of that element. The mole is a counting unit; a mole is 6.02×10^{23} things. One mole of an element has a mass in grams numerically equal to the atomic weight of the element.

The nuclei of some atoms decompose radioactively to produce alpha particles, beta particles, or gamma rays. These reactions can be shown by equations. The rate of decay of radioactive isotopes is measured in half-lives. Radioactivity has important applications in the health and biological sciences. Nuclear reactions can be used to produce energy.

Key Terms

alpha particle (4.2D2)
atomic mass unit (amu) (4.1C)
atomic number (4.2A)
atomic weight (4.3)
Avogadro's number (4.4)
beta particle (4.5B2)
electrical forces (4.2D1)
electron (4.1A)
gamma rays (4.5B3)
gravity (4.2D1)
half-life (4.6B)
ionizing radiation (4.5B)
ions (4.5B1)
isotopes (4.2C)

magnetic forces (4.2D1)
mass number (4.2B)
mole (4.4)
neutron (4.1C)
nuclear fission (4.8A)
nuclear forces (4.2D3)
nuclear fusion (4.8B)
nucleus of an atom (4.2D3)
proton (4.1B)
radioactivity (4.5A)
radioisotope (4.5A)
Rutherford's experiment (4.2D2)
subatomic particles (4.1)
tracers (4.7A)

Multiple-Choice Questions

For Questions 1–3 the possible answers are:

    1. electron   **2.** proton   **3.** neutron
    4. alpha particle   **5.** No answers are correct.

MC1. Which particle has the smallest mass?

MC2. Which particle carries no charge?

MC3. Which particles are found in the nucleus of an atom?

    a. 1 and 2   **b.** 1 and 3   **c.** 1 and 4   **d.** 2 and 3
    e. 2 and 4

MC4. Uranium exists as many isotopes, one of which is $^{235}_{92}U$. Which of the following might also be isotopes of uranium?

    1. atoms with atomic number 92, mass number 239
    2. atoms with atomic number 91, mass number 235
    3. atoms with atomic number 92, mass number 241
    4. atoms with atomic number 92, mass number 227
    5. atoms with atomic number 93, mass number 237
    a. 1, 3, and 4   **b.** 1, 2, and 5   **c.** 2 and 5   **d.** none   **e.** 1

MC5. Rutherford's experiment is diagrammed below. Which path is followed by most of the alpha particles that hit the gold foil: A, B, C, D, or E?

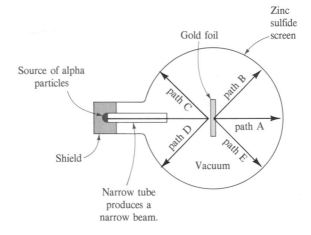

MC6. Why is Rutherford's gold-foil experiment so important in the history of atomic structure?

a. It showed that gold can be hammered into a thin foil only a few atoms thick.

b. It showed that atoms were composed of protons, electrons, and neutrons.

c. It showed that alpha particles have a relative charge of $+2$.

d. It showed that most of the atom was empty space surrounding a tiny, dense nucleus.

e. It showed that particles with like charges repel each other.

MC7. The atomic weight of gallium is 69.72. It occurs as gallium-69 and gallium-71. What is the approximate percent of gallium-69 in naturally occurring gallium?

a. 10%   b. 40%   c. 50%   d. 60%   e. 90%

MC8. A sample of iron weighs 6.93 g. How many moles of iron are contained in this sample?

a. 8.06 mol   b. 0.0806 mol   c. 0.124 mol
d. 0.0546 mol   e. 0.365 mol

MC9. What is the mass of 1.55×10^{-2} mol copper?

a. 9.85 g   b. 0.985 g   c. 91.3 g   d. 0.931 g
e. none of these

MC10. A sample of lead contains 3.01×10^{21} atoms. A sample of another metal contains the same number of atoms and weighs 1.01 g. What might the second metal be?

a. nickel   b. tin   c. iron   d. platinum   e. mercury

Problems

4.1 Subatomic Particles

4.12. Make a table showing the relative mass and charge of the following particles:

a. proton   b. alpha particle
c. neutron   d. electron

4.2 Atomic Structure

*4.13. Complete the following table:

| Element | Atomic number | Mass number | Number of | | |
			Electrons	Protons	Neutrons
———	11	23	——	——	——
sulfur	——	34	——	——	——
barium	——	——	56	——	81
———	20	40	——	——	——
———	——	——	——	8	8

$$\frac{3.01 \times 10^{21} \text{ Atoms}}{6.02 \times 10^{23} \text{ }} = \frac{1.01 \text{ gr}}{X} \quad 2.02 \times 10^{2}$$

4.14. Complete the following table:

Element	Atomic number	Mass number	Number of		
			Electrons	Protons	Neutrons
potassium	——	39	——	——	——
arsenic	——	——	33	——	75
——	17	37	——	——	——
——	——	——	——	13	14
——	——	40	18	——	——

4.15. Write symbols showing the mass number and the atomic number for the atoms listed in problems 4.13 and 4.14.

4.16. Describe the difference in atomic composition between:

chromium-53 and manganese-53
uranium-238 and neptunium-238

4.17. Describe the difference in atomic composition between:

antimony-121 and antimony-123
gallium-69 and gallium-71

*4.18. Name and give the composition of atoms having the following atomic number and mass number. Are they metals or nonmetals?

	Atomic number	Mass number
a.	28	58
b.	27	59
c.	26	58

4.19. Name and give the atomic composition of atoms having the following atomic and mass numbers:

	Atomic number	Mass number
a.	35	79
b.	80	200
c.	88	226

4.20. Describe the apparatus, the process studied, and the results of Rutherford's gold-foil experiment.

4.21. Suppose that the atom did not have a nucleus but did contain electrons, protons, and neutrons in a random arrangement. Can you predict the variety of paths that the alpha particles would follow?

4.3 Atomic Weights

*4.22. Samples of naturally occurring gallium contain 60.4% gallium-69, with an atomic mass of 68.93, and 39.6% gallium-71, with an atomic mass of 70.93. Calculate the atomic weight of gallium based on these data.

4.23. Iron occurs as the isotopes

$^{54}_{26}$Fe, $^{56}_{26}$Fe, $^{57}_{26}$Fe, and $^{58}_{26}$Fe

The atomic weight of iron is 55.85. Which of these four isotopes accounts for 92% of naturally occurring iron?

4.24. Naturally occurring silver contains 51.8% silver-107, with an atomic mass of 106.91, and 48.2% silver-109, with an atomic mass of 108.91. Calculate the approximate atomic weight of naturally occurring silver.

4.25. The naturally occurring isotopes of indium are

$^{113}_{49}$In and $^{115}_{49}$In

The atomic weight of indium is 114.8. Which of the indium isotopes is more abundant in naturally occurring samples?

4.4 The Mole

*4.26. Complete the following table:

Mass of sample	Moles of sample	Atoms in sample
_____	0.20 mol sodium	_____
5.0 g barium	_____	_____
_____	_____	1.0×10^{23} atoms calcium
_____	0.42 mol potassium	_____
9.3 g lithium	_____	_____

4.27. Calculate the mass of 0.15 mol of the following elements:
a. carbon   b. helium   c. zinc
d. lead   e. phosphorus
f. calcium   g. boron   h. iron

4.28. Calculate the number of atoms in 2.50 g of the following elements:
a. fluorine   b. sulfur   c. arsenic
d. tin   e. mercury

*4.29. Calculate the volume occupied by 1.00 mol of the following gases. All densities are given at 0°C and 1 atm.
a. neon ($d = 0.9002$ g/L)
b. krypton ($d = 3.733$ g/L)
c. radon ($d = 9.73$ g/L)
d. argon ($d = 1.783$ g/L)

4.30. Calculate the volume of 1.00 mol of the two liquid elements:
a. bromine ($d = 3.12$ g/mL at 20°C)
b. mercury ($d = 13.55$ g/mL at 20°C)

4.31. Calculate the number of moles of element in each of the following samples:
a. 16.3 g nickel   b. 56.3 g antimony
c. 109.01 g boron
d. 0.00546 g platinum

4.6 Characteristics of Nuclear Reactions

4.32. Complete the following equations.

$$^{103}_{46}\text{Pd} + ^{0}_{-1}\beta \longrightarrow \text{_____}$$

$$^{210}_{84}\text{Po} \longrightarrow ^{4}_{2}\alpha + \text{_____}$$

4.33. What is the half-life of an isotope if 6.00 g of the isotope decays to 0.75 g in 27 days?

4.34. Lead-210 has a half-life of 22 years.
a. If a 1.0-g sample of lead-210 is buried in 1988, how much of it will remain as lead-210 in 2076?
b. Lead-210 decays with the loss of an alpha particle. Write the equation for this decay.

4.35. What happens to the atomic number and the mass number of an atom if it loses:
a. an alpha particle?
b. a gamma ray?
c. a beta particle?

In which cases will the identity of the element change?

4.36. Why is radioactivity measured in half-lives rather than in the time required for total decay?

Review Problems

4.37. When gold costs $375 per ounce, what is the value of one atom of gold?

4.38. If copper is selling for $0.755 per pound, what is its cost per mole?

4.39. A normal, healthy body contains 34 parts per million (ppm) by weight of zinc. How many zinc atoms would be found in the body of a healthy person weighing 52.3 kg?

4.40. If regulations permit no more than 2 μg vanadium per cubic meter of air occupied by humans, how many atoms of vanadium would be contained in a room 3.0 m $\times$ 2.5 m $\times$ 4.0 m that contained the maximum allowable amount of vanadium?

****4.41.** According to the USFDA, fish sold for human consumption can contain no more than 5 ppm by weight of mercury. How many atoms of mercury would be in 1 lb of fish that contained this amount of mercury?

4.42. At a busy street intersection, the lead concentration in the air might be 9 μg per cubic meter. If 40% of the lead passing through the lungs is absorbed and if a typical adult breathes 20 m^3 of air per day, how many moles of lead are absorbed during an 8-hr working day by a newspaper vendor at this corner?

▪ 5 ▪

Electrons and the Properties of Elements

An atom contains protons, neutrons, and electrons. Rutherford showed in 1911 that the more massive particles—the protons and the neutrons—were inside, and the electrons outside, the nucleus of the atom. The radius of the nucleus is only 10^{-5} times the radius of the atom. The extranuclear space of the atom is empty except for the constantly moving, tiny electrons it contains. Research has led to a description of the properties of these electrons and explained how their number and individual properties are related to the properties of the elements. In this chapter we will describe such points as:

1. The properties of light and how the properties of electrons resemble the properties of light.
2. The configuration of electrons in an atom.
3. How the electron configuration of an atom is related to the location of the element in the periodic table.
4. What properties of an element can be predicted from its location in the periodic table.

5.1   Radiant Energy

To some extent, the properties of electrons can be compared with the properties of a closely related phenomenon, radiant energy or light. **Radiant energy,** also known as **electromagnetic radiation,** travels through a vacuum in waves at a constant speed of 3.0×10^8 m/sec.

In many ways the waves of electromagnetic radiation are like waves in water. You have probably seen closely spaced, choppy waves on a small lake. You may also have seen, on larger bodies of water like the ocean, waves that are farther apart. The difference between these two types of waves is in their **wavelength** λ (lambda), which is the distance from crest to crest. Three waves of different wavelength are shown in Figure 5.1.

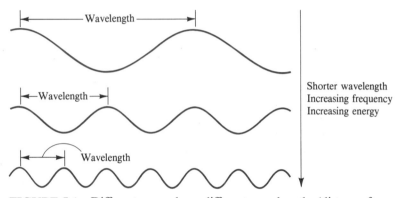

FIGURE 5.1   Different waves have different wavelengths (distance from crest to crest).

A wave can also be characterized by its frequency — that is, the number of wave crests that pass a given point in a unit time. The **frequency** ν (nu) of a wave is related to its wavelength λ by the equation

$$c = \lambda \nu$$

where c is the constant speed of light, 3.0×10^8 m/sec. From this equation you can see that as wavelength increases, the frequency of the wave decreases. The energy associated with a wave is directly proportional to its frequency. Hence, the higher the frequency, the shorter the wavelength and the higher the energy of the wave.

Figure 5.2 shows the wide range of electromagnetic radiation, from AM radio waves with a wavelength of 10^4 m to gamma waves with a wavelength of 10^{-12} m. This range is called the **electromagnetic spectrum.** The common names of the other kinds of electromagnetic radiation and their wavelengths are also given. Notice that **visible light,** electromagnetic radiation with wavelengths between 4×10^{-7} and 7×10^{-7} m, comprises only one small part of the elec-

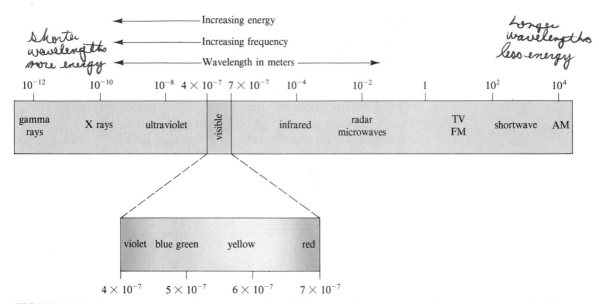

FIGURE 5.2 The electromagnetic spectrum. The visible range has been expanded to show the individual colors.

tromagnetic spectrum. In Figure 5.2 you can see that red light has a longer wavelength than blue light. Red light, then, has a lower frequency and is associated with less energy than blue light. Infrared light, microwaves, television waves, and radio waves are invisible forms of electromagnetic radiation; their wavelengths are greater than that of visible light, and thus their energies are lower than that of visible light.

Ultraviolet light, X rays, and gamma rays — all of which have wavelengths shorter and energies higher than those of visible light (see Figure 5.2) — are also invisible forms of radiant energy.

When an object is heated, it radiates energy, often in the form of visible light. Our sun is probably the most familiar example of a heated body giving off light. The "white" light from the sun is a collection of light of all wavelengths and is called **continuous light.** When light passes through a prism, it is separated into its various wavelengths. You have probably seen how sunlight, when passed through a prism, separates into all the colors and wavelengths of the rainbow. However, if a gaseous sample of a single element is heated, the light emitted is not continuous and is only of a few wavelengths. When this light is passed through a prism, instead of a rainbow we see a series of brightly colored lines, each line corresponding to a particular wavelength of the emitted light. The pattern of wavelengths (or lines) of light thus produced is unique for each element and is called its **emission spectrum** (plural, *spectra*). Because the pattern is characteristic, it can be used to show the presence of that element in even the tiniest amounts. Figure 5.3 shows the emission spectra of hydrogen, neon, and sodium in the visible range. The light from sodium-vapor street lamps is

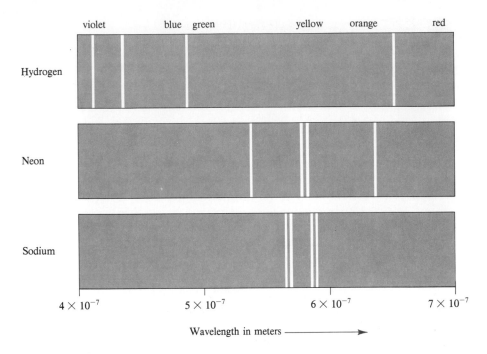

FIGURE 5.3   Emission spectra of hydrogen, neon, and sodium in the visible range.

that of the yellow-orange lines of the sodium spectrum. Neon signs use the red lines of the neon spectrum to produce their color. The spectra of these elements, as well as those of all the other elements, show other lines in the invisible parts of the electromagnetic spectrum.

The spectra shown in Figure 5.3 are called **emission spectra** because they show the light (energy) given off (emitted) by an unusually energetic atom. Atoms can also absorb energy. If continuous light is passed through the vapor of an element, some wavelengths of light are absorbed. Analysis of the emerging light shows a rainbow interspersed with some black lines. This spectrum, called an **absorption spectrum,** shows that some energy (measured by the wavelength at which the black lines appear) has been absorbed by the vaporized element. If the vaporized element is hydrogen, the black lines will appear at exactly the same places in the absorption spectrum as did the bright lines in the emission spectrum of hydrogen shown in Figure 5.3. Similarly for sodium, neon, and other elements, the bright lines in their emission spectra are of the same wavelengths as the black lines in their absorption spectra.

We conclude from these observations that, first, atoms can lose or gain energy in amounts similar to the energy of light and, second, the energy of an electron can change only by certain increments and not by random amounts.

5.2 The Energy of an Electron

The energy of an electron is of the same order of magnitude (is in the same range) as the energy of light. The lines in the spectrum of an element represent changes in the energy of electrons within the atoms of that element. By studying these spectra, scientists have drawn various conclusions about the behavior of electrons in atoms.

1. The energy of an electron depends on its location with respect to the nucleus of an atom. The higher the energy of an electron in an atom, the farther is its most probable location from the nucleus. Notice that we say *probable location.* Because of the electron's small size and high energy, we are limited in how precisely we can mark its position at any instant. We can only describe regions around the atom's nucleus within which the electron may be found.

2. In describing these regions of space, we also recognize that the energy of an electron is quantized. What does this statement mean? A property is **quantized** if it is available only in multiples of a set amount. If you are pouring a soft drink from a can, you can pour out as much or as little as you like. However, if you are buying a soft drink from a machine, you can buy only a certain amount. You cannot buy a half or a third of a can of soda; you can buy only a whole can or several cans. Soft drinks dispensed by a machine are available only in multiples of a set volume, or quantum. Thus, the dispensing of soft drinks by machine has been quantized.

Energy can also be quantized. If you are climbing a ladder, you can stop only on the rungs; you cannot stop between them. The energy needed to climb the ladder is used in finite amounts to lift your body from one rung to the next. To move upward, you must use enough energy to move your feet to the next higher rung. If the available energy is only enough to move partway up to the next rung, you cannot move at all because you cannot stop between rungs. Thus, in climbing the ladder, your expenditure of energy is quantized. If you are going up a hill instead of a ladder, your energy expenditure is not quantized. You can go straight up the hill or you can zigzag back and forth, going up gradually. You can take big steps or little steps; no limitations are placed on where you can stop or on how much energy you must use.

Let us apply the analogy of the ladder and its rungs to an atom and its electrons. In climbing the ladder, you can place your feet only on the rungs. Similarly, an atom has only certain places, set distances from the nucleus called **energy levels,** where electrons may be found. Unlike a ladder, which has a limited length, the energy levels of an atom extend infinitely out from the nucleus and the energy levels are not evenly spaced. As the distance from the nucleus increases, the levels get closer together and contain more-energetic electrons (Figure 5.4). The energy of an electron in one of the levels at a considerable distance from the nucleus is greater than that of an electron in a closer level.

For an electron to move from one energy level to the next higher level, it must gain the right amount of energy. If less than that amount is available, the

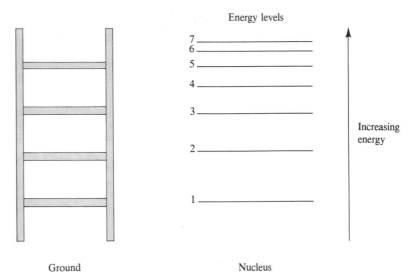

FIGURE 5.4   Energy levels. The energy levels in an atom are similar to the rungs of a ladder, but they get closer together as they get farther from the nucleus.

electron stays where it is. Electrons always move from one level to another; they cannot stop in between. Thus there are certain regions of space within an atom where an electron can be and other regions where an electron cannot be.

Example 5.1   The following diagram shows the energy levels of an atom. Each arrow shows a possible electron transition from a level of higher energy to one of lower energy. These transitions would release energy.

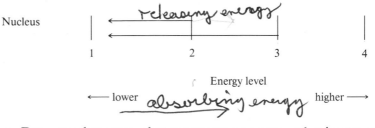

a. Draw another arrow that represents an energy-releasing transition.

b. Draw two arrows that show energy-absorbing transitions.

Solution   a. The arrows show electron transitions from the second to the first energy level and from the third to the first energy level. The transition of an electron from the fourth to the second energy level would also release energy.

b. When an electron moves from a lower to a higher energy level, it absorbs energy. An arrow from the second to the fourth energy level and an arrow from the first to the third energy level would show energy-absorbing transitions.

Problem 5.1 Draw a diagram of the energy levels of an electron, showing five energy levels. Draw three lines that represent energy-absorbing transitions and three lines that show energy-emitting transitions.

5.3 An Atomic Model

Our present model of the atom is based on the concept of energy levels for electrons within an atom and on the mathematical interpretation of detailed atomic spectra. The requirements for our model are:

1. Each electron in a particular atom has a unique energy that depends on the relationship between the negatively charged electron and both the positively charged nucleus and the other negatively charged electrons in the atom.

2. The energy of an electron in an atom can increase or decrease, but only by specific amounts, or quanta.

A. Energy Levels

We picture an atom as a small nucleus surrounded by a much larger volume of space containing the electrons. This space is divided into regions called **principal energy levels,** numbered 1, 2, 3, 4, . . . , ∞ outward from the nucleus. Early theories on the structure of the atom called these levels **electron shells** and designated them by letter. Occasionally, this designation is still encountered. With these designations, the first energy level is the K shell, the second the L shell, and so on through the alphabet.

Each principal energy level can contain up to $2n^2$ electrons, where n is the number of the level. Thus, the first level can contain up to 2 electrons, $2(1^2) = 2$; the second up to 8 electrons, $2(2^2) = 8$; the third up to 18, $2(3^2) = 18$; and so on. Only seven energy levels are needed to contain all the electrons in an atom of any of those elements now known.

As stated earlier, the energy associated with an energy level increases as the distance from the nucleus increases. An electron in the seventh energy level has more energy associated with it than does one in the first energy level.

The lower the number of the principal energy level, the closer the negatively charged electron in it is to the positively charged nucleus and the more difficult it is to remove this electron from the atom.

B. Sublevels and Orbitals

When an electron is in a particular energy level, it is more likely to be found in some parts of that level than in others. These parts are called **orbitals.** Orbitals of equivalent energy are grouped in **sublevels.** Each orbital can contain a maximum of two electrons. When in a magnetic field, the two electrons in a particular orbital differ very slightly in energy because of a property called **electron spin.** The theory of electron spin states that the two electrons in a single orbital spin in opposite directions on their axes, causing an energy difference between them. (Like many models, this explanation is an oversimplification, but for the purpose of this course it is a useful description.)

Each principal energy level has one sublevel containing one orbital, an *s* orbital, that can contain a maximum of two electrons. Electrons in this orbital are called *s* electrons and have the lowest energy of any electrons in that principal energy level. The first principal energy level contains only an *s* sublevel; therefore, it can hold a maximum of two electrons.

Each principal energy level above the first contains one *s* orbital and three *p* orbitals. A set of three *p* orbitals, called the *p* sublevel, can hold a maximum of six electrons. Therefore, the second level can contain a maximum of eight electrons—that is, two in the *s* orbital and six in the three *p* orbitals.

Each principal energy level above the second contains, in addition to one *s* orbital and three *p* orbitals, a set of five *d* orbitals, called the *d* sublevel. The five *d* orbitals can hold up to 10 electrons. Thus, the third level holds a maximum of 18 electrons: 2 in the *s* orbital, 6 in the three *p* orbitals, and 10 in the five *d* orbitals.

The fourth and higher levels also have an *f* sublevel, containing seven *f* orbitals, which can hold a maximum of 14 electrons. Thus, the fourth level can hold up to 32 electrons: 2 in the *s* orbital, 6 in the three *p* orbitals, 10 in the five *d* orbitals, and 14 in the seven *f* orbitals. The sublevels of the first four principal energy levels and the maximum number of electrons that the sublevels can contain are summarized in Table 5.1.

To distinguish which *s, p, d,* or *f* sublevel we are talking about, we precede the letter by the number of the principal energy level. For example, the *s* sublevel of the second principal energy level is designated 2*s*; the *s* sublevel of the third principal energy level is designated 3*s*; and so on. The number of electrons occupying a particular sublevel is shown by a superscript after the letter of the sublevel. The notation

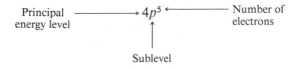

means that five electrons are contained in the *p* sublevel of the fourth energy level.

TABLE 5.1   Sublevels of the first four energy levels

Principal energy level	Sublevel	Number of orbitals in sublevel	Total possible occupying electrons
1	s	1	2
2	s	1	2 ⎫ 8
	p	3	6 ⎭
3	s	1	2 ⎫
	p	3	6 ⎬ 18
	d	5	10 ⎭
4	s	1	2 ⎫
	p	3	6 ⎪ 32
	d	5	10 ⎬
	f	7	14 ⎭

Example 5.2   For the third energy level, list the different kinds of orbitals (sublevels), how many of each there are, and the total number of electrons the energy level can hold.

Solution   The third energy level contains the following:

> one s orbital that can contain a total of 2 electrons
>
> three p orbitals that can contain a total of 6 electrons
>
> five d orbitals that can contain a total of 10 electrons

giving, for the entire third energy level, a maximum capacity of 18 electrons.

Problem 5.2   For the fourth energy level, list the different kinds of orbitals present, how many of each there are, and the total number of electrons this energy level can contain.

1. Orbital shapes and sizes

Each orbital has a unique shape and size. The shapes of s and p orbitals are shown in Figure 5.5. In these diagrams, the nucleus is at the origin of the axes. The s orbitals are spherically symmetrical about the nucleus and increase in size as distance from the nucleus increases. The $2s$ orbital is a larger sphere than the $1s$ orbital, the $3s$ orbital is larger than the $2s$ orbital, and so on (see Figure 5.6).

The three p orbitals are more or less dumbbell-shaped, with the nucleus at the center of the dumbbell. They are oriented at right angles to one another along the x, y, and z axes, hence we denote them as p_x, p_y, and p_z. Like the s

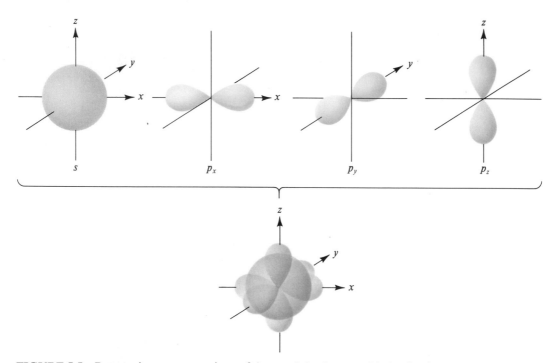

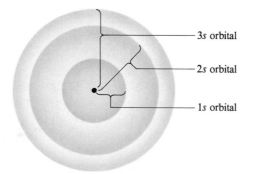

FIGURE 5.5   Perspective representations of the *s* and the three *p* orbitals of a single energy level. The clouds show the space within which the electron is most apt to be. The lower sketch shows how these orbitals overlap in the energy level.

orbitals, the *p* orbitals increase in size as the number of the principal energy level increases; thus a 4*p* orbital is larger than a 3*p* orbital.

The shapes of *d* orbitals are shown in Figure 5.7. The five *d* orbitals are denoted by d_{xy}, d_{yz}, d_{xz}, $d_{x^2-y^2}$, and d_{z^2}. Notice that these shapes are more complex than those of *p* orbitals, and recall that the shapes of *p* orbitals are more complex than those of *s* orbitals. Clearly, the shape of an orbital becomes more

3*s* orbital

2*s* orbital

1*s* orbital

FIGURE 5.6   Cross-sectional view of the *s* orbitals of an atom showing their relative sizes and overlap.

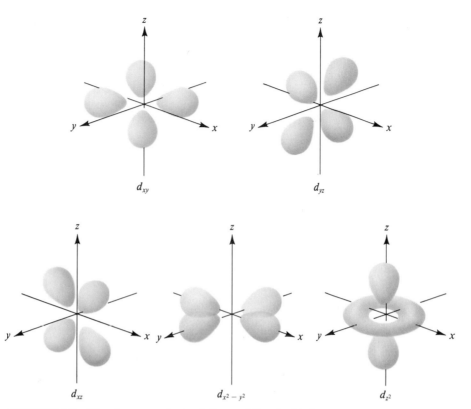

FIGURE 5.7   The shapes and orientations of the *d* orbitals.

complex as the energy associated with that orbital increases. We can predict that the shapes of *f* orbitals will be even more complex than those of the *d* orbitals.

One further, important note about **orbital shapes:** These shapes do *not* represent the path of an electron within the atom; rather, they represent the region of space in which an electron of that sublevel is most apt to be found. Thus, a *p* electron is most apt to be within a dumbbell-shaped space in the atom, but we make no pretense of describing its path.

2. The energy of an electron versus its orbital

Within a given principal energy level, electrons in *p* orbitals are always more energetic than those in *s* orbitals, those in *d* orbitals are always more energetic than those in *p* orbitals, and electrons in *f* orbitals are always more energetic than those in *d* orbitals. For example, within the fourth principal energy level, we have:

$$4s < 4p < 4d < 4f$$
$\xrightarrow{\quad\text{Increasing energy}\quad}$

In addition, the energy associated with an orbital increases as the number of the principal energy level of the orbital increases. For instance, the energy associated with a $3p$ orbital is always higher than that associated with a $2p$ orbital, and the energy of a $4d$ orbital is always higher than that associated with a $3d$ orbital. The same is true of s orbitals:

$$1s < 2s < 3s < 4s < 5s$$
Increasing energy

Each orbital is not a region of space separate from the space of other orbitals. This is implicit in Figures 5.5, 5.6, and 5.7. If all those orbitals were superimposed on one another, you would see that a great deal of space is included in more than one orbital. For example, a $3p$ electron can be within the space assigned to a $3d$ or $3s$ orbital as well as within its own $3p$ space.

There is also an interweaving of energy levels. Figure 5.8 shows, in order of increasing energy, all the orbitals of the first four energy levels. Notice that the energy of a $3d$ orbital is slightly higher than that of a $4s$ orbital, and that of a $4d$ orbital is a little higher than that of a $5s$ orbital. Note especially the overlap of orbitals in the higher principal energy levels.

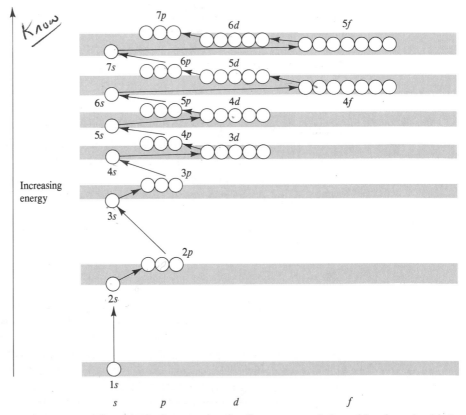

FIGURE 5.8   The principal energy levels of an atom and the sublevels and orbitals each contains. The arrows show the order in which the sublevels fill.

Example 5.3   Arrange the following orbitals in order of increasing energy:

4f 1s 3d 4p 5s

Solution   Using Figure 5.8, we determine the order as:

1s 3d 4p 5s 4f

Problem 5.3   Arrange the following orbitals in order of increasing energy:

3s 6s 4d 4f 5d

C. Our Model and the Spectra of Different Elements

According to our model of the atom, electrons are distributed among the energy levels and orbitals of the atom according to certain rules, and each electron has a unique energy determined by the position of its orbital. When an atom absorbs the right amount of energy, an electron moves from its original orbital to a higher-energy orbital that has a vacancy. Similarly, when an atom emits energy, the electron drops to a lower-energy orbital that has a vacancy. For example, an electron in a 3s orbital can drop to the 2p orbital, the 2s orbital, or the 1s orbital. The energy emitted by an electron in dropping to a lower-energy orbital is released in the form of radiation and determines the lines in the spectrum of the element.

When all the electrons of an atom are in the lowest possible energy states (meaning that the energy levels have been filled in order of increasing energy), the atom and its electrons are in the **ground state.** If one of these electrons moves to a higher energy level, the atom is in an **excited state.**

We know that each element has a unique spectrum. These spectra show that the energy differences among the electrons in an atom vary from one element to another. What causes this variation?

Recall that the nucleus of an atom is positively charged, that electrons carry a negative charge, and that oppositely charged bodies attract one another. The atoms of one element differ from those of another element in the number of protons in the nucleus and, consequently, in the charge on the nucleus. The attraction for an electron, and therefore its energy, will differ from one element to the next according to differences in nuclear charge. In addition, the atoms of one element contain a different number of electrons than do atoms of any other element. The energy of each electron within the atom depends not only on its interaction with the positively charged nucleus, but also on its interaction with the other electrons in the atom. Therefore, the energies of the electrons of one element will differ from the energies of the electrons of another element. Considering these two variables—nuclear charge and number of electrons—we can see that each element must have a unique spectrum derived from its unique set of electron energy levels.

The Electron Configurations of Atoms

The **electron configuration** of an atom shows the number of electrons in each sublevel in each energy level of the ground-state atom. To determine the electron configuration of a particular atom, start at the nucleus and add electrons one by one until the number of electrons equals the number of protons in the nucleus. Each added electron is assigned to the lowest-energy sublevel available. The first sublevel filled will be the $1s$ sublevel, then the $2s$ sublevel, the $2p$ sublevel, the $3s$, $3p$, $4s$, $3d$, and so on. This order is difficult to remember and often hard to determine from energy-level diagrams such as Figure 5.8.

A more convenient way to remember the order is to use Figure 5.9. The principal energy levels are listed in columns, starting at the left with the $1s$ level. To use this figure, read along the diagonal lines in the direction of the arrow. The order is summarized under the diagram.

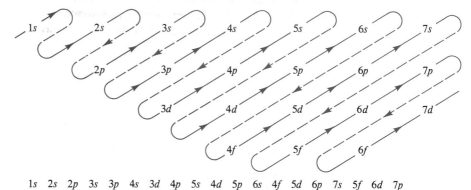

$1s$ $2s$ $2p$ $3s$ $3p$ $4s$ $3d$ $4p$ $5s$ $4d$ $5p$ $6s$ $4f$ $5d$ $6p$ $7s$ $5f$ $6d$ $7p$

FIGURE 5.9   The arrow shows a second way of remembering the order in which sublevels fill.

An atom of hydrogen (atomic number 1) has one proton and one electron. The single electron is assigned to the $1s$ sublevel, the lowest-energy sublevel in the lowest-energy level. Therefore, the electron configuration of hydrogen is written:

$$\text{H:}\quad 1s^1 \longleftarrow \text{Number of electrons}$$

Principal energy level ⟋   ⟍ Sublevel

For helium (atomic number 2), which has two electrons, the electron configuration is:

He: $1s^2$

Two electrons completely fill the first energy level. Because the helium nucleus is different from the hydrogen nucleus, neither of the helium electrons will have exactly the same energy as the single hydrogen electron, even though all are in the $1s$ sublevel.

The element lithium (atomic number 3) has three electrons. In order to write its electron configuration, we must first determine (from Figure 5.9) that the $2s$ sublevel is next higher in energy after the $1s$ sublevel. Therefore, the electron configuration of lithium is:

Li: $1s^2 2s^1$

Boron (atomic number 5) has five electrons. Four electrons fill both the $1s$ and $2s$ orbitals. The fifth electron is added to a $2p$ orbital, the sublevel next higher in energy (Figure 5.9). The electron configuration of boron is:

B: $1s^2 2s^2 2p^1$

Table 5.2 shows the electron configurations of the elements with atomic numbers 1 through 18. The electron configurations of elements with higher atomic numbers can be written by following the orbital-filling chart in Figure 5.9.

TABLE 5.2   Electron configurations of the first 18 elements

Element	Atomic number	Electron configuration
hydrogen	1	$1s^1$
helium	2	$1s^2$
lithium	3	$1s^2 2s^1$
beryllium	4	$1s^2 2s^2$
boron	5	$1s^2 2s^2 2p^1$
carbon	6	$1s^2 2s^2 2p^2$
nitrogen	7	$1s^2 2s^2 2p^3$
oxygen	8	$1s^2 2s^2 2p^4$
fluorine	9	$1s^2 2s^2 2p^5$
neon	10	$1s^2 2s^2 2p^6$
sodium	11	$1s^2 2s^2 2p^6 3s^1$
magnesium	12	$1s^2 2s^2 2p^6 3s^2$
aluminum	13	$1s^2 2s^2 2p^6 3s^2 3p^1$
silicon	14	$1s^2 2s^2 2p^6 3s^2 3p^2$
phosphorus	15	$1s^2 2s^2 2p^6 3s^2 3p^3$
sulfur	16	$1s^2 2s^2 2p^6 3s^2 3p^4$
chlorine	17	$1s^2 2s^2 2p^6 3s^2 3p^5$
argon	18	$1s^2 2s^2 2p^6 3s^2 3p^6$

Example 5.4   Write the electron configuration of vanadium.

Solution   The atomic number of vanadium is 23; we must therefore account for 23 electrons. The first sublevels to fill are $1s$, $2s$, $2p$, $3s$, $3p$, and $4s$, which together can hold 20 electrons. The remaining 3 electrons must be in the $3d$ orbital, giving the configuration:

$$\text{V:}\quad 1s^2 2s^2 2p^6 3s^2 3p^6 4s^2 3d^3$$

This is often written in order of principal energy levels

$$\text{V:}\quad 1s^2 2s^2 2p^6 3s^2 3p^6 3d^3 4s^2$$

Both ways are correct.

Problem 5.4   Write the electron configuration of arsenic (atomic number 33).

A. Box Diagrams of Electron Configuration

If an atom has a partially filled sublevel, it may be important to know how the electrons of that sublevel are distributed among the orbitals. Research has shown that unpaired electrons (a single electron in an orbital) are in a lower energy configuration than are paired electrons (two electrons in an orbital). The energy of the electrons in a sublevel would then be lower with half-filled orbitals than with some filled and some empty. We can show the distribution of electrons by using **box diagrams,** where each box represents an orbital and the arrows within the boxes represent the electrons in that orbital. The direction of the arrow represents the spin of the electron. (Recall from Section 5.3B that two electrons in an orbital spin in opposite directions on their axes.) Therefore, if an orbital contains two electrons, its box will contain two arrows, one pointing up and the other down.

Using a box diagram, we show the electron configuration of nitrogen as:

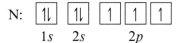

Notice that the $2p$ electrons are shown as

rather than

which would mean that, of the three p orbitals, one is filled, one is half-filled, and one is empty.

Example 5.5   Using a box diagram, show the electron configuration of sulfur.

Solution   The atomic number of sulfur is 16, so we must show 16 electrons. In simplest notation the electron configuration is:

S: $1s^2 2s^2 2p^6 3s^2 3p^4$

To show this configuration as a box diagram, we draw one box for each orbital. Place two electrons in each box until you get to the $3p$ orbitals. In filling the $3p$ boxes, first put one arrow in each $3p$ box and then add one more to each until the available electrons are used up (in this case, only one more arrow is needed).

S: ⌷1↓⌷ ⌷1↓⌷ ⌷1↓⌷⌷1↓⌷⌷1↓⌷ ⌷1↓⌷ ⌷1↓⌷⌷1⌷ ⌷1⌷

1s   2s   2p   3s   3p

Problem 5.5   Using a box diagram, show the electron configuration of phosphorus.

5.5 The Periodic Table

In Section 3.3B, the periodic table was introduced as a list of the elements. We also pointed out that the design of the periodic table separates the metals from the nonmetals. In this section we will show how the various features of the table relate to the electron configuration of the different elements and to their position in the table. First let us point out those features using the complete periodic table shown in Figure 5.10.

In the table, the elements are placed in rows and columns of varying length. Seven rows are used to show all of the elements now known. These rows are called **periods** and each period is numbered. Notice that the display of elements labeled "lanthanides" and placed below the table belongs in period 6 between element 57 (lanthanum) and element 72 (hafnium). In some periodic tables, lanthanum is the first member of the lanthanide series. Similarly, the display labeled "actinides" belongs in period 7 between element 89 (actinium) and element 104 (unnilquadium). Again, in some tables actinium is the first member of the actinide series. These two displays are customarily put below the table so that the table will fit into a reasonable space, as on a book page. The columns of the periodic table vary in length. Some are numbered and some have been given a letter A or B. That system of numbering and assigning letters differs depending on whether you are using an American or a European periodic table. In this book we have classified the columns by length and numbered the long columns with Roman numerals. The short columns, those in the middle of the table, have not been numbered. We have chosen not to use any letters in designating the columns.

Figure 5.10 also shows a column notation that was adopted by IUPAC in 1985 but is still fairly controversial and as such has not yet come into wide

Period																		
Numbering system used in this text	I	II											III	IV	V	VI	VII	VIII
IUPAC system	1	2	3	4	5	6	7	8	9	10	11	12	13	14	15	16	17	18
Old European system	IA	IIA	IIIB	IVA	VA	VIA	VIIA	VIIIA	VIIIA	VIIIA	IB	IIB	IIIB	IVB	VB	VIB	VIIB	0
Old U.S.A. system	IA	IIA	IIIB	IVB	VB	VIB	VIIB	VIII	VIII	VIII	IB	IIB	IIIA	IVA	VA	VIA	VIIA	0
1	1 H 1.008																	2 He 4.00
2	3 Li 6.94	4 Be 9.01											5 B 10.81	6 C 12.01	7 N 14.01	8 O 16.00	9 F 19.00	10 Ne 20.18
3	11 Na 22.99	12 Mg 24.31											13 Al 26.98	14 Si 28.09	15 P 30.97	16 S 32.06	17 Cl 35.45	18 Ar 39.95
4	19 K 39.10	20 Ca 40.08	21 Sc 44.96	22 Ti 47.88	23 V 50.94	24 Cr 52.00	25 Mn 54.94	26 Fe 55.85	27 Co 58.93	28 Ni 58.69	29 Cu 63.55	30 Zn 65.38	31 Ga 69.72	32 Ge 72.59	33 As 74.92	34 Se 78.96	35 Br 79.90	36 Kr 83.80
5	37 Rb 85.47	38 Sr 87.62	39 Y 88.91	40 Zr 91.22	41 Nb 92.91	42 Mo 95.94	43 Tc (98)	44 Ru 101.07	45 Rh 102.91	46 Pd 106.42	47 Ag 107.87	48 Cd 112.41	49 In 114.82	50 Sn 118.69	51 Sb 121.75	52 Te 127.60	53 I 126.90	54 Xe 131.29
6	55 Cs 132.91	56 Ba 137.33	57 La 138.91	72 Hf 178.49	73 Ta 180.95	74 W 183.85	75 Re 186.21	76 Os 190.2	77 Ir 192.22	78 Pt 195.08	79 Au 196.97	80 Hg 200.59	81 Tl 204.38	82 Pb 207.2	83 Bi 208.98	84 Po (209)	85 At (210)	86 Rn (222)
7	87 Fr (223)	88 Ra 226.03	89 Ac 227.03	104 Unq[a] (261)	105 Unp[a] (262)	106 Unh[a] (263)	107 Uns[a] (262)	108	109 Une[a]									

Period

Lanthanides	58 Ce 140.12	59 Pr 140.91	60 Nd 144.24	61 Pm (145)	62 Sm 150.36	63 Eu 151.96	64 Gd 157.25	65 Tb 158.92	66 Dy 162.50	67 Ho 164.93	68 Er 167.26	69 Tm 168.93	70 Yb 173.04	71 Lu 174.97
Actinides	90 Th 232.04	91 Pa 231.04	92 U 238.03	93 Np 237.05	94 Pu (244)	95 Am (243)	96 Cm (247)	97 Bk (247)	98 Cf (251)	99 Es (252)	100 Fm (257)	101 Md (258)	102 No (259)	103 Lr (260)

Note: Atomic masses shown here are 1981 IUPAC values rounded off to two decimals. [a]Symbols based on IUPAC systematic names.

FIGURE 5.10 Periodic table of the elements. This figure shows the confusion that has existed in designating the columns of the periodic table. In this text, only the long columns are numbered (with Roman numerals). The A and B designations are not used.

usage. The IUPAC system uses Arabic numbers from 1 to 18 to number the columns. You should be aware of this system because you may encounter it in advanced chemistry classes.

The elements in a column make up a family of elements. A family is also known as a group. Thus the elements in column VIII are known as the family or group of noble gases.

A. Electron Configuration and the Periodic Table

Figure 5.11 again shows the periodic table but without the symbols of the elements. Instead it shows the last sublevels filled in describing the electron configurations of the elements in each section. We will use Figures 5.11 and 5.8 to relate the electron configuration of an element to its position in the periodic table.

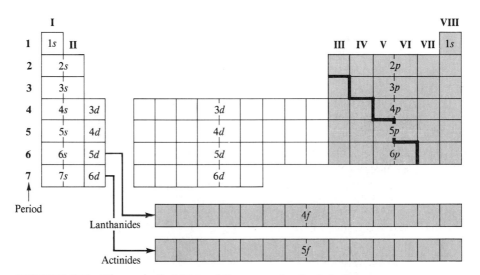

FIGURE 5.11   The periodic table and the energy level subshells.

In period 1, there are two boxes. In the usual table, these boxes would contain the symbols for hydrogen and helium, the elements in this period. In Figure 5.11 we show instead the letter *s* indicating that the last added electron for the elements in these boxes is in the 1*s* sublevel. In period 2, there are eight boxes. Instead of symbols for eight elements, Figure 5.11 shows *s* in the first two boxes and *p* in the last six boxes, showing that the 2*s* and 2*p* sublevels are being filled as the electron configurations of the elements in these boxes are completed. Period 3 also has eight boxes, which would correspond to the electrons needed to fill the 3*s* and 3*p* sublevels.

Look back now to Figure 5.8, which shows the order in which the sublevels fill. Notice that the 4*s* sublevel is filled immediately after the 3*p* sublevel. Figure 5.11 shows that elements whose last added electron goes into an *s* sublevel are in columns I and II. So we must start here a new period, period 4, and put boxes for the elements formed by filling the 4*s* sublevel in those columns. Figure 5.8 shows that the next sublevel to fill is the 3*d* sublevel. These are the first *d* electrons added, so we start new columns for the elements formed by their addition. Ten electrons are needed to fill the five *d* orbitals, so we start ten columns in this fourth period, placing the columns next to column II and between it and column III. The 4*p* sublevel is filled next, after the 3*d* sublevel. The boxes for the elements formed by filling the *p* orbitals are in place under the boxes for the elements formed by adding the 3*p* electrons.

By consulting Figure 5.8, we see that the next sublevels filled are in the order: 5*s*, 4*d*, and 5*p*. Boxes for the elements formed by filling the orbitals of these sublevels are arranged as were those in period 4. Just as period 4 contains more elements than period 3, period 6 contains more elements than period 5. Period 6 starts with elements whose last added electron is in the 6*s* sublevel. The next step is where period 6 differs from period 5. Look again at Figure 5.8 and note that the 4*f* sublevel is filled after the 6*s* sublevel and before the 5*d* sublevel. We will need 14 boxes to contain the electrons needed to fill the seven *f* orbitals. These are the boxes of the lanthanide series, shown below the table. There is some evidence that these orbitals do not fill before one electron is in a 5*d* orbital, so we have shown in Figure 5.11 the lanthanide series coming after the first *d* column. After the 4*f* orbitals are filled, boxes are shown for the rest of the elements formed by adding 5*d* and 6*p* electrons. The seventh period contains boxes for the elements formed by filling the 7*s*, the 5*f* (the actinide series shown below the table), and finally the 6*d* sublevels.

Figure 5.11 thus shows the close relationship that exists between the electron configuration of an element and its location in the periodic table. This relationship is further expressed by the following names sometimes given to parts of the table:

columns I and II   *s* block
columns III–VIII   *p* block
short columns   *d* block
lanthanides and actinides   *f* block

The groups of elements found in these blocks are also known by other names.

B. Categories of Elements in the Periodic Table

1. The representative elements

Elements in the *s* and *p* blocks are known as **representative elements** or main group elements. The term *representative* dates from early times, when chemists believed that the chemistry of these elements was representative of all elements. Group VIII is not always included in the representative elements because the chemistry of the noble gases is unique to them. In period 7 there are no elements in the p block.

The p block of period 7 would contain elements with atomic numbers greater than 112; such elements have not yet been found in the Earth's crust nor have they been prepared by nuclear reaction.

In the s and p blocks, the period in which the element occurs has the same number as the highest energy level that contains electrons in a ground-state atom. The number of the column in which an element is found is the same as the number of s and p electrons in that level. Sodium is a representative element with 11 electrons. Its electron configuration is:

$$1s^2 2s^2 2p^6 3s^1$$

Sodium is in column I of the third period. In a sodium atom, the highest-energy principal energy level containing electrons is the third energy level, and that energy level contains one electron.

Example 5.6   Relate the electron configuration of arsenic to its location in the periodic table.

Solution   The atomic number of arsenic is 33. Its electron configuration is:

$$1s^2 2s^2 2p^6 3s^2 3p^6 3d^{10} 4s^2 4p^3$$

The highest occupied energy level is the fourth, and arsenic is in the fourth period of the table. There are five s and p electrons in this fourth energy level. Arsenic is in column V.

Problem 5.6   Relate the electron configuration of chlorine to its location in the periodic table.

---

2. The transition elements

The **transition elements** (or transition metals, for they are all metals) are those elements found in the short columns of the d block. Many of these elements are probably familiar to you. The coinage metals—gold, silver, and copper—are here. So is iron, the principal ingredient of steel, as well as those elements that are added to iron to make particular kinds of steel: chromium, nickel, and manganese. In period 7, the d block is not filled. The reason is the same as the reason why the p section of period 7 is empty: these elements do not occur naturally and have not yet been found as the product of a nuclear reaction. Many of the properties of the transition elements are related to the fact that, in their electron structures, the occupied s and d sublevels of highest energy are very close in energy.

3. The inner transition elements

The **inner transition elements** are those found in the f block of the periodic table (in the two rows below the main body of the table). The elements in this block are chemically very much alike, which will seem reasonable when you consider that they have the same electron configurations in the two outermost energy levels. The differences occur in the next further-in energy level. For example, the electron configuration of cerium (Ce, #58) is:

$$1s^2 2s^2 2p^6 3s^2 3p^6 4s^2 3d^{10} 4p^6 5s^2 4d^{10} 5p^6 6s^2 4f^2$$

and that of praseodymium (Pr, #59) is:

$$1s^2 2s^2 2p^6 3s^2 3p^6 4s^2 3d^{10} 4p^6 5s^2 4d^{10} 5p^6 6s^2 4f^3$$

The only difference between these two configurations is in the number of $4f$ electrons. Both the fifth and sixth energy levels contain electrons.

The elements in the **lanthanide series** are also known as the rare earths. They are used extensively in producing monitors for color television.

The elements in the **actinide series** are all radioactive, and only three are found in appreciable concentration in the Earth's crust. Of the others, only some have been found in trace amounts in the Earth or in the stars. All have been produced in laboratories as products of nuclear reactions.

C. The Electron Configuration of the Noble Gases; Core Notation

We have established a relationship between the electron configuration of an element and its location in the periodic table. Let us look closer now at the electron configurations of the **noble gases,** those elements in Group VIII of the periodic table. The electron configurations of these elements are shown in Table 5.3.

TABLE 5.3   Electron configurations of the noble gases (Group VIII elements)

Element	Atomic number	Electron configuration
He	2	$1s^2$
Ne	10	$1s^2 2s^2 2p^6$
Ar	18	$1s^2 2s^2 2p^6 3s^2 3p^6$
Kr	36	$1s^2 2s^2 2p^6 3s^2 3p^6 3d^{10} 4s^2 4p^6$
Xe	54	$1s^2 2s^2 2p^6 3s^2 3p^6 3d^{10} 4s^2 4p^6 4d^{10} 5s^2 5p^6$
Rn	86	$1s^2 2s^2 2p^6 3s^2 3p^6 3d^{10} 4s^2 4p^6 4d^{10} 4f^{14} 5s^2 5p^6 5d^{10} 6s^2 6p^6$

A careful examination of these configurations shows that none has any partially filled sublevels. The symbol of a noble gas enclosed in brackets is used to represent those filled sublevels. As an example, consider the electron configuration of bromine:

Br: $1s^2 2s^2 2p^6 3s^2 3p^6 3d^{10} 4s^2 4p^5$

The first 18 electrons are in the same orbitals as those of an atom of argon (see Table 5.3). If we use the symbol [Ar] to represent those 18 electrons, we can write the electron configuration of bromine as

Br: $[Ar]3d^{10}4s^2 4p^5$

This device is useful because we can write electron configurations more quickly. More importantly, this notation emphasizes the electron configurations in the higher energy levels, where the differences are important in determining the chemistry of an element. This use of the noble gases to represent certain configurations is known as **core notation.** The symbol of a noble gas enclosed in brackets represents the inner, filled orbitals of an element. Additional electrons are shown outside the brackets in the standard way. Note that *only* the noble gases can be used in core notation. When using this method, remember that, even though the inner configuration of an element may be written the same as that of a noble gas, the energies of these inner electrons are slightly different.

Example 5.7   Using core notation, write the electron configuration of lead (atomic number 82).

Solution   Lead has 82 electrons. The nearest noble gas of fewer electrons is xenon (atomic number 54). Let [Xe] represent the first 54 electrons of lead. Xenon is at the end of period 5 of the table, therefore its outermost electrons must be $5p$ electrons. The configuration of the next 28 electrons can be written by using Figure 5.9 and following the arrows after $5p$. Filling the sublevels in order of energy, we obtain:

Pb: $[Xe]6s^2 4f^{14} 5d^{10} 6p^2$

Expressing this configuration in order of principal energy level, we get:

Pb: $[Xe]4f^{14} 5d^{10} 6s^2 6p^2$

Recall that either notation is correct.

Problem 5.7   Using core notation, write the electron configuration of: **a.** strontium and **b.** arsenic.

Table 5.4 shows, in core notation, the electron configurations of the elements in Groups I and VI of the periodic table. Notice how this method emphasizes the similar structure of the elements in a single column.

TABLE 5.4   Electron configurations of elements in Groups I and VI, using core notations

Group I		Group VI	
H	$1s^1$		
Li	$[He]2s^1$	O	$[He]2s^22p^4$
Na	$[Ne]3s^1$	S	$[Ne]3s^23p^4$
K	$[Ar]4s^1$	Se	$[Ar]4s^23d^{10}4p^4$
Rb	$[Kr]5s^1$	Te	$[Kr]5s^24d^{10}5p^4$
Cs	$[Xe]6s^1$	Po	$[Xe]6s^24f^{14}5d^{10}6p^4$
Fr	$[Rn]7s^1$		

D. Valence Electrons

In discussing the chemical properties of an element, we often focus on electrons in the outermost occupied energy level. These outer-shell electrons are called **valence electrons,** and the energy level they occupy is called the valence shell. Valence electrons participate in chemical bonding and chemical reactions. The valence electrons of an element are shown by using a representation of the element called an **electron-dot structure** or **Lewis structure,** named after G. N. Lewis, the twentieth-century American chemist who first pointed out the importance of outer-shell electrons. A Lewis structure shows the symbol of the element surrounded by a number of dots equal to the number of electrons in the outer energy level of the atoms of the element.

You may have noticed in writing electron configurations that the s sublevel of a principal energy level n is always occupied before d electrons are added to the principal energy level numbered $n - 1$. Immediately after filling the d sublevel of principal level $n - 1$, the p sublevel of principal level n is filled, and the next sublevel filled will be the s sublevel of the $n + 1$ principal energy level. This order of filling is illustrated in the configurations of krypton, xenon, and radon in Table 5.3 and of selenium, tellurium, and polonium in Table 5.4. The significance of these observations is that, in the electron configuration of any atom, the principal energy level with the highest number that contains any electrons cannot contain more than eight electrons. This also means that the valence electrons of an atom are the s and p electrons in the occupied principal energy level of highest number. Consequently, no atom can have more than eight valence electrons.

In drawing the Lewis structure of an atom, we imagine a four-sided box around the symbol of the atom and consider that each side of that box corresponds to an orbital. We represent each valence electron as a dot. The first two valence electrons will be s electrons; they would be represented by two dots on a side (it doesn't matter which side) of the symbol. The valence electrons that are in the p subshell are placed first, one on each of the remaining sides of the symbol, and then a second one is added to each side. This method of filling is similar to the one used in drawing box diagrams of electron configurations. As an example, consider the Lewis structure of sodium.

Looking back at Table 5.4, we see that the core notation for sodium is $[Ne]3s^1$. This tells us that a sodium atom has one electron in its outer shell, so its Lewis structure is Na·. The core notation for selenium is $[Ar]3d^{10}4s^24p^4$. Its Lewis structure is ·S̈e:. The ten $3d$ electrons of selenium are not shown because they are not in the outer shell, which is the principal energy level 4. Lewis structures for the elements in the first three periods and Group II of the periodic table are shown in Table 5.5.

TABLE 5.5 Lewis structures for the elements of the first three periods and Group II

Period	I	II	III	IV	V	VI	VII	VIII
1	H·							He:
2	Li·	Be:	B̈:	·C̈:	·N̈:	·Ö:	:F̈:	:N̈e:
3	Na·	Mg:	Äl:	·S̈i:	·P̈:	·S̈:	:C̈l:	:Ä̈r:
4		Ca:						
5		Sr:						
6		Ba:						
7		Ra:						

Example 5.8 Give the Lewis structure of: **a.** gallium and **b.** bromine.

Solution

a. Gallium is in Group III. Its atoms have three electrons in the highest energy level. The Lewis structure is Ġa·.

b. Bromine is in Group VII. Its atoms have seven electrons in the highest energy level. Its Lewis structure is :B̈r:.

Problem 5.8 Give the Lewis structure of: **a.** potassium and **b.** arsenic.

5.6 Historic Classification of the Elements

Our discussion has implied that the arrangement of elements in the periodic table followed knowledge of electron configuration. That implication is incorrect. In the late eighteenth and early nineteenth centuries, many elements were discovered and their properties described; those elements with similar properties were grouped in families and the families arranged in a periodic table. Figure 5.12 shows the location of these families in the periodic table.

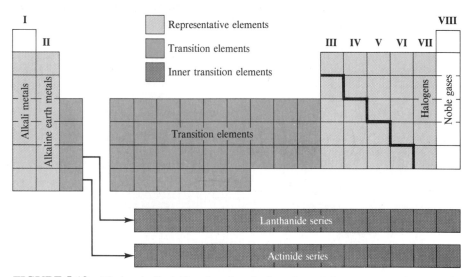

FIGURE 5.12   The periodic table, showing the location of the representative, transition, and inner transition elements.

A. Families of Elements

One family of elements is the **alkali metals:** lithium, sodium, potassium, rubidium, cesium, and francium. These elements, found in column I of the periodic table, have a single valence electron. They are all soft, silvery gray solids with a clearly metallic luster. They are all very reactive, and their reactions with water are vigorous. These elements do not occur free in nature but are found combined with other elements, often with chlorine. The most common of these compounds is sodium chloride (table salt). Sodium chloride, NaCl, is the substance that makes seawater salty; it is also found in huge underground deposits (salt mines).

The **alkaline earth metals,** another family, are the elements in column II of the periodic table. Their atoms have two valence electrons. These metals are harder and stronger than the alkali metals. They, too, are found in combination with other elements. Although all alkaline earth metals have similar chemical reactivity, they have a wider range of reactivity than do the alkali metals. Beryllium and magnesium are unaffected by water, calcium reacts slowly with boiling water, and barium reacts violently with cold water.

The elements of Group VII, with seven valence electrons, are known as the **halogens.** Chemically, these elements are very similar; physically, they are less alike. The lightest halogen, fluorine, is a pale yellow gas; iodine is a shiny, black solid. The heaviest halogen, astatine, is quite rare and is found in uranium ores. The total amount of astatine in the Earth's crust is probably less than 1 g. The longest-lived isotope of this element, astatine-210, has a half-life of only 8.3 hours. Several characteristic properties of halogens are shown in Table 5.6. Notice how these properties change as the atomic number increases.

TABLE 5.6 **Properties of halogens**

	Symbol	*Atomic number*	*Lewis structure*	*Atomic weight*	*Radius* $(\times 10^{-8}$ cm)	*Melting point, °C*
fluorine	F	9	$:\ddot{F}:$	19.0	0.72	-219.6
chlorine	Cl	17	$:\ddot{C}l:$	35.5	0.99	-101.0
bromine	Br	35	$:\ddot{B}r:$	79.9	1.14	-7.2
iodine	I	53	$:\ddot{I}:$	126.9	1.33	113.7
astatine	At	85	$:\ddot{A}t:$	(210)	1.45	302

The elements in column VIII are the noble gases. Recall from Chapter 3 that all these elements are monatomic gases and occur free and uncombined. They are all singularly unreactive, so much so that they were known earlier as the "inert gases." Only in 1960 were any of them shown to take part in chemical reactions. Even now, only krypton and xenon are known to form chemical compounds with other elements.

The elements in other columns of the table do not show striking similarities of properties. These columns are crossed by the stair-step line that separates metals from nonmetals, so they have nonmetals at the top and metals at the bottom of the column. For example, Group IV is headed by the nonmetal carbon and has lead, a typical metal, as its heaviest member. Group V starts with the nonmetal nitrogen and ends with the metal bismuth.

B. Historical Development of the Periodic Table

By the middle of the nineteenth century, scientists knew the atomic weights of all the then-known elements and had observed that these weights progressed from 1 amu (hydrogen) to over 200 amu (lead). Efforts were made to find a systematic arrangement of all the elements that would place them both in order of increasing atomic weight and in groups with similar properties. Such an arrangement, to be known as the periodic table, was conceived in 1869 both by Julius Lothar **Meyer** (1830–1895) in Germany and by Dmitri **Mendeleev**

(1834–1907) in Russia. Mendeleev is usually given most of the credit for constructing the periodic table because he not only described the ordering of the known elements but also used this arrangement to predict the existence of other elements not yet discovered.

Mendeleev's arrangement of the elements is shown in Figure 5.13. It resembles today's periodic table but has fewer columns and a different overall shape. Our current Group VIII is missing because the noble gases had not yet been discovered. The shorter columns of the *d* block are missing; instead, the transition elements are placed as subgroups in the long columns. Nevertheless, Mendeleev's table is the ancestor of our periodic table.

Series	Group I	Group II	Group III	Group IV	Group V	Group VI	Group VII	Group VIII
1	H = 1							
2	Li = 7	Be = 9.1	B = 11	C = 12	N = 14	O = 16	F = 19	
3	Na = 23	Mg = 24.4	Al = 27	Si = 28	P = 31	S = 32	Cl = 35.5	
4	K = 39.1	Ca = 40	— = 44	Ti = 48.1	V = 51.2	Cr = 52.3	Mn = 55	Fe = 56, Ni = 58.5, Co 59.1, Cu 63.3.
5	(Cu) = 63.3	Zn = 65.4	— = 68	— = 72	As = 75	Se = 79	Br = 80	
6	Rb = 85.4	Sr = 87.5	Y = 89	Zr = 90.7	Nb = 94.2	Mo = 95.9	— = 100	Rh = 103, Ru = 103.8, Pd = 108, Ag = 107.9.
7	(Ag) = 107.9	Cd = 112	In = 113.7	Sn = 118	Sb = 120.3	Te = 125.2	I = 126.9	
8	Cs = 132.9	Ba = 137	La = 138.5	Ce = 141.5	Di = 145	—	—	— — —
9	(—)	—	—	—	—	—	—	
10	—	—	Yb = 173.2	—	Ta = 182.8	W = 184	—	Ir = 193.1, Pt = 194.8, Os = 200, Au = 196.7.
11	(Au) = 196.7	Hg = 200.4	Tl = 204.1	Pb = 206.9	Bi = 208	—	—	
12	—	—	—	Tb = 233.4	—	U = 239	—	— — —

FIGURE 5.13   Mendeleev's periodic table (1871). The formulas of the oxide and the hydride of an element determined which column it would be in. Notice the spaces left empty but filled later as new elements were discovered. The atomic weights shown are those accepted in 1871; many have since been changed.

In making this table, Mendeleev followed most closely his aim to put similar elements in the same column. This aim had two important results. First, if the elements did not match, spaces were left and properties predicted for the elements yet to be discovered. Germanium, one of those for which a space was left, was discovered in 1886 by using Mendeleev's predicted properties as a guide. Second, if two neighboring elements were very close in atomic weight and did not match their respective columns when placed in consecutive order, the order was reversed. Iodine and tellurium are two such elements. Although tellurium has a higher atomic weight, it is clearly not a halogen, but iodine is; thus, tellurium was placed in Group VI and iodine in Group VII.

Mendeleev's arrangement of the elements held until about 1912 as an illustration of his statement of the periodic law: The properties of the elements are periodic functions of their atomic weights. In 1912 H. G. J. **Moseley** (1887–1915), a student of Rutherford's, assigned to each element an atomic number that was equal to the charge on its nucleus and showed that this charge increased by one from one element to the next. Moseley's work suggested a restatement of the periodic law to read: The properties of the elements are periodic functions of their atomic numbers. This restatement resolved the tellurium–iodine problem as well as other similar problems. It recognized that atomic weights are based on isotopic distribution and that the real distinction between elements is the difference in the number of protons in their atoms. The periodic table we use today came into wide acceptance during the 1940s as the importance of electron configuration in predicting properties was recognized. Today we use the table as a summary of knowledge about the elements, whereas Mendeleev's table was held to be an interesting but not particularly useful artifact.

5.7   Properties That Can Be Predicted from the Periodic Table

A. Atomic Radius

Figure 5.14 shows the relative sizes of the atoms of the representative elements. Notice that atom size increases from top to bottom in a column and from right to left across a row. This trend is related to electron configuration. As we look at the elements in column I, for example, we see that the single valence electron for each successive element is in a higher principal energy level than the last, and the electron is thus farther away from the positively charged nucleus; hence, the **atomic radius** increases going from top to bottom. This same regular increase in size can be observed in each column of the periodic table.

Atoms decrease in size going across a period from left to right. For elements within a period, electrons are being added one by one to the same principal energy level. At the same time, protons are also being added one by one to the nucleus, increasing its positive charge. This increasing positive charge increases the attraction of the nucleus for all electrons and pulls them all closer to the nucleus, decreasing the atom's radius. Thus, atomic size is a periodic property that increases from top to bottom within a column and from right to left across a period.

Example 5.9   Arrange the following elements in order of increasing atomic radius: calcium, sulfur, magnesium.

Solution   Magnesium and calcium are both in column II of the periodic table. Because calcium is further down, its atoms will be larger than those of magnesium. Sulfur and magnesium are both in period 3. Sulfur is further to the right, so its atoms will be smaller. The order is: sulfur, magnesium, calcium.

Problem 5.9   Arrange the following elements in order of increasing atomic radius: fluorine, sodium, chlorine.

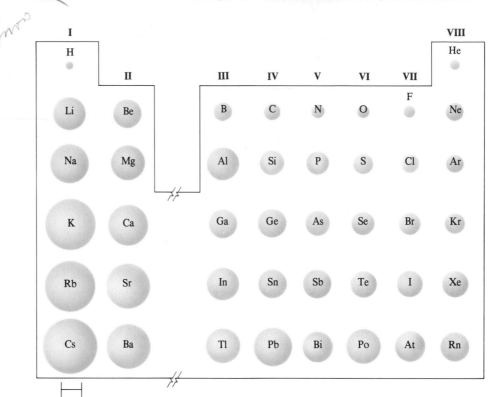

$(2 \times 10^{-10}$ m)

FIGURE 5.14   The relative sizes of the atoms of the representative elements.

B. Ionization Energy

The **ionization energy** of an element is the minimum energy required to remove an electron from a gaseous atom of that element, leaving a positive ion. An equation expressing the ionization of sodium would be:

$$Na \longrightarrow Na^+ + e^-$$

Electrons are held in the atom by the attractive force of the positively charged nucleus. The farther the outermost electrons are from the nucleus, the less tightly they are held. Thus, the ionization energy within a group of elements decreases as the elements increase in atomic number. Among the atoms of naturally occurring alkali metals, the single valence electron of cesium is farthest from the nucleus (in the sixth principal energy level), and we can correctly predict that the ionization energy of cesium is the lowest of all the alkali metals. (Recall that francium is not naturally occurring.)

From left to right across a period, the ionization energy of the elements tends to increase. The number of protons in the nucleus (the nuclear charge) increases, yet the valence electrons of the elements are in the same energy level.

It becomes increasingly more difficult to remove an electron from the atom. The ionization energy of chlorine is much greater than that of sodium, an element in the same period.

The ionization energies of elements 1 through 36 are plotted versus their atomic numbers in Figure 5.15. The peaks of the graph are the high ionization energies of the noble gases. The height of the peaks decreases as the number of the highest occupied energy level increases. The low points of the graph are the ionization energies of the alkali metals, which have only one electron in their valence shell. These points, too, decrease slightly as the number of the highest occupied energy level increases. The graph shows that ionization energy is periodically related to atomic number. Even within a row of the periodic table, the variations in ionization energy are closely related to electron configuration.

The ionization energy of an element is a measure of its metallic nature. From Figure 5.15, we see that each alkali metal has the lowest ionization energy of the elements in its period. Therefore, alkali metals are the most metallic elements. From bottom to top in the periodic table and from left to right across it, the metallic nature of the elements decreases.

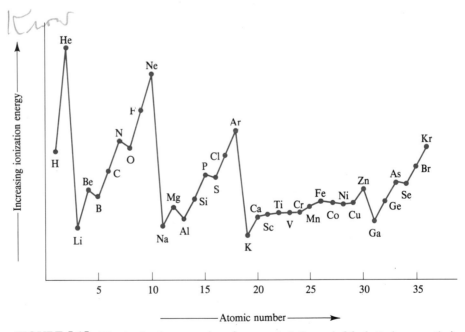

FIGURE 5.15 The ionization energies of elements 1 through 36 plotted versus their atomic numbers.

Nonmetals, located in the upper-right section of the periodic table, have high ionization energies. Except for the noble gases, fluorine has the highest ionization energy. Therefore, excluding the noble gases, fluorine is the least metallic (or most nonmetallic) element. From top to bottom in a column or to the left of fluorine, elements become more metallic. In summary, ionization energy increases from bottom to top of a column and from left to right across a period.

Example 5.10   **a.** Write the equation for the ionization (loss of an electron) of potassium.

b. Arrange the following elements in order of increasing ionization energy: potassium, calcium, fluorine.

Solution   **a.** The equation for the loss of an electron by potassium is:

$$K \longrightarrow K^+ + e^-$$

b. Fluorine, being in the upper-right corner of the periodic table, will have the highest ionization energy of these elements. Calcium is to the left and below fluorine; it will have a smaller ionization energy. Potassium is even further to the left and below calcium; its ionization energy will be even lower. The order is: potassium, calcium, fluorine.

Problem 5.10   **a.** Write the equation for the ionization of lithium.

b. Arrange the following elements in order of increasing ionization energy: chlorine, sodium, lithium.

Electron affinity is closely related to but the opposite of ionization energy. **Electron affinity** is the energy change that occurs when an electron is added to a neutral atom. For a nonmetal this change is usually a release of energy. The equation showing this reaction for chlorine is:

$$Cl_2 + 2\ e^- \longrightarrow 2\ Cl^-$$

Electron affinities are fairly difficult to measure. Accurate values have been determined for only a few elements. In general the values become increasingly negative from left to right across a period. Consequently, a halogen will have the most-negative electron affinity of all the representative elements in its period (remember that the noble gases are not representative elements). Electron affinity does not change with the same regularity as does atomic radius or ionization energy, and relative values cannot be predicted as easily.

Figure 5.16 summarizes how atomic radii, ionization energies, and metallic properties change within the periodic table.

C. The Formation of Ions

Atoms are electrically neutral. The number of positively charged protons in the nucleus of an atom equals the number of negatively charged electrons outside the nucleus. If electrons are added or lost as an atom reacts, the atom acquires a charge and becomes an **ion.**

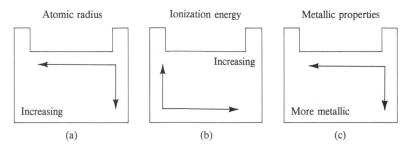

FIGURE 5.16 Trends of various atomic properties as related to position in the periodic table.

1. The octet rule

We have already observed that the noble gases are very unreactive (Section 5.6A). This lack of reactivity is attributable to a stable electron configuration. Looking back to Section 5.5C (Table 5.3), you can see that all the noble gases but helium have eight electrons (two s and six p) in the highest occupied energy level. When atoms of the other representative elements react, they lose, gain, or share enough electrons to attain the noble-gas electron structure — a complete octet, eight electrons, in their outer shell. This tendency is expressed by the **octet rule:** An atom generally reacts in ways that give it an octet of electrons in its outer shell. Hydrogen and lithium are exceptions; they react in ways that give them the same electron configuration as helium, with two outer-shell electrons.

An atom with one, two, or three valence electrons usually reacts by losing these electrons to acquire the electron configuration of the noble gas next below it in atomic number. An atom with six or seven valence electrons will usually react by adding enough electrons to acquire the electron configuration of the noble gas next above it in atomic number. Other atoms may attain a complete octet by sharing electrons with a neighboring atom (discussed in Section 7.1).

2. Positive ions, or cations

When a neutral atom loses an electron, it forms a positively charged ion, called a **cation** (pronounced "cát-i-on"). In general, metals lose electrons to form cations. The atom thereby attains the electron configuration of the noble gas next below it in atomic number.

For example, an alkali metal loses one electron to form a cation with a single positive charge. Sodium loses its single $3s$ valence electron to form the ion Na^+, which has the electron configuration of neon:

$$Na\cdot \longrightarrow Na^+ + e^-$$

An alkaline earth metal loses two electrons to form a cation with a charge of $+2$. In forming the magnesium ion, Mg^{2+}, a magnesium atom loses its two valence electrons:

$$Mg\colon \longrightarrow Mg^{2+} + 2\ e^-$$

Aluminum loses its three valence electrons to form a cation with a charge of $+3$:

$$\overset{\cdot}{Al}: \longrightarrow Al^{3+} + 3\ e^-$$

The names of these cations are the same as the metals from which they are formed (see Table 5.7).

TABLE 5.7   Cations of metals

Alkali metal cations		Alkaline earth metal cations		Other metal cations	
Symbol	Name	Symbol	Name	Symbol	Name
Li^+	lithium ion	Mg^{2+}	magnesium ion	Al^{3+}	aluminum ion
Na^+	sodium ion	Ca^{2+}	calcium ion		
K^+	potassium ion	Sr^{2+}	strontium ion		
Rb^+	rubidium ion	Ba^{2+}	barium ion		
Cs^+	cesium ion				

Transition elements and the metals to their right do not always follow the octet rule; frequently they form more than one cation. For example, iron forms Fe^{2+} and Fe^{3+}; cobalt forms Co^{2+} and Co^{3+}. The names of these ions must indicate the charge they carry. The preferred system of nomenclature (naming) is that recommended by the International Union of Pure and Applied Chemistry (IUPAC). In this system, the name of the metal is followed by a Roman numeral (in parentheses) showing the charge on the ion. No extra space is left between the name and the number. Thus, Fe^{2+} is iron(II) (pronounced "iron two"), and Fe^{3+} is iron(III). In the older system, the name of the cation of lower charge ends in *ous,* and the name of the cation of higher charge ends in *ic.* Examples of both systems of naming are given in Table 5.8.

TABLE 5.8   Naming of cations

Symbol	IUPAC name	Older name	Symbol	IUPAC name	Older name
Co^{2+}	cobalt(II)	cobaltous	Cr^{2+}	chromium(II)	chromous
Co^{3+}	cobalt(III)	cobaltic	Cr^{3+}	chromium(III)	chromic
Cu^+	copper(I)	cuprous	Fe^{2+}	iron(II)	ferrous
Cu^{2+}	copper(II)	cupric	Fe^{3+}	iron(III)	ferric

Notice that none of the cations discussed here have a charge greater than $+3$. When ions are formed, electrons are pulled off one by one from the atom. Thus, the first electron is removed from a neutral atom, the second electron from an ion of charge $+1$, the third electron from an ion of charge $+2$, and so on. The amount of energy necessary to remove an electron increases dramatically as the positive charge of the ion increases. To remove a fourth electron and form an ion of charge $+4$ is energetically unlikely.

3. Negative ions, or anions

When a neutral atom gains an electron, it forms a negatively charged ion, called an **anion** (pronounced "án-i-on"). Typically, nonmetals form anions, gaining enough electrons to acquire the electron configuration of the noble gas of next higher atomic number. Elements of group VI, with six valence electrons, form anions by gaining two electrons; the halogens, with seven valence electrons, form anions by gaining one electron. The names of these anions include the root name of the element and the ending *ide*. Table 5.9 lists several common anions and their names; in each case, the root of the name is italicized.

TABLE 5.9 Anions of nonmetals

Symbol	Name	Symbol	Name
F^-	*fluor*ide ion	O^{2-}	*ox*ide ion
Cl^-	*chlor*ide ion	S^{2-}	*sulf*ide ion
Br^-	*brom*ide ion		
I^-	*iod*ide ion		

4. Polyatomic ions

The ions described in the preceding paragraphs are monatomic ions; that is, each contains only one atom. Many polyatomic ions are also known. **Polyatomic ions** are groups of atoms bonded together that carry a charge due to an excess or deficiency of electrons. Table 5.10 lists the formulas and names of several common polyatomic ions. The symbols in the formula show which elements are present. The subscripts ("1" is understood) tell how many atoms of each element are present in the ion.

TABLE 5.10 Common polyatomic ions

Charge	Formula	Name	Charge	Formula	Name
+1	NH_4^+	ammonium ion	−2	CO_3^{2-}	carbonate ion
−1	OH^-	hydroxide ion		SO_4^{2-}	sulfate ion
	NO_3^-	nitrate ion	−3	PO_4^{3-}	phosphate ion
	HCO_3^-	bicarbonate ion			

Example 5.11   Show how the following elements follow the octet rule in forming ions: **a.** magnesium and **b.** sulfur.

Solution   **a.** The electron configuration of magnesium is:

$$1s^2 2s^2 2p^6 3s^2 \quad \text{or} \quad [\text{Ne}]3s^2$$

Magnesium, having two valence electrons, loses two electrons to form the ion Mg^{2+}. The electron configuration of this ion is:

$1s^2 2s^2 2p^6$ or   [Ne]

a structure with a complete octet in the outer energy level.

b. The electron configuration of sulfur is:

$1s^2 2s^2 2p^6 3s^2 3p^4$ or $[Ne]3s^2 3p^4$

Sulfur has six valence electrons and therefore forms the ion S^{2-} by adding two electrons. This ion has the electron configuration:

$1s^2 2s^2 2p^6 3s^2 3p^6$ or   [Ar]

The outer occupied energy level contains a complete octet.

Problem 5.11   Show how the following elements follow the octet rule in forming ions: **a.** potassium and **b.** bromine.

D. Metals and Nonmetals; Acids and Bases

So far we have shown that metals usually have one, two, or three valence electrons. They have low ionization energies and are found to the left in the periodic table. Nonmetals have four, five, six, or seven valence electrons, have high ionization energies, and are in the upper-right section of the periodic table. All these properties are closely related to electron configurations, and we have used electron configurations in discussing them. However, before electron configurations were known, people did know the differences between metals and nonmetals.

Early chemists observed how the physical properties of a metal (malleability, luster, and conductivity), described in Section 3.3B1, contrasted with those of nonmetals. These chemists also identified differences in chemical properties. The compounds formed when oxides of metals react with water are very much alike and very different from those formed when oxides of nonmetals react with water.

When a **metallic oxide** reacts with water, a hydroxide is formed:

Metal oxide	Equation	Hydroxide
sodium oxide	$Na_2O + H_2O \longrightarrow 2\ NaOH$	sodium hydroxide
magnesium oxide	$MgO + H_2O \longrightarrow Mg(OH)_2$	magnesium hydroxide
aluminum oxide	$Al_2O_3 + 3\ H_2O \longrightarrow 2\ Al(OH)_3$	aluminum hydroxide

When a **nonmetallic oxide** reacts with water, an acid is formed:

Nonmetal oxide	Equation	Acid
carbon dioxide	$CO_2 + H_2O \longrightarrow H_2CO_3$	carbonic acid
sulfur trioxide	$SO_3 + H_2O \longrightarrow H_2SO_4$	sulphuric acid
oxide of phosphorus	$P_4O_{10} + 6\ H_2O \longrightarrow 4\ H_3PO_4$	phosphoric acid

Table 5.11 lists several common hydroxides and acids.

TABLE 5.11   Common hydroxides and acids

Common hydroxides		Common acids	
sodium hydroxide	NaOH	hydrochloric acid	HCl
potassium hydroxide	KOH	acetic acid	$HC_2H_3O_2$
calcium hydroxide	$Ca(OH)_2$	nitric acid	HNO_3
aluminum hydroxide	$Al(OH)_3$	sulfuric acid	H_2SO_4
ammonium hydroxide	NH_4OH	carbonic acid	H_2CO_3
		phosphoric acid	H_3PO_4

Hydroxides are a subset of a larger group called **bases,** although not all bases are hydroxides. In all but ammonium hydroxide, the cation of a hydroxide is a metallic ion. A hydroxide dissolves in water to yield hydroxide ion. The solution of a hydroxide feels slippery because of the action of these ions on the skin. (You may have noticed this property in household ammonia, a dilute solution of ammonium hydroxide.) The solution of a hydroxide gives a class of compounds called **indicators** characteristic colors (see Table 5.12).

TABLE 5.12   Properties of acids and hydroxides

	Acids	Hydroxides
In aqueous solutions	release H^+	release OH^-
Indicators		
litmus	red	blue
phenolphthalein	colorless	red
methyl orange	red	yellow
Other properties	taste sour	feel slippery

Most common **acids** contain hydrogen, a nonmetal, and frequently oxygen. Acids dissolve in water to yield hydrogen ions. The solution of an acid tastes sour because of the hydrogen ions it contains. Indicators in acid solutions show different colors than they do in solutions of hydroxides.

Example 5.12

Write balanced equations to show:

a. the reaction of iron with oxygen to form iron(III) oxide, Fe_2O_3, and the reaction of the oxide with water to form iron(III) hydroxide, $Fe(OH)_3$

b. the reaction of sulfur with oxygen to form sulfur dioxide, SO_2, and the reaction of the oxide with water to form sulfurous acid, H_2SO_3.

Solution

a. Iron is a metal; its oxide reacts with water to form a hydroxide.

$$4\ Fe + 3\ O_2 \longrightarrow 2\ Fe_2O_3 \qquad Fe_2O_3 + 3\ H_2O \longrightarrow 2\ Fe(OH)_3$$

b. Sulfur is a nonmetal; its oxides react with water to form acids.

$$S + O_2 \longrightarrow SO_2 \qquad SO_2 + H_2O \longrightarrow H_2SO_3$$

Problem 5.12

Write balanced equations to show:

a. the reaction of calcium with oxygen to form calcium oxide, CaO, and the reaction of that oxide with water to form calcium hydroxide

b. the reaction of chlorine with oxygen to form dichloropentoxide, Cl_2O_5, and the reaction of that oxide with water to form chloric acid, $HClO_3$.

5.8   Summary

Our knowledge of the electron configuration of atoms is based on studies of the properties of light and of the spectra of atoms. Because of the electrons' tiny mass and high energies, their properties are similar to those of light.

Each electron in a ground-state atom has unique energy properties. In describing these properties, we assume that the nucleus of an atom is surrounded by energy levels; the energy associated with these levels increases as their distance from the nucleus increases. Each energy level is divided into sublevels that differ slightly from one another in energy. The sublevels are divided into orbitals, each of which can contain two electrons spinning in opposite directions on their axes. The electron configuration of a particular atom is described by stating how many electrons are in each sublevel of a particular principal energy level. If electron spin is important, box diagrams are used. Core notation emphasizes the outer-level electrons. If only the valence electrons are of interest, Lewis structures are used.

Although the periodicity of elements was clearly recognized in 1869 when the periodic table was first proposed, it is only in this century that the relationship of an element's electron configuration to its location in the periodic table has been clearly defined. Elements in the same long columns of the table have the same valence-shell configuration and have very similar chemical properties. Elements in the same row of the table show a reasonably regular progression of properties, particularly with respect to atomic radius, metallic nature, and ionization potential.

Elements lose or gain electrons to become ions. Metals become cations by losing electrons, thus gaining a positive charge. Nonmetals become anions by gaining electrons, thus attaining a negative charge. Polyatomic ions are groups of atoms with an excess or deficiency of electrons.

The oxides of metals react with water to yield hydroxide ions; hydroxides are a subset of a larger group called bases. Oxides of nonmetals typically react with water to form acids. The solution of an acid contains an excess of hydrogen ions; that of a base, an excess of hydroxide ions.

Key Terms

absorption spectrum (5.1)
acids (5.7D)
actinide series (5.5B3)
alkali metals (5.6A)
alkaline earth metals (5.6A)
anion (5.7C3)
atomic radius (5.7A)
bases (5.7D)
box diagrams (5.4A)
cation (5.7C2)
continuous light (5.1)
core notation (5.5C)
electromagnetic radiation (5.1)
electromagnetic spectrum (5.1)
electron affinity (5.7B)
electron configuration (5.4)
electron-dot structure (5.5D)
electron shells (5.3A)
electron spin (5.3B)
emission spectrum (5.1)
energy levels (5.2)
excited state (5.3C)
ground state (5.3C)
halogens (5.6A)
hydroxides (5.7D)
indicators (5.7D)

inner transition elements (5.5B3)
ionization energy (5.7B)
ions (5.7C)
lanthanide series (5.5B3)
Lewis structure (5.5D)
Mendeleev (5.6B)
metallic oxides (5.7D)
Meyer (5.6B)
Moseley (5.6B)
noble gases (5.5C, 5.6A)
nonmetallic oxides (5.7D)
octet rule (5.7C1)
orbital (5.3B)
orbital shapes (5.3B1)
periods (of periodic table) (5.5)
polyatomic ions (5.7C4)
principal energy levels (5.3A)
quantized (5.2)
radiant energy (5.1)
representative elements (5.5B1)
sublevels (5.3B)
transition elements (5.5B2)
valence electrons (5.5D)
visible light (5.1)
wave frequency, v (5.1)
wavelength, λ (5.1)

Multiple-Choice Questions

MC1. The electromagnetic spectrum contains the following regions:
1. X rays   **2.** yellow light   **3.** blue light   **4.** infrared
In order of increasing energy, they would be arranged:
a. 1, 2, 3, 4   **b.** 4, 3, 2, 1   **c.** 4, 2, 3, 1   **d.** 1, 3, 2, 4
e. None of these arrangements is in order of increasing energy.

MC2. Which of the following statements are always true of an electron?
1. It has wave-like properties.
2. There are restrictions on the amounts by which its energy can change.
3. Its location in the atom can be exactly pinpointed.
4. If it is a *p* electron, it moves in a figure-8 path around the nucleus of the atom.
5. All electrons in the third principal energy level of an atom have identical energies.
a. 1 and 2   **b.** 1, 2, and 4   **c.** 1, 2, and 5   **d.** 2, 4, and 5
e. 3, 4, and 5

MC3. When an electron makes the transition shown in the figure,

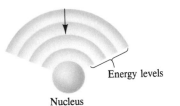

Nucleus

1. energy is emitted.   **2.** energy is absorbed.
3. the electron loses energy.   **4.** the electron gains energy.
5. The electron cannot make this transition.
a. 1 and 4   **b.** 1 and 3   **c.** 2 and 3   **d.** 2 and 4   **e.** 5

MC4. For an atom with 16 electrons, how many principal energy levels in the ground-state atom will contain electrons?
a. 0   **b.** 1   **c.** 2   **d.** 3   **e.** 4

MC5. An atom of magnesium in the unexcited (ground) state would have no electrons in an orbital shaped:

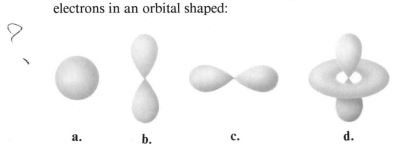

a.   **b.**   **c.**   **d.**

     e. It would have electrons in orbitals of all these shapes.

MC6. What is the electron configuration of calcium?
     a. $1s^2 2s^2 2p^6 3s^2 3p^6 4s^2$ **b.** $1s^2 2s^2 2p^6 2d^{10}$
     c. $1s^2 2s^2 2p^6 3s^2 3p^6 3d^2$ **d.** $1s^2 2s^3 2p^6 3s^3 3p^3 4s^2$
     e. $1s^2 2s^3 2p^5 3s^4 3p^5 4s^1$

MC7. Which of the following statements is not true of germanium?
     a. Its atoms are larger than those of arsenic.
     b. Its Lewis structure shows five valence electrons.
     c. It is more metallic than carbon.
     d. It has a lower ionization energy than selenium.
     e. It is less metallic than tin.

MC8. When a sodium atom becomes an ion, it loses an electron. Which of the following are true?
     1. The sodium ion then has a charge of $+1$.
     2. The sodium ion then has a charge of -1.
     3. The electron comes from within the nucleus.
     4. The electron comes from outside the nucleus.
     5. The mass number of the sodium atom changes.
     a. 1 and 3   **b.** 1 and 4   **c.** 2 and 3   **d.** 2 and 4
     e. 1, 4, and 5

MC9. Which of these oxides will dissolve in water to form a solution that turns litmus red?
     1. CO_2 **2.** NO_2 **3.** SO_2 **4.** P_2O_5 **5.** Cl_2O_7
     a. none   **b.** all   **c.** 1, 2, and 3   **d.** 4 and 5   **e.** 1

MC10. Which of the following elements is a transition element?
     a. iron   **b.** magnesium   **c.** silicon   **d.** sulfur   **e.** iodine

Problems

5.3 An Atomic Model

*5.13. Draw a diagram of an atom showing its nucleus and its first six energy levels. Draw arrows showing:
     a. three energy-emitting transitions to the third energy level.
     b. two energy-absorbing transitions to the fourth energy level.

5.14. Arrange the following sublevels in order of increasing energy in a ground-state atom:
     a. $2s$, $4d$, $1s$, $5p$
     b. $2p$, $6s$, $4f$, $3p$, $5s$

5.4 The Electron Configurations of Atoms

*5.15. Write the complete electron configuration of the following elements:
     a. magnesium   **b.** phosphorus
     c. argon   **d.** oxygen
     e. chlorine   **f.** sodium

5.16. Write the complete electron configuration of the following elements:
     a. silicon   **b.** tin
     c. vanadium   **d.** lead
     e. aluminum   **f.** sulfur
     g. iodine   **h.** barium

*5.17. Using box diagrams, show the electron configuration of the elements in Problem 5.16.

5.5 The Periodic Table

*5.18. What characterizes the electron configuration of:
   a. elements in the same column of the periodic table?
   b. elements in the same period of the table?
   c. the noble gases?

5.19. Use core notation to give the electron configuration of the elements in Problem 5.15.

5.20. Classify the following elements as representative, transition, or inner transition elements: iron, manganese, selenium, strontium, silicon, and neodymium.

5.21. Classify all those elements whose name begins with "C" as representative, transition, or inner transition elements.

5.22. Write the complete electron configuration of the following elements. Tell how this configuration relates to the element's position in the periodic table.
   a. calcium   **b.** antimony
   c. boron   **d.** radium
   e. potassium   **f.** gallium
   g. germanium

5.23. Write the complete electron configuration of each of the alkaline earth metals.

*5.24. For the elements in period 3 of the periodic table,
   a. give the complete electron configuration.
   b. give the electron configuration using core notation.

5.25. Draw the Lewis (electron-dot) structures of the alkali metals.

*5.26. Draw the Lewis structure of an atom of each of the following elements:

   a. sulfur   **b.** chlorine
   c. magnesium   **d.** tellurium
   e. carbon   **f.** boron
   g. lithium   **h.** barium

5.27. Draw the Lewis structure of the elements in Problem 5.22.

5.28. Draw the Lewis structure of the elements in Problem 5.15.

5.29. Draw the Lewis structure of the members of the halogen family.

5.7 Properties That Can Be Predicted from the Periodic Table

*5.30. a. Indicate how each of the following properties is related to position in the periodic table: metallic properties, nonmetallic properties, ionization energy, atomic radius.
   b. How is each property in part **a** above related to electron configuration?

5.31. Classify the following elements as metals or nonmetals:
   a. vanadium   **b.** palladium
   c. selenium   **d.** sulfur
   e. zinc   **f.** fluorine

*5.32. Element number 117 has not yet been discovered. When (and if) it is discovered, we expect it will be in column VII of the periodic table; explain why. Predict the following properties for this element, comparing them to those of a known element: metal or nonmetal, atomic radius, ionization energy, Lewis structure.

*5.33. Which element in each of the following pairs has a higher ionization energy?
   a. cesium, cerium
   b. arsenic, bismuth
   c. aluminum, silicon
   d. iodine, bromine

5.34. Which element in each of the following pairs has a larger atomic radius?

a. potassium, rubidium
b. nitrogen, arsenic
c. aluminum, sulfur
d. carbon, oxygen

5.35. Arrange the following elements in order of increasing ionization energy:
a. Be, Mg, Sr   b. Na, Al, S
c. Bi, Cs, Ba

5.36. Write the complete electron configuration of the following ions:
a. Se^{2-} b. Br^-
c. Rb^+ d. Sr^{2+}
Which noble gas has the same electron configuration as these ions?

***5.37.** The following ions are known to exist:
a. Cs^+ b. Ga^+
c. Te^{2+} d. Bi^{3+}
e. Pb^{2+}
Which are exceptions to the octet rule?

5.38. Name each of the following ions in two different ways:
a. Fe^{2+}, Fe^{3+} b. Cr^{2+}, Cr^{3+}
c. Cu^+, Cu^{2+} d. Ni^{2+}, Ni^{3+}

5.39. Write the formulas of the following cations:
a. iron(III)   b. tin(II)
c. platinum(II)   d. osmium(III)
e. lead(II)   f. mercury(II)

***5.40.** Write the formula and give the name of the monatomic anion formed by the following elements:
a. iodine   b. oxygen
c. bromine   d. sulfur
e. fluorine

5.41. Write the formula of the monatomic cation formed by the following elements:
a. aluminum   b. strontium
c. cesium   d. lithium
e. magnesium   f. potassium

5.42. Complete the following table.

Name of ion	Formula
	NH_4^+
Nitrate	NO_3^-
Bicarbonate	HCO_3^-
Sulfate	SO_4
Carbonate	CO_3^{2-}
Phosphate	PO_4^3

***5.43.** Write balanced equations showing the formation of the oxides of the elements in Group II by the reaction of the element with oxygen. Write equations that show the reactions of these oxides with water.

Review Problems

5.44. Given the following data, calculate the density of the nucleus of an atom of sodium-23.

mass of proton = 1.007 amu
mass of neutron = 1.008 amu
mass of electron = 5.45×10^{-4} amu
volume of sphere = $(4/3)\pi r^3$
1 amu = 1.67×10^{-24} g
diameter of nucleus = 1.16×10^{-14} m
diameter of atom = 3.08×10^{-10} m

5.45. Plutonium is very harmful to humans. The Atomic Energy Commission recommends that the concentration of plutonium in the air be no greater than 3.00×10^{-11} g/m³. How many atoms of plutonium are contained in a cubic meter at this concentration? (The atomic weight of plutonium is 244.)

5.46. The density of copper is 8.96 g/cm³. What is the mass of a block of copper that measures 1.65 cm $\times$ 1.02 cm $\times$ 0.921 cm? How many atoms of copper are in this sample? What mass of lead will contain the same number of atoms?

5.47. The addition of one part per million of fluorine (1 g fluorine/10^6 g water) to drinking water has caused a dramatic decrease in dental cavities in the population served by such water supplies. Evanston, Illinois, is one of the communities that add fluorine to their drinking water. If you drink one glass (250 mL) of Evan-

ston water, how many atoms of fluorine do you imbibe? Assume the density of Evanston's drinking water is the same as pure water.

5.48. The mass of an electron is 5.45×10^{-4} amu. In a uranium atom of mass 238 amu, what percent of the total mass is contributed by the 92 electrons in the atom? What percent of the mass is contributed by the 92 protons? What percent of the mass is contributed by the neutrons in the atom?

$$\bar{e} = 5.014 \times 10^{-2}$$

$$\rho =$$

▪ 6 ▪

Compounds I: Names and Formulas

Elements combine to form compounds. From various combinations of the hundred or so elements, millions of compounds are formed each of which is unique. In this chapter we will discuss how compounds are named and their formulas determined. This discussion will include:

1. The various categories of compounds.
2. How compounds are named.
3. Calculation of the formula weight and percent composition of a compound.
4. How the formula of a compound can be determined from its percent composition.
5. The difference between the empirical and molecular formulas of a compound.

Categories of Compounds

A compound is a chemical combination of elements. It has a constant composition and a unique set of properties. Its composition is represented by its formula, which lists the symbols of the elements it contains, with each symbol followed by a subscript that tells how many atoms of that element are contained in the simplest unit of the compound.

Some compounds exist as **molecules.** These **molecular compounds** usually contain only nonmetals. Table 6.1 lists several molecular compounds, their common names, and their melting points.

TABLE 6.1   Some molecular compounds

Common name	Formula	Melting point, °C
acetone	C_3H_6O	-95
ammonia	NH_3	-77
cane sugar	$C_{12}H_{22}O_{11}$	185
chloroform	$CHCl_3$	-63
ethyl alcohol	C_2H_5OH	-117
hydrazine	N_2H_4	1.4
phosgene	Cl_2CO	-118
water	H_2O	0

Other compounds are ionic. These compounds are formed by the combination of ions. (The names and formulas of ions were discussed in Section 5.7C.) **Ionic compounds** include hydroxides (Section 5.7D) and salts. A **salt** is an ionic compound that contains a cation other than hydrogen and an anion other than hydroxide. Many acids exist as ions in solution but as molecules in the pure state.

The formula of an ionic compound is neutral. The ratio of the cations and anions it contains is such that there is no excess charge. Thus, the combination of a sodium ion, Na^+, with a sulfate ion, SO_4^{2-}, to form sodium sulfate must be in a $2:1$ ratio so that the resulting compound is neutral, Na_2SO_4. The combination of an aluminum ion, Al^{3+}, with a chloride ion, Cl^-, to form aluminum chloride must be in a $1:3$ ratio, giving the formula $AlCl_3$, which is neutral. The combination of an ammonium ion, NH_4^+, with a phosphate ion, PO_4^{3-}, to form ammonium phosphate must be in a $3:1$ ratio, giving the neutral formula $(NH_4)_3PO_4$. Notice in this last formula that the ammonium ion is enclosed in parentheses and followed by the subscript 3. This notation means that the whole ion is taken three times. When a polyatomic ion is taken more than once in a formula, it is enclosed in parentheses and the number of ions contained in the formula is indicated by a subscript following the parentheses. Monatomic ions and polyatomic ions taken only once (for example, the sulfate ion in sodium sulfate) are not enclosed in parentheses.

Example 6.1   Write the formulas of the following compounds:

a. potassium sulfide   **b.** magnesium nitrate   **c.** aluminum sulfate

Solution   **a.** The potassium ion is K^+ (Table 5.7), and the sulfide ion is S^{2-} (Table 5.9). Their neutral combination must have two K^+ to one S^{2-}, so potassium sulfide is written K_2S.

b. Magnesium ion is Mg^{2+}; nitrate ion is NO_3^- (Table 5.10). Their neutral combination must combine one Mg^{2+} with two NO_3^-, so magnesium nitrate is $Mg(NO_3)_2$.

c. Aluminum ion is Al^{3+}; sulfate ion is SO_4^{2-}. The smallest number that is divisible by both 2 and 3 is 6. If we combine two Al^{3+} ions with three SO_4^{2-} ions, the charges will be balanced and the compound neutral, as required. The formula of aluminum sulfate is $Al_2(SO_4)_2$.

Problem 6.1   Write the formulas of:

a. calcium chloride   **b.** barium nitrate   **c.** magnesium phosphate

Notice two things about Example 6.1: (1) In both **b** and **c**, the polyatomic ion was enclosed in parentheses because it was taken more than once; the monatomic ion was not enclosed in parentheses regardless of its subscript. (2) If the charge on the two ions differs in magnitude, the number of times the cation is taken equals the magnitude of the charge on the anion. Similarly, the number of times the anion is taken equals the magnitude of the charge on the cation. For example,

$$K_2S \qquad Mg(NO_3)_2 \qquad Al_2(SO_4)_3$$

$$K^+ \quad S^{2-} \qquad Mg^{2+} \quad NO_3^- \qquad Al^{3+} \quad SO_4^{2-}$$

6.2 Naming Compounds

Each compound has a name. Ideally, this name should indicate the composition of the compound and perhaps something of its properties. Such names are called **systematic names** and are based on a set of rules drawn up by IUPAC. Although all compounds have systematic names, many also have trivial, or common, names. Table 6.1 lists the common (trivial) names of some molecular compounds. Several ionic compounds are listed in Table 6.2, with both their common and systematic names.

TABLE 6.2   Names and formulas of some common ionic compounds

Common name	Systematic name	Formula
bleach	sodium hypochlorite	NaOCl
chalk	calcium carbonate	$CaCO_3$
lime	calcium oxide	CaO
milk of magnesia	magnesium hydroxide	$Mg(OH)_2$

A. Oxidation Numbers

Many of the rules by which names are assigned are based on the concept of oxidation numbers. The **oxidation number** of an element represents the positive or negative character (nature) of an atom of that element in a particular bonding situation. Oxidation numbers are assigned according to the following rules:

1. The oxidation number of an uncombined element is 0. In the equation

$$Zn + 2 HCl \longrightarrow H_2 + ZnCl_2$$

   the oxidation number of zinc (Zn) as an uncombined atom is 0, and the oxidation number of hydrogen in H_2 is 0.

2. The oxidation number of a monatomic ion is the charge on that ion. In $ZnCl_2$, the oxidation number of chlorine as Cl^- is -1 and that of zinc as Zn^{2+} is $+2$. In Ag_2S, the oxidation number of silver as Ag^+ is $+1$ and that of sulfur as S^{2-} is -2.

3. Hydrogen in a compound usually has the oxidation number $+1$. An exception to this rule occurs when hydrogen is bonded to a metal.

4. Oxygen in a compound usually has the oxidation number -2. Peroxides are an exception to this rule: In hydrogen peroxide, H_2O_2, for example, the oxidation number of oxygen is -1.

5. The sum of the oxidation numbers of the atoms in a compound is 0. For example, in the compound $ZnCl_2$, the oxidation number of the zinc ion is $+2$ and that of each chloride ion is -1. The sum of these oxidation numbers ($+2$ for zinc and -2 for the two chloride ions) is 0.

6. In a polyatomic ion, the net charge on the ion is the sum of the oxidation numbers of the atoms in the ion. We can use this rule to calculate the oxidation number of nitrogen in the nitrate ion, NO_3^-, by setting up the following equation:

   Oxidation number of nitrogen
   $$+ 3 \text{ (oxidation number of oxygen)} = -1$$
   Oxidation number of oxygen $= -2$

By substituting, we get:

Oxidation number of nitrogen $+ 3(-2) = -1$

By rearranging, this equation becomes:

Oxidation number of nitrogen $= -1 - 3(-2)$

$$= -1 + 6 = +5$$

Example 6.2   Assign an oxidation number to each atom in the following compounds and polyatomic ions:

   a. CO_2 **b.** SO_4^{2-} **c.** NH_4^+

Solution   **a.** CO_2: oxygen $= -2$ (rule 4), carbon $= +4$ (rule 5)

   b. SO_4^{2-}: oxygen $= -2$ (rule 4), sulfur $= +6$ (rule 6)

   c. NH_4^+: hydrogen $= +1$ (rule 3), nitrogen $= -3$ (rule 6)

Problem 6.2   Assign an oxidation number to each atom in the following compounds:

   a. Fe_2O_3 **b.** $NaMnO_4$ **c.** NO_2

B. Binary Compounds

Many chemical compounds are **binary;** that is, they contain two elements. Binary compounds are of several varieties.

1. Binary compounds containing a metal and a nonmetal

Binary compounds of a metal and a nonmetal contain a metallic cation and a nonmetallic anion. The names and formulas of cations and anions were introduced in Section 5.7 (Tables 5.7–5.9). Recall that the alkali metals form only ions with a $+1$ charge, the alkaline earth metals form only ions with a $+2$ charge, and aluminum forms only the ion Al^{3+}. For these ions, the name of the element followed by the term *ion* is an unambiguous name. For example, the sodium ion can only be Na^+, the calcium ion only Ca^{2+}. According to IUPAC rules, the names of all other metallic cations contain the name of the element followed by its oxidation state (in parentheses) in that ion. This rule prevents ambiguity. This particular rule is frequently called the Stock System of nomenclature. The name *chromium ion* does not say whether the ion is Cr^{2+} or Cr^{3+}; the proper names for these ions are chromium(II) and chromium(III). The anions in binary compounds are named by using the root name of the element, followed by the suffix *ide;* for example, *brom*ide ion is Br^-, the *sulf*ide ion is S^{2-}, and the *ox*ide ion is O^{2-}. In these examples, the root name of the element is italicized. For binary compounds, the cation is named first and the anion second. Thus,

$NiCl_2$ is nickel(II) chloride

K_2S is potassium sulfide

$CaBr_2$ is calcium bromide

ZnO is zinc(II) oxide

Example 6.3

Name the following compounds:

a. $CrCl_3$ **b.** MgI_2 **c.** Fe_2O_3

Solution

a. $CrCl_3$: The anion is the chloride ion, Cl^-. Note the ending of the name has changed from the *ine* of the element to the *ide* of the ion. There are three chloride ions for each chromium cation; therefore the chromium ion has a $+3$ charge and is named chromium(III). The compound is chromium(III) chloride.

b. MgI_2: The anion is iodide, I^-. Again, note the change in the ending from *ine* to *ide*. Magnesium is an alkaline earth metal; therefore, its oxidation number of $+2$ need not be included in its name. The compound is magnesium iodide.

c. Fe_2O_3: The anion is the oxide ion O^{2-}, which has a -2 charge. Note the change from *oxygen* to *oxide*. There are three oxide ions, giving a total negative charge of -6. The neutral compound contains two iron cations; together they have a charge of $+6$. Each iron cation then must have a charge of $+3$. Therefore, the cation is iron(III) and the compound is iron(III) oxide.

Problem 6.3

Name the following compounds:

a. CuO **b.** FeS **c.** $SrCl_2$

Before leaving this group of compounds, we should mention again the second and less-preferred method of naming cations of the same element in different oxidation states. This older method gives the ending *ous* to the ion of lower oxidation state and the ending *ic* to the ion of higher oxidation state. Often this system also uses the Latin root of the name of the element. Thus, in this system, Fe^{2+} is ferrous and Fe^{3+} is ferric; Pb^{2+} is plumbous and Pb^{4+} is plumbic. Those elements that use Latin roots are shown in Table 6.3.

TABLE 6.3 Some elements with non-English root names (root is italicized)

Element	Latin name	Element	Latin name
copper	*cup*rum	lead	*plumb*um
gold	*aur*um	silver	*argent*um
iron	*ferr*um	tin	*stann*um

Example 6.4

Name each of the following compounds in two different ways:

    a. Cu_2O, CuO **b.** $CoBr_2$, $CoBr_3$

Solution

a. According to the Stock System, these compounds would be copper(I) oxide and copper(II) oxide. According to the older method, they would be cuprous oxide and cupric oxide. (The Latin root of copper is *cupr.*)

b. According to the Stock System, these compounds would be cobalt(II) bromide and cobalt(III) bromide. Cobalt does not have a Latin root; by the older method, these compounds would be cobaltous bromide and cobaltic bromide.

Problem 6.4

Name the following compounds by the two different methods of nomenclature:

    a. NiS, Ni_2S_3 **b.** $PtCl_2$, $PtCl_4$

2. Binary compounds containing two nonmetals but not hydrogen

Binary compounds of two nonmetals, neither of which is hydrogen, are molecular rather than ionic. They do not contain cations and anions. Carbon dioxide (CO_2) and phosphorus trichloride (PCl_3) are examples of such compounds. Although oxidation states can be used in naming these compounds, the preferred method uses prefixes to state how many atoms of an element are in one molecule of the compound. (The prefixes are listed in Table 6.4.) The name

TABLE 6.4 Prefixes used in naming binary compounds of two nonmetals

Number of atoms	Prefix	Number of atoms	Prefix	Number of atoms	Prefix
1	mono-	5	penta-	9	nona-
2	di-	6	hexa-	10	deca-
3	tri-	7	hepta-	11	hendeca-
4	tetra-	8	octa-	12	dodeca-

of the second element is modified to the root of its name followed by the ending *ide*. In both the formula and the name of these compounds, the most nonmetallic element comes first (see Figure 5.16 in Chapter 5). The prefix *mono* is often omitted for the first element but never omitted for the second. Thus,

CO is carbon monoxide

SF_6 is sulfur hexafluoride

N_2O is dinitrogen monoxide

Example 6.5   Name the following compounds:

a. SO_3 **b.** N_2O_3 **c.** XeO_2

Solution   In all of these compounds, both elements are nonmetals. Therefore, prefixes are used in their names.

a. The prefix *mono* can be omitted for the sulfur. There are three oxygens, therefore we use the prefix *tri* and change the name *oxygen* to *oxide*. The compound is sulfur trioxide (two words).

b. There are two nitrogens per molecule, therefore the first part is *dinitrogen*. There are three oxygens, therefore the second part is *trioxide*. The compound is dinitrogen trioxide.

c. There is only one atom of xenon, therefore the name is unchanged. There are two oxygens, thus the name *dioxide*. The compound is xenon dioxide.

Problem 6.5   Name the following compounds:

a. Cl_2O **b.** N_2O_4 **c.** $AsCl_3$

3. Binary acids

The binary compound formed when a halogen or any element, except oxygen, from Group VI of the periodic table combines with hydrogen can be named as were the binary nonmetallic compounds discussed in the preceding section. However, when these compounds are dissolved in water, the solution contains hydrogen ions. Because this property identifies an acid (Section 5.7D), these compounds must also be named as acids. Therefore, these compounds have two sets of names, one for the pure state and one for the compound dissolved in water (see Table 6.5).

Two points should be noted: (1) The acid name has the prefix *hydro* and the suffix *ic*. (2) These formulas are always written with hydrogen first. Other nonmetals form compounds with hydrogen, but they are *not* acids; their formulas are written with hydrogen last. Methane, CH_4, ammonia, NH_3, and arsine, AsH_3, are examples.

TABLE 6.5 **Nomenclature for binary acids**

Formula	Name in pure state	Name in water solution
HCl	hydrogen chloride	hydrochloric acid
H_2S	hydrogen sulfide	hydrosulfuric acid
HBr	hydrogen bromide	hydrobromic acid

4. Pseudo-binary compounds

Several polyatomic ions act so much like monatomic ions that they are classified as such. These ions are called pseudo-binary ions. They include the ammonium ion, NH_4^+, the hydroxide ion, OH^-, the cyanide ion, CN^-, and others. Compounds containing these ions are **pseudo-binary compounds.**

The properties of the ammonium ion are much like those of the alkali-metal ions.

Compounds containing the hydroxide ion are bases. A general definition of a base is that its aqueous solution contains more hydroxide than hydrogen ions. (Bases were introduced in section 5.7D.)

The cyanide ion behaves very much like a halogen ion. Many compounds containing the cyanide ion are extremely toxic.

C. Ternary Compounds

Ternary compounds are those compounds containing three elements. Ionic ternary compounds are formed by the combination of a monatomic cation with a polyatomic (containing several atoms) anion, as in sodium nitrate, $NaNO_3$. A **polyatomic anion** is derived from a **ternary acid.**

1. Ternary acids and their anions

When a ternary compound contains hydrogen and a polyatomic anion (for example, HNO_3), its name in the pure state is *hydrogen* followed by the name of the anion. Pure HNO_3 has the name hydrogen nitrate. When this compound is dissolved in water, it is an acid and is named as such. HNO_3 in water solution is named nitric acid. Table 6.6 lists the formulas of some of these compounds, the names they carry when in water solution, the oxidation number of the non-metal other than oxygen that they contain, and the name and formula of their anions. The rules for naming these compounds as acids follow the table. Be sure to study the table as you read the rules and notice the pattern shown in the names and formulas.

The rules for naming ternary acids are as follows:

1. The name of the most common oxyacid for a particular nonmetal is the root of the element's name plus the suffix *ic.* It has no prefix. The name of the anion of this acid is the root of the element's name plus the suffix

ate. The oxidation number of the nonmetal in this acid is high but not necessarily the highest possible. These acids are sometimes referred to as *ic–ate* acids. Of the oxyacids in Table 6.6, nitric, sulfuric, phosphoric, and chloric are in the category of "most common." In the table, the formulas marked with an asterisk are the most common acids for a particular nonmetal.

2. The name of the acid in which the nonmetal has the next lower oxidation number is the element's root plus the suffix *ous*. The name of its anion is the root plus *ite*. These acids may be called *ous–ite* acids. Of the acids in Table 6.6, nitrous, sulfurous, and chlorous fall into this group. Their formulas can be predicted if you have learned the formulas in the first group. The anion of an *ous–ite* acid contains one less oxygen atom than that of the *ic–ate* acid.

3. As with the halogens, if there is an oxyacid in which the nonmetal has an even lower oxidation number, that acid is named using the prefix–suffix *hypo–ous*, and its anion using *hypo–ite*. Of the acids in Table 6.6, only hypochlorous is in this category. Its formula can be predicted if you know the formula of chloric acid. The anions of these acids contain two less oxygen atoms than the anions of the *ic–ate* acids.

4. Again as with the halogens, if there is an oxyacid in which the nonmetal has a higher oxidation number than in the most common acid, that acid is named *per–ic* acid and its anion *per–ate*. Of the acids in Table 6.6, only perchloric is in this group. Its formula can be predicted from the formula of chloric acid. The anion will contain one more oxygen atom than the anion of the *ic–ate* acid.

TABLE 6.6   **Compounds that are acids in water solution and their anions**

Acid formula	Oxidation number of nonmetal	Name in aqueous solution	Name and formula of anion
*HNO_3	+5	nitric acid	nitrate, NO_3^-
HNO_2	+3	nitrous acid	nitrite, NO_2^-
*H_2SO_4	+6	sulfuric acid	sulfate, SO_4^{2-}
H_2SO_3	+4	sulfurous acid	sulfite, SO_3^{2-}
*H_3PO_4	+5	phosphoric acid	phosphate, PO_4^{3-}
H_2CO_3	+4	carbonic acid	carbonate, CO_3^{2-}
$HClO_4$	+7	perchloric acid	perchlorate, ClO_4^-
*$HClO_3$	+5	chloric acid	chlorate, ClO_3^-
$HClO_2$	+3	chlorous acid	chlorite, ClO_2^-
**HClO	+1	hypochlorous acid	hypochlorite, ClO^-
HCl	−1	hydrochloric acid	chloride, Cl^-

* These acids are the most common for a particular nonmetal.
** Although only chlorine is shown, similar compounds are formed by the other halogens and would be named the same way as are these chlorine-containing compounds.

The formulas of the salts of these acids are neutral combinations of ions (discussed in Section 6.1). To name them, first name the cation according to the rules given in Section 6.2B1. The names of the anions are given in Table 6.6.

2. Ternary acids containing carbon

Many acids contain only carbon, hydrogen, and oxygen. Acetic acid is an example. Its formula can be written as

$$HC_2H_3O_2 \quad \text{or} \quad HCH_3CO_2 \quad \text{or} \quad CH_3COOH \quad \text{or} \quad CH_3CO_2H$$

Regardless of how it is written, there is only *one* acidic hydrogen in acetic acid; the other three hydrogens do not separate as hydrogen ions in aqueous solution. Notice how the acidic hydrogen is placed by itself in each of the formulas to signify this difference. Many acids, like acetic acid, contain a group of atoms bonded to a —COOH group. Only the hydrogen of the —COOH group is an acidic hydrogen. For example,

$$C_6H_5COOH \quad \text{benzoic acid} \qquad C_2H_3COOH \quad \text{acrylic acid}$$

These acids are called carboxylic acids. They are discussed more fully in Chapter 15.

In naming the anions of these acids, the *ic* of the acid is replaced by *ate*. Thus,

acetic acid, $HC_2H_3O_2$, yields acetate ion, $C_2H_3O_2^-$

benzoic acid, C_6H_5COOH, yields benzoate ion, $C_6H_5COO^-$

acrylic acid, C_2H_3COOH, yields acrylate ion, $C_2H_3COO^-$

3. Salts containing more than one cation

Occasionally you will encounter a salt containing more than one cation. If both cations are metals, they are named in the order in which they are written according to the rules already given. If one of the cations is hydrogen, the salt can be named either by calling the cation *hydrogen* or by adding *bi* as a prefix to the name of the anion. Thus, $NaHCO_3$ can be named sodium hydrogen carbonate or sodium bicarbonate. Salts with more than one cation, one of which is hydrogen, are sometimes called acid salts.

Example 6.6

Write the formulas of the following compounds:

    a. potassium nitrite   **b.** iron(III) sulfate
    c. magnesium perchlorate   **d.** bromous acid

Solution

    a. Potassium nitrite: Potassium ion is K^+. The nitrite ion is NO_3^-. The formula is KNO_3.

b. Iron(III) sulfate: Iron(III) is Fe^{3+}. The sulfate ion is SO_4^{2-}. To balance charges, we need two Fe^{3+} (6 positive charges) and three SO_4^{2-} (6 negative charges). The formula is $Fe_2(SO_4)_3$.

c. Magnesium perchlorate: Magnesium ion is Mg^{2+}. The perchlorate ion is ClO_4^-. The formula is $Mg(ClO_4)_2$. The charge of $+2$ from the magnesium ion requires two perchlorate ions, each with a single negative charge, to balance.

d. Bromous acid: Bromine forms compounds similar to those of chlorine. Therefore, bromic acid would be $HBrO_3$ and bromous acid $HBrO_2$.

Problem 6.6   Write the formulas of the following compounds:

         a. ammonium sulfite   **b.** iron(II) phosphate
         c. calcium carbonate   **d.** iodic acid

Example 6.7   Name the following compounds:

         a. $Fe(NO_3)_2$ **b.** $CuSO_3$ **c.** Na_2HPO_4 **d.** $LiClO$

Solution

a. Each nitrate ion has a single negative charge. There are two nitrate ions, so the cation is iron(II). The compound is iron(II) nitrate.

b. The sulfite ion has a charge of -2, therefore the cation is copper(II). The compound is copper(II) sulfite.

c. The prefix *di* is needed on the sodium to specify that there are two sodium ions. The compound is disodium hydrogen phosphate.

d. Lithium forms only one ion (Section 5.7C2); therefore the cation is named lithium. ClO^- is the hypochlorite ion (see Table 6.6). The name of $LiClO$ is lithium hypochlorite.

Problem 6.7   Name the following compounds:

         a. Ag_2S **b.** $NiCO_3$ **c.** $KHSO_3$ **d.** $NaClO_3$

 6.3   **Formula Weights**

A. Calculation of Formula Weights

The formula weight of a compound or ion is the sum of the atomic weights of all elements in the compound or ion, with each element's atomic weight being multiplied by the number of atoms of that element occurring in the formula.

Example 6.8   **a.** The formula of sulfuric acid is H_2SO_4. Calculate its formula weight.

b. What is the formula weight of the nitrate ion, NO_3^-?

c. What is the formula weight of ammonium carbonate, $(NH_4)_2CO_3$?

Solution   In each of these examples, the mass contributed by each element in the compound is calculated by multiplying its atomic weight by the number of its atoms occurring in the formula. The sum of these contributions is the formula weight of the compound or ion.

a. The formula weight of sulfuric acid is calculated as follows:

	Atomic weight (amu)		*No. of atoms in the formula*		*Mass contributed by element (amu)*
hydrogen	1.008	×	2	=	2.016
sulfur	32.06	×	1	=	32.06
oxygen	16.00	×	4	=	64.00
					98.076 or
					98.08 amu (answer)

b. The formula weight of the nitrate ion is calculated as

nitrate	14.01	×	1	=	14.01
oxygen	16.00	×	3	=	48.00
					62.01 amu (answer)

c. The formula weight of ammonium carbonate is the sum of the formula weight of the carbonate ion plus twice the formula weight of the ammonium ion. First calculate the formula weight of the ammonium ion:

nitrogen	14.01	×	1	=	14.01
hydrogen	1.008	×	4	=	4.032
					18.04 amu (for NH_4^+)

Next calculate the formula weight of the carbonate ion:

carbon	12.01	×	1	=	12.01
oxygen	16.00	×	3	=	48.00
					60.01 amu (for CO_3^{2-})

Finally, calculate the formula weight of ammonium carbonate:

$(NH_4^+)_2$	18.04	×	2	=	36.08
CO_3^{2-}	60.01	×	1	=	60.01
					96.09 amu (answer)

Problem 6.8   Calculate the formula weight of:

a. potassium sulfide   **b.** the phosphate ion

c. magnesium nitrate

B. Moles of Compounds

In Section 4.4 the mole was defined as 6.02×10^{23} items. In that section, we talked about moles of atoms and you learned that one mole of a particular element has a mass in grams equal to the element's atomic weight. We can also have moles of compounds or of ions. One **mole** of a compound or an ion will have a mass in grams equal to the formula weight of that compound or ion. In Example 6.8 we calculated the formula weight of sulfuric acid to be 98.08 amu, thus it follows that one mole of sulfuric acid has a mass of 98.08 g. The formula of a compound also tells how many moles of a particular element are contained in one mole of the compound. For example, one mole of sulfuric acid contains:

2 mol hydrogen atoms, weighing 2(1.008 g) or 2.016 g

1 mol sulfur atoms, weighing 1(32.06 g) or 32.06 g

4 mol oxygen atoms, weighing 4(16.00 g) or 64.00 g

Note that these individual masses add up to our calculated formula weight for the acid. This relationship between mass and moles of a compound is often used as a conversion factor in solving problems.

Example 6.9

An experimental procedure requires 1.76 mol glucose, $C_6H_{12}O_6$. What mass of glucose is required?

Solution

Use the steps developed in Section 2.3B.

Wanted

? g glucose

Given

1.76 mol glucose. The formula weight of glucose is:

carbon	$6 \times 12.01 =$	72.06
hydrogen	$12 \times 1.008 =$	12.10
oxygen	$6 \times 16.00 =$	96.00
		180.16 amu (for glucose)

Conversion factor

1 mol glucose = 180.16 g

Equation

$$? \text{ g glucose} = 1.76 \text{ mol glucose} \times \frac{180.16 \text{ g glucose}}{1 \text{ mol glucose}}$$

Answer

317 g glucose (The answer is given to three significant figures because that is the number of significant figures in 1.76 mol glucose.)

Problem 6.9

Calculate the mass of 0.875 mol carbon dioxide.

Example 6.10   How many atoms of oxygen are in 0.262 g carbon dioxide, CO_2?

Solution

Wanted
? atoms oxygen

Given
0.262 g carbon dioxide

Conversion factors
The formula weight of carbon dioxide is:

carbon $1 \times 12.01 = 12.01$

oxygen $2 \times 16.00 = \underline{32.00}$

44.01 amu   (for CO_2)

1 mol CO_2 = 44.01 g
1 mol CO_2 contains 2 mol O atoms
1 mol atoms = 6.02×10^{23} atoms

Equation

$$? \text{ atoms O} = 0.262 \text{ g } CO_2 \times \frac{1 \text{ mol } CO_2}{44.01 \text{ g } CO_2} \times \frac{2 \text{ mol O atoms}}{1 \text{ mol } CO_2}$$

$$\times \frac{6.02 \times 10^{23} \text{ atoms}}{1 \text{ mol atoms}}$$

Answer
7.17×10^{21} atoms O

Problem 6.10   Calculate the number of hydrogen atoms in 5.32×10^{-3} g ammonia, NH_3.

In Example 6.10, notice that each factor involving a mole states the chemical composition of the mole: "1 mol CO_2" and "2 mol O atoms." As problems become increasingly complex, this bookkeeping habit becomes especially important. Notice also that the example deals with atoms of oxygen; we are not concerned here with the fact that oxygen exists in nature as a diatomic molecule, O_2.

Example 6.11   A solution of glucose contains 9.00 g glucose per 100 mL of solution. How many moles of glucose are contained in a liter of this solution?

Solution

Wanted
? mol glucose per liter of solution

Given
9.00 g glucose per 100 mL of solution

Conversion factors

1 mol glucose = 180.16 g   (from Example 6.9)

1 L = 1000 mL

Equation

$$\frac{? \text{ mol glucose}}{1 \text{ L solution}} = \frac{9.00 \text{ g glucose}}{100 \text{ mL solution}} \times \frac{1 \text{ mol glucose}}{180.16 \text{ g glucose}}$$

$$\times \frac{1000 \text{ mL solution}}{1 \text{ L solution}}$$

Answer

0.500 mol glucose per liter of solution

Problem 6.11   Calculate the number of moles of sulfuric acid, H_2SO_4, in 1 L of solution if 200 mL of solution contain 6.23 g of acid.

C. Percent Composition

Percent means parts per hundred. The **percent composition** of a compound is the number of grams of each element or group of elements in 100 g of the compound, expressed as a percent.

$$\text{Percent element} = \frac{\text{g element}}{\text{g compound}} \times 100\%$$

For example, the percent composition of sodium chloride, NaCl, can be calculated from the atomic weights of sodium and chlorine and the formula weight of NaCl.

Formula weight of NaCl: 23.00 g + 35.45 g = 58.45 g

Percent sodium: $\dfrac{23.00 \text{ g Na}}{58.45 \text{ g NaCl}} \times 100\% = 39.35\%$ sodium

Percent chlorine: $\dfrac{35.45 \text{ g Cl}}{58.45 \text{ g NaCl}} \times 100\% = 60.65\%$ chlorine

Example 6.12   Calculate the percent composition of carbon tetrachloride, CCl_4.

Solution   The formula weight of carbon tetrachloride is:

Formula weight of CCl_4:   carbon   12.01 × 1 =   12.01

chlorine 35.45 × 4 = 141.8

153.8 amu

Percent carbon: $\dfrac{12.01 \text{ g C}}{153.8 \text{ g CCl}_4} \times 100\% = 7.809\%$ carbon

Percent chlorine: $\dfrac{141.8 \text{ g Cl}}{153.8 \text{ g CCl}_4} \times 100\% = 92.19\%$ chlorine

It is always wise to check these percent calculations by assuring yourself that they add up to 100%, as they do here.

Problem 6.12   Calculate the percent composition of magnesium carbonate.

Example 6.13   Calculate the percent nitrogen in the fertilizer ammonium sulfate, $(NH_4)_2SO_4$.

Solution   **1.** Calculate the formula weight of $(NH_4)_2SO_4$:

Formula weight of NH_4^+:   nitrogen $14.01 \times 1 = 14.01$
hydrogen $1.008 \times 4 = \underline{4.032}$
18.04 amu

Formula weight of SO_4^{2-}:   sulfur $32.06 \times 1 = 32.06$
oxygen $16.00 \times 4 = \underline{64.00}$
96.06 amu

Formula weight of $(NH_4)_2SO_4$: $(NH_4^+)_2$ $18.04 \times 2 = 36.08$
SO_4^{2-} $96.06 \times 1 = \underline{96.06}$
132.14 amu

2. Calculate the percent nitrogen. Each formula unit contains two nitrogen atoms; therefore, there are $2(14.01 \text{ g})$, or 28.02 g, of nitrogen in 132.14 g ammonium sulfate.

Percent nitrogen $= \dfrac{28.02 \text{ g N}}{132.14 \text{ g } (NH_4)_2SO_4} \times 100\% = 21.20\%$ nitrogen

Problem 6.13   Calculate the percent nitrogen in the fertilizer ammonium phosphate $(NH_4)_3PO_4$.

Example 6.14   When 6.932 g silver oxide is decomposed, the silver residue weighs 5.351 g. What is the percent composition of silver oxide?

Solution   The percent silver can be calculated directly:

Percent silver $= \dfrac{5.351 \text{ g silver}}{6.932 \text{ g silver oxide}} \times 100\% = 77.19\%$ silver

The percent oxygen can be calculated by subtraction:

Percent oxygen $= 100\% - 77.19\% = 22.81\%$

Problem 6.14   When a red oxide of mercury is heated, oxygen is driven off, leaving a silver-colored pool of pure mercury. After a 5.00-g sample of this oxide has been heated, 4.63 g mercury remain. Calculate the percent mercury in this compound.

Example 6.15   When 0.652 g of a compound that contains only carbon and hydrogen is burned, 1.974 g carbon dioxide and 1.010 g water are formed. What is the percent composition of this hydrocarbon?

Solution   Consider the statement of the problem. All the carbon in the compound went into carbon dioxide; all the hydrogen in the compound went into water. We can calculate the amount of carbon in 1.974 g carbon dioxide and the amount of hydrogen in 1.010 g water. Knowing these weights, we can calculate the percent composition of the original compound. Use the following steps:

1. Calculate the mass of carbon in 1.974 g CO_2. The formula weight of carbon dioxide is:

$$12.01 + 2(16.00) = 44.01 \text{ amu}$$

The mass of carbon is:

$$? \text{ g C} = 1.974 \text{ g } CO_2 \times \frac{1 \text{ mol } CO_2}{44.01 \text{ g } CO_2} \times \frac{1 \text{ mol C}}{1 \text{ mol } CO_2} \times \frac{12.01 \text{ g C}}{1 \text{ mol C}}$$

$$= 0.5387 \text{ g C}$$

2. Calculate the mass of hydrogen in 1.010 g water. The formula weight of water is:

$$2(1.008) + 16.00 = 18.01$$

The mass of hydrogen is:

$$? \text{ g H} = 1.010 \text{ g } H_2O \times \frac{1 \text{ mol } H_2O}{18.01 \text{ g } H_2O} \times \frac{2 \text{ mol H}}{1 \text{ mol } H_2O} \times \frac{1.008 \text{ g H}}{1 \text{ mol H}}$$

$$= 0.1131 \text{ g H}$$

3. Calculate the percent composition. The percent carbon is:

$$\% \text{ carbon} = \frac{0.5387 \text{ g C}}{0.652 \text{ g compound}} \times 100\% = 82.6\% \text{ carbon}$$

The percent hydrogen is:

$$\% \text{ hydrogen} = \frac{0.1131 \text{ g H}}{0.652 \text{ g compound}} \times 100\% = 17.3\% \text{ hydrogen}$$

Problem 6.15   When 0.873 g of a hydrocarbon is burned, 2.74 g carbon dioxide and 1.12 g water are formed. What is the percent composition of the hydrocarbon?

6.4   Empirical Formulas

The **empirical formula** of a compound expresses a ratio between the numbers of atoms of different elements present in a molecule of the compound. This ratio is a **mole ratio** as well as a ratio between numbers of atoms. From the formula it is possible to calculate the percent composition of a compound. Going in the opposite direction from the composition of a compound, it is possible to calculate its empirical formula.

Consider the compound chloroform. The percent composition of chloroform is 10.06% carbon, 0.85% hydrogen, and 89.09% chlorine. We know then that 100 g chloroform contain 10.06 g carbon, 0.85 g hydrogen, and 89.09 g chlorine. This weight relationship can be converted to a mole ratio by the following calculations:

carbon $\qquad 10.06 \text{ g C} \times \dfrac{1 \text{ mol C}}{12.01 \text{ g C}} = 0.838 \text{ mol C}$

hydrogen $\qquad 0.85 \text{ g H} \times \dfrac{1 \text{ mol H}}{1.008 \text{ g H}} = 0.84 \text{ mol H}$

chlorine $\qquad 89.09 \text{ g Cl} \times \dfrac{1 \text{ mol Cl}}{35.45 \text{ g Cl}} = 2.513 \text{ mol Cl}$

These calculations show that the mole ratio between the elements in chloroform is 0.84 mol C to 0.84 mol H to 2.51 mol Cl. This ratio can be expressed by the formula:

$$C_{0.84}H_{0.84}Cl_{2.51}$$

However, formulas by definition can contain only whole numbers of atoms. The ratio can be changed to whole numbers by dividing each subscript by the smallest subscript, giving the formula of chloroform as:

$$C_{0.84/0.84}H_{0.84/0.84}Cl_{2.51/0.84} \quad \text{or} \quad CHCl_3$$

Example 6.16   Calculate the empirical formula of a compound that contains 36.8% nitrogen and 63.25% oxygen.

Solution
1. Assume that you have 100 g of the compound, which will contain 36.8 g nitrogen and 63.25 g oxygen.

2. Convert these weights to moles.

$$? \text{ mol N} = 36.8 \text{ g N} \times \dfrac{1 \text{ mol N}}{14.0 \text{ g N}} = 2.63 \text{ mol N}$$

$$? \text{ mol O} = 63.25 \text{ g O} \times \dfrac{1 \text{ mol O}}{16.0 \text{ g O}} = 3.95 \text{ mol O}$$

This calculation gives the formula:

$$N_{2.63}O_{3.95}$$

3. Change this ratio to whole numbers:

$$N_{2.63/2.63}O_{3.95/2.63} = NO_{1.5}$$

But the ratio is not yet whole numbers as wanted. If each subscript is multiplied by 2, the subscripts then become whole numbers, giving

$$N_2O_3$$

the correct empirical formula for the compound.

Problem 6.16   Calculate the empirical formula of another oxide of nitrogen that contains 63.6% nitrogen and 36.4% oxygen.

Example 6.16 presents a situation often encountered in calculating empirical formulas. When the result of a calculation of an empirical formula contains a subscript more than 0.1 away from a whole number, it must not be rounded off. Rather, the whole formula must be multiplied by a factor that will make that subscript a whole number. In general, when the subscript is 1.5, multiply by 2 as was done in Example 6.16. When the subscript is 1.3 or 1.7, multiply by 3.

We have calculated the formula of compounds from their percent composition. The following example shows how to determine the formula of a compound when its composition is given not in percent but in grams.

Example 6.17   Analysis of 3.23 g of a compound shows that it contains 0.728 g of phosphorus and 2.50 g chlorine. What is the empirical formula of the compound?

Solution   **1.** Calculate the number of moles of each compound in 3.23 g of the compound.

$$\text{phosphorus} \qquad 0.728 \text{ g P} \times \frac{1 \text{ mol P}}{30.97 \text{ g P}} = 0.0235 \text{ mol P}$$

$$\text{chlorine} \qquad 2.50 \text{ g Cl} \times \frac{1 \text{ mol Cl}}{35.45 \text{ g Cl}} = 0.0705 \text{ mol Cl}$$

This calculation gives the formula:

$$P_{0.0235}Cl_{0.0705}$$

2. Change this ratio to whole numbers by dividing through by the smallest number of moles:

$$P_{0.0235/0.0235}Cl_{0.0705/0.0235} = PCl_3$$

Problem 6.17   Analysis of a 1.16 g sample of a compound shows that it contains 0.51 g of iron, with the rest being chlorine. What is the empirical formula of this compound?

6.5 Empirical versus Molecular Formulas

The formulas we have calculated in the preceding section express the simplest atomic ratio between the elements in the compound. Such formulas are called empirical formulas. An empirical formula does not necessarily represent the actual numbers of atoms present in a molecule of a compound; it represents only the ratio between those numbers. The actual numbers of atoms of each element that occur in the smallest freely existing unit or molecule of the compound is expressed by the **molecular formula** of the compound. The molecular formula of a compound may be the empirical formula, or it may be a multiple of the empirical formula. For example, the molecular formula of butene, C_4H_8, shows that each freely existing molecule of butene contains four atoms of carbon and eight atoms of hydrogen. One molecule of ethylene (molecular formula C_2H_4) contains two atoms of carbon and four atoms of hydrogen. Both butene and ethylene contain two hydrogen atoms for each carbon atom. They have the same empirical formula, yet they are different compounds with different molecular formulas. Butene is C_4H_8, or four times the empirical formula; ethylene is C_2H_4, or twice the empirical formula.

Table 6.7 shows three groups of compounds. Within each group, the compounds have the same empirical formula and percent composition but different molecular formulas. That they are different compounds is shown by their different boiling points.

TABLE 6.7 Compounds with the same empirical formula but different molecular formulas

Empirical formula	Compound	Molecular formula	Boiling point, °C
CH (92.2% C; 7.8% H)	acetylene	C_2H_2	−84
	benzene	C_6H_6	80
CH_2 (85.6% C; 14.4% H)	ethylene	C_2H_4	−103
	butene	C_4H_8	−6.3
	cyclohexane	C_6H_{12}	80.7
CH_2O (40.0% C; 6.7% H; 53.3% O)	formaldehyde	CH_2O	−21
	acetic acid	$C_2H_4O_2$	117
	glyceraldehyde	$C_3H_6O_3$	140

The molecular formula of a compound can be determined from the empirical formula if the molecular weight is known.

Example 6.18   The empirical formula of hexane is C_3H_7. Its molecular weight is 86.2 amu. What is the molecular formula of hexane?

Solution   The molecular formula of a compound is a multiple of its empirical formula. The molecular formula weight is the same multiple of the empirical formula weight. We know the empirical formula and thus can calculate the empirical formula weight. We can calculate what multiple the molecular formula weight is of the empirical formula weight. As stated, the molecular formula is the same multiple of the empirical formula.

1. Calculate the formula weight of C_3H_7:

$$\begin{array}{lrl} \text{carbon} & 12.01 \times 3 = & 36.03 \\ \text{hydrogen} & 1.008 \times 7 = & \underline{7.056} \\ & & 43.09 \text{ amu} \end{array}$$

2. Calculate the ratio between the molecular weight and the empirical weight:

$$\frac{\text{molecular weight}}{\text{empirical weight}} = \frac{86.2}{43.09} = 2$$

3. The molecular formula must be twice the empirical formula:

$$(C_3H_7)_2 \quad \text{or} \quad C_6H_{14}$$

Problem 6.18   The empirical formula of the sugar ribose is CH_2O, and its molecular weight is 150 amu. What is the molecular formula of ribose?

Combining the information in Sections 6.4 and 6.5, we see that two kinds of data are needed to determine the molecular formula of a compound: (1) its composition, from which we can calculate its empirical formula, and (2) its molecular weight. The molecular weight will be a multiple of the empirical formula weight. The molecular formula is the same multiple of the empirical formula.

Example 6.19   The compound ethylene glycol is often used as an antifreeze. It contains 38.70% carbon, 9.75% hydrogen, and the rest oxygen. The molecular weight of ethylene glycol is 62.07 g. What is the molecular formula of ethylene glycol?

Solution   1. Calculate the empirical formula as was done in Example 6.16. Assume 100 g of the compound, which will contain 38.70 g carbon, 9.75 g hydrogen, and the rest oxygen.

$$? \text{ g O} = 100 \text{ g ethylene glycol} - 38.70 \text{ g C} - 9.75 \text{ g H} = 51.55 \text{ g O}$$

2. Calculate the moles of each element present:

$$? \text{ g mol C} = 38.70 \text{ g C} \times \frac{1 \text{ mol C}}{12.01 \text{ g C}} = 3.22 \text{ mol C}$$

$$? \text{ mol H} = 9.75 \text{ g H} \times \frac{1 \text{ mol H}}{1.008 \text{ g H}} = 9.67 \text{ mol H}$$

$$? \text{ mol O} = 51.55 \text{ g O} \times \frac{1 \text{ mol O}}{16.00 \text{ g O}} = 3.22 \text{ mol O}$$

This calculation gives the formula:

$$C_{3.22}H_{9.67}O_{3.22} \quad \text{or} \quad C_{3.22/3.22}H_{9.67/3.22}O_{3.22/3.22} = CH_3O$$

3. Next calculate the ratio of molecular weight to empirical formula weight. The molecular weight is given. The empirical formula is CH_3O, so the empirical formula weight is $12.01 + 3(1.008) + 16.00 = 31.03$:

$$\frac{\text{Molecular weight}}{\text{Empirical formula weight}} = \frac{62.07}{31.03} = 2$$

Therefore the molecular formula is twice the empirical formula: $C_2H_6O_2$.

Problem 6.19 A compound known as fumaric acid has a molecular weight of 116.1 g. It contains 41.4% carbon, 3.5% hydrogen, and the rest oxygen. What is the molecular formula of fumaric acid?

Example 6.20 The compound dioxane contains only carbon, hydrogen, and oxygen. When 0.956 g dioxane is burned, 1.91 g carbon dioxide and 0.782 g water are formed. In another experiment, it was determined that 6.04×10^{-3} mol dioxane weighs 0.532 g. What is the molecular formula of dioxane?

Solution

1. Using the method shown in Example 6.15, calculate the mass of carbon, hydrogen, and oxygen in 0.956 g dioxane.

$$? \text{ g C} = 1.91 \text{ g } CO_2 \times \frac{1 \text{ mol } CO_2}{44.0 \text{ g } CO_2} \times \frac{1 \text{ mol C}}{1 \text{ mol } CO_2} \times \frac{12.0 \text{ g C}}{1 \text{ mol C}} = 0.521 \text{ g C}$$

$$? \text{ g H} = 0.782 \text{ g } H_2O \times \frac{1 \text{ mol } H_2O}{18.0 \text{ g } H_2O} \times \frac{2 \text{ mol H}}{1 \text{ mol } H_2O} \times \frac{1.0 \text{ g H}}{1 \text{ mol H}} = 0.087 \text{ g H}$$

$$? \text{ g O} = 0.956 \text{ g dioxane} - 0.521 \text{ g C} - 0.087 \text{ g H} = 0.348 \text{ g O}$$

2. Using the data from step 1 above, calculate the empirical formula of dioxane.

$$? \text{ mol C} = 0.521 \text{ g C} \times \frac{1 \text{ mol C}}{12.01 \text{ g C}} = 0.043 \text{ mol C}$$

$$? \text{ mol H} = 0.087 \text{ g H} \times \frac{1 \text{ mol H}}{1.0 \text{ g H}} = 0.087 \text{ mol H}$$

$$? \text{ mol O} = 0.348 \text{ g O} \times \frac{1 \text{ mol O}}{16.0 \text{ g O}} = 0.022 \text{ mol O}$$

The empirical formula is:

$$C_{0.043}H_{0.087}O_{0.022} \quad \text{or} \quad C_2H_4O$$

3. Calculate the molecular weight of dioxane:

$$\text{Molecular weight} = \frac{\text{grams}}{\text{mole}} = \frac{0.532 \text{ g dioxane}}{6.04 \times 10^{-3} \text{ mol dioxane}} = 88.1 \text{ g/mol}$$

4. Calculate the molecular formula. The empirical formula weight is $2(12.0) + 4(1.01) + 16.0 = 44.0$. The molecular weight is 88.08. The ratio of molecular weight to empirical weight is:

$$\frac{\text{Molecular weight}}{\text{Empirical weight}} = \frac{88.0}{44.0} = 2$$

Thus the molecular formula is: $C_4H_8O_2$.

Problem 6.20   The compound resorcinol contains only carbon, hydrogen, and oxygen. When 2.63 g resorcinol is burned, 6.30 g carbon dioxide and 1.29 g water are formed. In another experiment, it was determined that 4.79 g resorcinol is 4.35×10^{-2} mol. What is the molecular formula of resorcinol?

6.6   **Summary**

Compounds fall into two large groups: molecular, in which the smallest unit of the compound is a molecule, and ionic. Ionic compounds are formed by the union of ions in a ratio such that the compound and its formula have no net charge. Each compound has a systematic name that identifies its composition. Many compounds also have common names. Compounds are given systematic names based on certain rules, many of which depend on the oxidation numbers of the elements in the compound. Another set of rules is used to assign oxidation numbers.

The formula of a compound states which elements and how many atoms of each are present in the smallest unit of the compound. It also represents the mole ratio between the elements in the compound. The formula weight of a compound can be calculated from the formula and the atomic weights of the elements present in the compound. The formula weight in grams is the mass of one mole of the compound. Using formula weights, the percent composition of a compound can be calculated.

If the percent composition of a compound is known, its empirical formula can be calculated. If both the molecular weight and the percent composition of a compound are known, the molecular formula of the compound can be calculated. Several compounds may have the same empirical formula but different molecular formulas.

Key Terms

binary acids (6.2B3)
binary compounds (6.2B)
binary compounds of a metal and a
   nonmetal (6.2B1)

binary compounds of two nonmetals
   (6.2B2)
empirical formula (6.4)
ionic compounds (6.1)

mole (6.3B)   *percent composition (6.3C)*
molecular compounds (6.1)   *polyatomic anion (6.2C1)*
molecular formula (6.5)   *pseudo-binary compounds (6.2B4)*
molecules (6.1)   *salt (6.1)*
mole ratio (6.4)   *systematic (Stock) names (6.2)*
oxidation number (6.2A)   *ternary acids (6.2C)*

Multiple-Choice Questions

For Questions 1 and 2, use the following formulas. The name of $HBrO_3$ is given to help you in choosing the correct answer.

      a. HBr **b.** $HBrO_2$ **c.** $HBrO_3$, bromic acid   **d.** $HBrO_4$

MC1. Which formula is perbromic acid?

MC2. In which compound does bromine have an oxidation number of $+7$?

MC3. What is the formula of nickel(II) bromite?
    a. $NiBr_2$ **b.** $NiBrO$ **c.** $NiBrO_2$ **d.** $Ni(BrO_2)_2$
    e. $NiBrO_3$

MC4. What is the formula of ammonium phosphate?
    a. $(NH_3)_2PO_4$ **b.** $(NH_4)_2PO_4$ **c.** NH_4PO_3
    d. $(NH_3)_4PO_3$ **e.** none of these

MC5. Which of the following is (are) correctly named?
    1. Fe_2O_3, iron(II) oxide
    2. $MgCl_2$, manganese(II) chloride
    3. $Cr(NO_3)_3$, chromium(III) nitrite
    4. PCl, potassium chloride
    a. none   **b.** all   **c.** 1 and 2   **d.** 2 and 4   **e.** 3

MC6. What is the formula weight of ammonium sulfate, $(NH_4)_2SO_4$?
    a. 114.1   **b.** 132.1   **c.** 228.1   **d.** 156.3   **e.** none of these

MC7. How many moles of calcium are in 15.5 g calcium chloride?
    a. 0.530   **b.** 0.274   **c.** 0.140   **d.** 0.205   **e.** none of these

MC8. A 1.60 g sample of a substance contains 0.380 g carbon. What percent carbon does it contain?
    a. 20.4%   **b.** 23.8%   **c.** 4.2%   **d.** 38.1%   **e.** none of these

MC9. A fluoride of uranium contains 24.2% fluorine and the rest uranium. What is the formula of this compound?
    a. UF_2 **b.** U_2F_3 **c.** UF_4 **d.** UF_6 **e.** U_2F_5

MC10. The empirical formula of a group of hydrocarbons is CH_2. One of these, eicosene, has a molecular weight of 280 g. How many carbons are there in a molecule of eicosene?
    a. 10   **b.** 2   **c.** 20   **d.** 15   **e.** no way of knowing

Problems

6.1 Categories of Compounds

6.21. State which elements and how many atoms of each are in a formula unit of:
   a. naphthalene, $C_{10}H_8$
   b. potassium sulfate, K_2SO_4
   c. lysergic acid diethylamide (LSD), $C_{20}H_{24}N_3O$

*6.22. Write the formula of the compound whose formula unit contains:
   a. 4 atoms phosphorus and 10 atoms oxygen
   b. 1 atom calcium, 2 atoms carbon, and 4 atoms oxygen
   c. 1 atom bromine, 6 atoms carbon, and 5 atoms hydrogen

6.23. Complete the following chart as done in line 1 for sodium.

	SO_4^{2-}	HCO_3^-	SO_3^{2-}	Br^-	PO_4^{3-}
Na^+	Na_2SO_4	$NaHCO_3$	Na_2SO_3	$NaBr$	Na_3PO_4
NH_4^+	———	———	———	———	———
Fe^{3+}	———	———	———	———	———
Mg^{2+}	———	———	———	———	———

6.2 Naming Compounds

*6.24. Determine the oxidation number of nitrogen in each of the following compounds:
   a. N_2O **b.** NH_4Cl **c.** $NaNO_2$
   d. HNO_3 **e.** N_2 **f.** N_2O_4

6,25. Determine the oxidation number of chromium in each of the following compounds:
   a. K_2CrO_4 **b.** Cr_2O_3
   c. $Cr(NO_3)_2$ **d.** $CrCl_3$
   e. $Na_2Cr_2O_7$ **f.** $(NH_4)_2CrO_4$

*6.26. Name the following compounds:
   a. N_2O_4 **b.** CCl_4 **c.** Cl_2O
   d. CO **e.** SO_3 **f.** P_2O_3
   g. $SiBr_4$ **h.** Cl_2O_7

*6.27. Give formulas for the following compounds:

a. silver(I) nitrate
b. ammonium nitrate
c. magnesium sulfite
d. calcium bromide
e. sodium perchlorate
f. copper(I) oxide
g. hydrogen iodide
h. lead(IV) chloride

6.28. Name the following compounds:
   a. $MnCl_2$ **b.** $NaBrO$ **c.** Ag_2Se
   d. $Cr(NO_3)_2$ **e.** KH_2PO_4 **f.** LiI
   g. $HBrO$ **h.** $(NH_4)_2SO_4$
   i. $AlCl_3$

6.29. Name the following compounds:
   a. $Fe(NO_3)_2$ **b.** $KHCO_3$
   c. Fe_2O_3 **d.** Na_2S **e.** $MgSO_4$
   f. LiF **g.** $Ca_3(PO_4)_2$ **h.** NH_4I
   i. $NaNO_2$ **j.** $Mn(NO_3)_2$
   k. $CaHPO_4$ **l.** $AlBr_3$

6.30. Write the formula for each of the following compounds:
   a. copper(II) chloride
   b. iron(III) nitrate
   c. ammonium sulfate
   d. cesium chloride
   e. lithium carbonate
   f. aluminum iodide
   g. potassium sulfite
   h. calcium sulfide
   i. tin(II) nitrate

6.3   Formula Weights

6.31. Calculate the formula weight of:
   a. nitric acid, HNO_3
   b. methane, CH_4

c. sulfur dioxide, SO_2
d. sodium phosphate, Na_3PO_4

6.32. Calculate the formula weight of:
   a. citric acid, $C_6H_8O_7$
   b. ascorbic acid, $C_6H_8O_6$ (vitamin C)
   c. cholesterol, $C_{27}H_{46}O$
   d. glucose, $C_6H_{12}O_6$

* 6.33. Calculate the formula weight of:
   a. chloroform, $CHCl_3$
   b. ammonium sulfate, $(NH_4)_2SO_4$
   c. magnesium oxide, MgO
   d. camphor, $C_{10}H_{16}O$
   e. aspirin, $C_8H_8O_4$

6.34. Calculate the percent carbon in each of the compounds in Problem 6.32.

6.35. What is the percent composition of:
   a. octane, C_8H_{18}
   b. sodium bicarbonate, $NaHCO_3$
   c. magnesium chloride, $MgCl_2$

*6.36. Calculate the moles present in each of the following samples:
   a. 6.85 g sodium iodide, NaI
   b. 437.1 g sulfur dioxide, SO_2
   c. 0.442 g nitrogen dioxide, NO_2
   d. 7.41 g lithium oxide, Li_2O
   e. 32.5 g glucose, $C_6H_{12}O_6$
   f. 0.27 g oxygen gas
   g. 32.5 g sucrose, $C_{12}H_{22}O_{11}$

6.37. Calculate the moles of sodium present in each of the following samples:
   a. 0.155 g sodium chloride, $NaCl$
   b. 0.155 g sodium sulfate, Na_2SO_4
   c. 0.155 g sodium phosphate, Na_3PO_4
   d. 0.155 g sodium bicarbonate, $NaHCO_3$
   e. 0.155 g sodium benzoate, $NaC_7H_5O_2$
   f. 0.155 g monosodium glutamate, $NaC_5H_8NO_4$

6.38. Calculate the percent sodium in each of the compounds in Problem 6.37.

6.39. Calculate the mass of 0.155 mol of each of the compounds in Problem 6.31.

6.40. Calculate the mass of 2.13 mol of each of the compounds in Problem 6.32.

6.4 Empirical Formulas

*6.41. A compound contains 53.5% carbon, 11.1% hydrogen, and 35.6% oxygen by weight. Its formula weight is 90.1 amu. What is the empirical formula? the molecular formula?

6.42. A sulfide of arsenic contains 60.9% by weight of arsenic and 39.1% sulfur. What is the empirical formula of this compound?

6.43. A compound has the empirical formula C_4H_6. Its formula weight is 162 amu. What is its molecular formula?

6.44. Which compounds in each of the following groups have the same percent composition?
   a. C_2H_4, C_6H_{12}, C_5H_8, $C_{16}H_{22}$
   b. N_2O_4, NO_2, N_2O_5
   c. $C_2H_4O_2$, $C_6H_8O_6$, $C_3H_6O_3$

*6.45. A compound contains 85.6% carbon and 14.4% hydrogen. Its formula weight is 84.18 amu. What is the molecular formula of this compound?

6.46. A compound contains 29.1% sodium, 40.5% sulfur, and the rest oxygen. What is the formula of this compound?

6.47. An oxide of phosphorus contains 43.6% phosphorus and the rest oxygen. What is the empirical formula of this compound?

6.5 Empirical versus Molecular Formulas

6.48. A compound having a molecular weight of 174.2 g contains 41.4% carbon, 8.1% hydrogen, 18.4% oxygen, and the rest nitrogen. What is the molecular formula of this compound?

6.49. When 3.10 g of a compound containing only carbon, hydrogen, and oxygen was burned, 4.4 g carbon dioxide and 2.7 g water were obtained. In another experiment, it was determined that 0.204 mol of the compound weighed 12.7 g. What is the molecular formula of this compound?

Review Problems

6.50. The molecular weight of saccharin is 183.2 g. How many molecules of saccharin will be contained in a sample weighing 0.916 g?

6.51. Calculate the formula of a compound that contains 32.8% chromium and the rest chlorine. Name the compound. Give the oxidation number of chromium in this compound.

6.52. Lutetium (atomic number 71) costs $26.00 per pound. What is the cost of 0.275 mol of this element?

*6.53. To fry an egg properly, one should have the frying pan at a temperature of about 125°C. If the frying pan weighs 815 g and is made completely of iron, how many joules must be supplied to raise its temperature from room temperature (25°C) to 125°C? The specific heat of iron is 0.447 J/g°C.

6.54. In some states, pollution control boards regulate the amount of nitrogen that can be applied to the fields. If a farmer can apply 200 lb nitrogen per year per acre per crop, how many moles of ammonia can be applied to a 40-acre field for each crop grown in a single year?

7

Compounds II: Bonding and Geometry

We can predict some properties of a compound if we know its composition—that is, if we know which elements and how many atoms of each are contained in each formula unit of the compound. We can predict other properties if we know other things about the compound—namely, the nature of the bonds between its atoms and how these atoms are arranged in space. This chapter discusses those aspects of compounds. In the discussion we will describe:

1. The various kinds of chemical bonds and how to predict the kind of bond that will be formed between two particular atoms.
2. How to show the distribution of electrons in a compound by drawing its Lewis structure.
3. How the electron distribution around an atom can be used to predict the geometry around that atom.
4. Why some molecules and not others dissolve in water.
5. How the bonding within a molecule of a compound can be used to predict some physical properties of that compound.

7.1   ## The Chemical Bond

The atoms of a compound are held together by **chemical bonds** formed by the interaction of electrons from each atom. According to the **octet rule** (Section 5.7C1), atoms bond together to form molecules in such a way that each atom participating in a chemical bond acquires an electron configuration resembling that of the noble gas nearest it in the periodic table. Thus the outer shell of each bonded atom will contain eight electrons (or two electrons for hydrogen and lithium).

The simplest chemical bond is that formed between two hydrogen atoms. Each hydrogen atom has one electron. As the two atoms approach each other, the nucleus of one atom attracts the electron of the other. Eventually the two orbitals overlap, becoming a single orbital containing two electrons (see Figure 7.1).

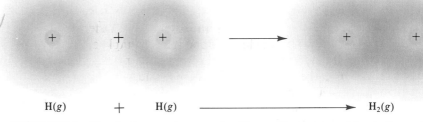

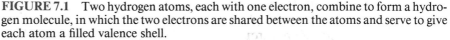

H(g)   +   H(g)   ⟶   H₂(g)

FIGURE 7.1   Two hydrogen atoms, each with one electron, combine to form a hydrogen molecule, in which the two electrons are shared between the atoms and serve to give each atom a filled valence shell.

This orbital encompasses space around both nuclei. Although the electrons may be in any part of this orbital, we can predict that they are most likely to be in the space between the nuclei, shielding one nucleus from the other and being attracted by both. In the resulting molecule, both atoms have two electrons and a filled outer (valence) shell. These shared electrons form a bond between the two atoms. This chemical bond is a **covalent bond,** a pair of electrons shared between two atoms. When this bond forms, energy is released. This release of energy shows that the molecule of hydrogen is more stable than the separate atoms.

A. Covalent, Polar Covalent, and Ionic Bonds

Because the hydrogen molecule contains two identical atoms, it can be assumed that the bonding electrons in this covalent bond are shared equally by these atoms.

Most chemical bonds are not between like atoms but form between atoms of different elements. These bonds are slightly different from that in a hydrogen molecule. Consider the bond between hydrogen and chlorine: Again both atoms require one more electron to satisfy the octet rule. As the atoms come together, their orbitals overlap and the two atoms share a pair of electrons. However, the hydrogen–chlorine bond differs from the hydrogen–hydrogen bond because the electrons are not shared equally between hydrogen and chlorine but are more strongly attracted to the chlorine. They are more apt to be found close to the chlorine than close to the hydrogen. Because of this unequal sharing, the chlorine atom assumes a slightly negative character and the hydrogen atom a slightly positive character. We say that the bond is **polar covalent,** meaning that the bond consists of electrons shared between two atoms (therefore covalent) but shared unequally, thus giving the bond a positive and a negative end, a condition described by the term **polar.** We can also say that the bond is a **dipole** or has a dipole moment, meaning that the bond has a positive end (the hydrogen) and a negative end (the chlorine). The bond between hydrogen atoms is **nonpolar** (has no positive and negative ends) **covalent** (electrons are shared).

An **ionic bond** is the extreme case of a polar covalent bond. In an ionic bond, the bonding atoms differ so markedly in their attraction for electrons that one or more electrons are essentially transferred from one atom to the other. The sodium–chlorine bond is an example of an ionic bond. The attraction of the chlorine atom for electrons is so much greater than that of a sodium atom that the 3s electron of sodium is said to be completely transferred from sodium to chlorine.

In summary, then, the three types of bonds are: (1) a covalent bond, in which the electrons are shared equally; (2) a polar covalent bond, in which the electrons are shared unequally; and (3) an ionic bond, in which electrons are transferred from one atom to the other. These bonds are illustrated in Figure 7.2.

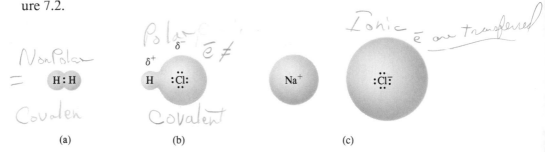

FIGURE 7.2 Electrons in nonpolar covalent, polar covalent, and ionic bonds: (a) the electrons are shared equally; (b) the electrons are held closer to the more-negative chlorine atom; (c) one electron has been tranferred from sodium to chlorine.

B. Predicting Bond Type; Electronegativity

It is possible to predict the type of bond that will form between two elements. The farther apart (left to right) the two elements are in the periodic table, the more ionic and the less covalent will be the bond between them. Thus, metals react with nonmetals to form ions joined predominantly by ionic bonds. Bonds with the highest degree of ionic character are formed by the reaction of alkali or alkaline earth metals with the halogens, particularly with fluorine or chlorine. Nonmetals react together to form covalent bonds. If the bond is between two different nonmetals, the bond will be polar covalent. If the two nonmetals are neighbors in the table, the bond will be less polar than if the nonmetals are separated by other elements. For example, carbon and nitrogen are in neighboring columns, and carbon and fluorine are in Groups IV and VII, respectively. A carbon–nitrogen bond will be less polar than a carbon–fluorine bond. Finally, if the two atoms are of the same element, as in a hydrogen molecule or a chlorine molecule, the bond will be essentially nonpolar.

The concepts in the previous paragraph have been quantified by the concept of **electronegativity.** The electronegativity (EN) of an element measures its attraction for the electrons in a chemical bond.

One scale of electronegativity was developed by the American chemist Linus Pauling (b. 1901). On this scale, fluorine, the most electronegative element, has an electronegativity of 4.0. Carbon has an electronegativity of 2.5, hydrogen 2.1, and sodium 0.9. Figure 7.3 shows the electronegativities of the elements with which we deal most often.

Increasing electronegativity →

	I	II											III	IV	V	VI	VII
1	H 2.1																
2	Li 1.0	Be 1.5											B 2.0	C 2.5	N 3.0	O 3.5	F 4.0
3	Na 0.9	Mg 1.2				Transition elements							Al 1.5	Si 1.8	P 2.1	S 2.5	Cl 3.0
4	K 0.8	Ca 1.0	Sc 1.4	Ti 1.5	V 1.6	Cr 1.7	Mn 1.6	Fe 1.8	Co 1.9	Ni 1.9	Cu 2.0	Zn 1.6	Ga 1.8	Ge 2.0	As 2.2	Se 2.6	Br 2.8
5	Rb 0.8	Sr 1.0	Y 1.2	Zr 1.3	Nb 1.6	Mo 2.2	Tc —	Ru 2.2	Rh 2.3	Pd 2.2	Ag 1.9	Cd 1.7	In 1.8	Sn 1.8	Sb 2.0	Te 2.1	I 2.5
6	Cs 0.79	Ba 0.9								Pt 2.3	Au 2.5	Hg 2.0	Tl 2.0	Pb 2.3	Bi 2.0	Po —	

Increasing electronegativity ↑

FIGURE 7.3   Electronegativities of some elements (Pauling scale).

Notice that the electronegativity of most metals is close to 1.0 and that the electronegativity of a nonmetal, although dependent on its location in the table, is always greater than 1.0. In general, electronegativity increases from bottom to top in a column and from left to right across a period.

Note that the noble gases, Group VIII, do not appear in this table. Electronegativity measures the relative attraction of atoms for electrons in chemical bonds. The noble gases react differently from the halogens and other nonmetals. The concepts of electronegativity do not apply to them.

When two atoms combine, the nature of the bond between them is determined by the difference between their electronegativities (denoted ΔEN). If the atoms forming the bond differ in electronegativity by more than 1.7 units, the bond will be at least 50% ionic (referred to as **percent ionic character**); we treat such a bond as wholly ionic. If the values differ by less than 0.4 units, we consider the bond to be wholly nonpolar. If the difference is between 0.4 and 1.7 electronegativity units, the bond is considered to be polar covalent. Remember that electronegativities have been calculated from fairly imprecise data for particular bonding situations. Electronegativity is useful in predicting the nature of a bond and for comparing bond types, but the prediction is only an approximation. Remember too that no sharp distinction exists between ionic, polar covalent, and nonpolar bonds; rather, they form a continuum. Even the most ionic bond (between cesium and fluorine) has some covalent character, and only bonds between atoms of the same element have no ionic character.

In these bonds, the atom with the higher electronegativity will be the negative end of the bond and, in extreme situations, will become the negative ion. To show these **partial charges** on a polar covalent bond, we mark the positive end of the bond with a δ^+ and the negative end of the bond with a δ^-. Table 7.1 summarizes these data.

TABLE 7.1 Guidelines for predicting bond type from electronegativity data

Difference in electronegativity (ΔEN)	Type of bond predominant	Example	ΔEN	More positive atom
>1.7	ionic	NaCl	2.1	sodium
0.4–1.7	polar covalent	C—Cl	0.5	carbon
<0.4	covalent	H—H	0.0	neither
		C—H	0.4	neither

Example 7.1

Predict the nature of the bond between the following pairs of atoms as predominantly nonpolar covalent, polar covalent, or ionic. For each polar covalent bond, use a small Greek letter δ to show which atom bears a partial positive charge (δ^+) and which a partial negative charge (δ^-).

a. S—O   **b.** C—O   **c.** Al—F

Solution

a. The electronegativity of oxygen is 3.5 and that of sulfur is 2.5. The difference (ΔEN) is 1.0 unit; we predict the S—O bond to be polar covalent. The oxygen is partially negative, and the sulfur is partially positive, so we write:

$$\overset{\delta^+}{\text{S}}-\overset{\delta^-}{\text{O}}$$

b. The electronegativity difference between oxygen and carbon is 1.0 unit (3.5 − 2.5). Therefore, we predict the C—O bond to be polar covalent. Because oxygen is the more electronegative of the two, it carries the negative charge.

$$\overset{\delta^+}{C}-\overset{\delta^-}{O}$$

c. The electronegativity difference between fluorine and aluminum is 2.5 units (4.0 − 1.5). Therefore, we predict the Al—F bond to be largely ionic. The aluminum forms a cation, the fluorine an anion.

Problem 7.1   Is the bond between carbon and nitrogen in a C—N bond nonpolar covalent, polar covalent, or ionic? In this pair of atoms, which bears the partial positive charge and which bears the partial negative charge?

C. Single, Double, and Triple Bonds

A covalent bond represents the sharing of electrons between two atoms. **Single bonds** result from the sharing of a single pair of electrons. The covalent bonds shown in Figure 7.2 are single bonds. Usually, as in the hydrogen molecule, each atom forming the bond contributes one electron to the bond. Sometimes, as in the reaction of ammonia, NH_3, with a hydrogen ion, H^+, to form the ammonium ion, NH_4^+, both electrons come from the same atom:

$$H:\overset{..}{N}:H + H^+ \longrightarrow \left[H:\overset{\overset{\textstyle H}{}}{\underset{\underset{\textstyle H}{}}{\overset{..}{N}}}:H \right]^+$$

It is common practice to use a dash to represent a pair of electrons. In this text we will use dashes for shared electrons and dots for unshared (lone-pair) electrons. With this notation, the above equation is written:

$$H-\overset{..}{\underset{\underset{\textstyle H}{|}}{N}}-H + H^+ \longrightarrow \left[H-\overset{\overset{\textstyle H}{|}}{\underset{\underset{\textstyle H}{|}}{N}}-H \right]^+$$

In the ammonia molecule, the nitrogen shares a pair of electrons with each of the three hydrogens. In each bond, one electron comes from nitrogen and one from hydrogen. The nitrogen still has an unshared pair of electrons.

A hydrogen ion has no electrons; the single hydrogen electron was lost when the atom became an ion and gained a positive charge. When the hydrogen ion bonds to the ammonia molecule, both electrons of the bond come from the nitrogen. A bond in which one atom has donated both electrons is often referred to as a **coordinate covalent bond.** It is most important to realize that the different name refers *only* to the method of formation. Once the ammonium ion is formed, all hydrogen–nitrogen bonds in the ion are equivalent. Notice, too, that the entire ammonium ion now carries a positive charge, denoted by placing brackets around the ion and writing a superscript +.

In addition to single bonds, there are double bonds and triple bonds. A **double bond** represents the sharing of four electrons by two atoms. The bond between carbon and oxygen is often a double bond, as in formaldehyde, CH_2O.

Here carbon is singly bonded to each of the hydrogens and doubly bonded to oxygen. Of this double bond, two electrons have come from carbon and two from oxygen. The single carbon–hydrogen bonds are nonpolar ($\Delta EN = 0.4$); the double carbon–oxygen bond is polar covalent ($\Delta EN = 1.0$). Note that each atom in the diagram of formaldehyde now follows the octet rule. Each hydrogen has two electrons; the carbon and the oxygen have eight electrons each. Notice too that the oxygen has two pairs of unshared electrons. Such an unshared pair is sometimes known as a **lone pair.** We will see that the negative end of a polar bond often holds unshared electron pairs.

A **triple bond** is formed when two atoms share six electrons (three pairs). The nitrogen molecule contains a triple bond. Its structure is

$$:N{\equiv}N:$$

Each nitrogen donates three electrons to the bond and retains a lone pair.

7.2   Lewis Structures

A compound contains two or more atoms of different elements joined by chemical bonds. The properties of the compound depend on the arrangement of atoms in the compound and the types of bonds between them. To help in gaining this information for covalently bonded molecules, we draw structures for the molecules such as those shown in Section 7.1 for hydrogen, ammonia, formaldehyde, and nitrogen. These structures are called **Lewis structures.** A Lewis structure shows each atom in the molecule or ion and its relationship to the other atoms. It also shows all bonding electrons as well as those valence electrons that are nonbonding.

A. The Arrangement of Atoms

Most often the formula of a compound is written in a way that predicts the arrangement of its atoms. Each covalently bonded molecule or ion has one central atom or a chain of central atoms. The central atom or chain of atoms will be the least electronegative element in the structure. This chain is often of carbon atoms, although it may be of nitrogen or some other nonmetal. Methane, CH_4, has one central carbon atom. Butane, C_4H_{10}, has a chain of four carbon atoms. The element symbols that directly follow the central atom in the formula indicate that these atoms are bonded to the central atom(s). Thus, methane, CH_4, has four hydrogen atoms bonded to its central carbon atom. Butane, C_4H_{10}, has ten hydrogen atoms bonded to the four carbon atoms. Formaldehyde, whose formula is written CH_2O or HCHO, has two hydrogen atoms and one oxygen atom bonded to the central carbon atom. Methyl amine, CH_3NH_2, has three hydrogen atoms bonded to the carbon atom. The carbon atom is also bonded to a nitrogen atom, and two more hydrogen atoms are also bonded to the nitrogen atom.

If more than one arrangement of atoms seems possible, we choose the one with the most **symmetry.** In carbon dioxide, CO_2, the atoms are arranged O—C—O, a more symmetrical arrangement than C—O—O. Similarly, sulfur trioxide, SO_3, is written as

$$O—S—O \quad \text{rather than} \quad S—O—O—O$$
$$\underset{\displaystyle O}{\overset{\displaystyle |}{}}$$

In showing the arrangement of atoms, keep in mind two things. Hydrogen rarely bonds to more than one atom. Halogens are usually bonded to only one atom. Only in polyatomic ions or molecules, such as bromate ion, BrO_3^-, or chloric acid, $HClO_3$, are halogens the central atom and thus bonded to more than one other atom.

Oxyacids like H_2SO_4, H_3PO_4, $HClO_3$, and acetic acid are a little different. The acid hydrogens, those that are written at the beginning of the formula, are not bonded to the central atom but are bonded to oxygen. Therefore, these compounds have the atomic arrangements:

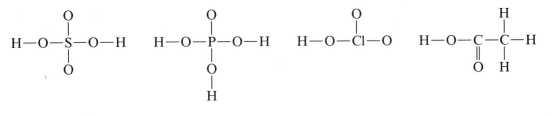

Example 7.2   Show the arrangement of atoms in the following compounds.

a. SO_2 **b.** C_2H_5Cl **c.** C_2H_2

Solution   **a.** From the formula you can predict that the two oxygens are bonded to the sulfur; or you can choose the most symmetrical arrangement with the least electronegative atom, the sulfur atom, at the center. Both arrangements are the same: O—S—O.

b. The two carbon atoms will be bonded together, forming a chain. Three hydrogen atoms are bonded to one carbon atom; two hydrogen atoms and the chlorine atom are bonded to the other carbon. The arrangement is

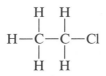

c. Again the two carbon atoms will be bonded together. The most symmetrical arrangement for the two hydrogen atoms is one on each carbon, giving the arrangement: H—C—C—H.

Problem 7.2   Show the arrangement of atoms in the following compounds.

      a. H_2S **b.** CH_3OH **c.** PCl_3

B. The Number and Placement of Electrons in Lewis Structures

The Lewis structure of a molecule or ion shows the arrangement of atoms and the distribution of electrons in that molecule or ion. In Section 7.2A, we showed how to predict the arrangement of atoms; in this section we show how to predict the number and distribution of electrons. Some of these electrons will be shared; some will be unshared. The number in each category can be determined using the following steps. Notice that in these examples all the atoms follow the octet rule. We will illustrate each step of this process using a molecule of formaldehyde, CH_2O.

1. Each atom in the molecule except hydrogen is assumed to require eight electrons; hydrogen will require only two. (See Section 5.7C1 on the octet rule exceptions.) Formaldehyde, CH_2O, will require electrons to fill 20 spaces.

$$\underset{\text{hydrogen}}{2(2)} + \underset{\text{carbon}}{8} + \underset{\text{oxygen}}{8} = 20 \text{ spaces to fill}$$

2. Each atom contributes its valence electrons toward filling these spaces. (Valence electrons were defined in Section 5.5D.) Formaldehyde has 12 available valence electrons:

$$\underset{\text{hydrogen}}{2(1)} + \underset{\text{carbon}}{4} + \underset{\text{oxygen}}{6} = 12 \text{ valence electrons available}$$

3. The difference between these numbers is the number of bonding electrons shared by two atoms. Formaldehyde will contain eight shared electrons:

$$20 - 12 = 8 \text{ shared electrons}$$

4. The number of unshared electrons is the difference between the number of valence electrons available and the number of shared or bonding electrons. Formaldehyde will have $12 - 8 = 4$ unshared electrons. Now we draw the Lewis structure. Having determined the arrangement of atoms, we place a pair of electrons between each set of neighboring atoms. For formaldehyde, we write

$$\begin{array}{c} \text{H} \\ | \\ \text{H} - \text{C} - \text{O} \end{array}$$

which uses six electrons and leaves two more electrons to be shared. These electrons will group with another pair to form a double bond. Carbon–hydrogen bonds are always single; therefore, the double bond will be between carbon and oxygen, giving the structure

$$\begin{array}{c} \text{H} \\ | \\ \text{H} - \text{C} = \text{O} \end{array}$$

5. The unshared electrons are then added where needed to give an octet. In the structure for formaldehyde, we have four unshared electrons. Oxygen still does not have an octet, so the unshared electrons are added to the oxygen, giving

$$\begin{array}{c} \text{H} \\ | \\ \text{H} - \text{C} = \overset{..}{\underset{..}{\text{O}}} \end{array}$$

This structure is the complete Lewis structure of formaldehyde.

Rules for drawing Lewis structures of molecules:

■ 1. Draw skeleton of molecule.
   Least electronegative atom will be at center.
   Hydrogen and halogens will be at perimeter.

■ 2. Calculate the following.
   Number of spaces to fill if each atom obeys octet rule (A)
   Total number of valence electrons available (V)
   Number of bonding electrons (B): $B = A - V$
   Number of nonbonding electrons (N): $N = V - B$

■ 3. Assign two bonding electrons to each bond. Assign any bonding electrons remaining to double or triple bonds between appropriate atoms (not hydrogen or halogen).

■ 4. Assign nonbonding electrons (N) where needed to complete octets.

Example 7.3   Draw the Lewis structure of the following:

a. ethylene, C_2H_4 **b.** phosphorus trichloride, PCl_3
c. methyl alcohol, CH_3OH

Solution   **a.** The most symmetrical arrangement for ethylene, C_2H_4, is

H H

H C C H

The number of electron spaces is $\underset{\text{hydrogen}}{4(2)}$ + $\underset{\text{carbon}}{2(8)}$ = 24.

The number of valence electrons is $\underset{\text{hydrogen}}{4(1)}$ + $\underset{\text{carbon}}{2(4)}$ = 12.

The number of bonding electrons is $24 - 12 = 12$.
The number of unshared electrons is $12 - 12 = 0$.
Using a pair of electrons for each bond predicted in the most symmetrical arrangement gives:

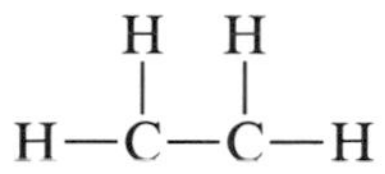

H H
| |
H—C—C—H

Two bonding electrons remain unused, and each carbon atom still needs two more electrons. Two electrons can fill this need by forming a double bond between the carbons, which gives the structure:

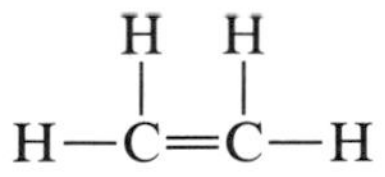

H H
| |
H—C=C—H

All atoms in the structures as shown satisfy the octet rule; therefore no nonbonding (unshared) electrons are needed.

b. The symmetrical atomic arrangement for phosphorus trichloride, PCl_3, is

Cl

Cl P Cl

The number of electron spaces is $\underset{\text{phosphorus}}{8}$ + $\underset{\text{chlorine}}{3(8)}$ = 32.

The number of valence electrons is $\underset{\text{phosphorus}}{5}$ + $\underset{\text{chlorine}}{3(7)}$ = 26.

The number of bonding electrons is $32 - 26 = 6$.
The number of unshared electrons is $26 - 6 = 20$.

Adding six bonding electrons to the predicted arrangement gives:

$$\begin{array}{c} Cl \\ | \\ Cl-P-Cl \end{array}$$

Adding the unshared electrons completes the octet of each atom and gives:

$$\begin{array}{c} :\overset{\cdot\cdot}{Cl}: \\ | \\ :\overset{\cdot\cdot}{\underset{\cdot\cdot}{Cl}}-P-\overset{\cdot\cdot}{\underset{\cdot\cdot}{Cl}}: \end{array}$$

c. Reading the formula for methyl alcohol, CH_3OH, we predict the atomic arrangement to be three hydrogen atoms and the oxygen atom bonded to the carbon atom, with another hydrogen bonded to the oxygen:

$$\begin{array}{c} H \\ \\ H \quad C \quad O \quad H \\ \\ H \end{array}$$

The number of electron spaces available is:

$$\underset{\text{carbon}}{8} \;+\; \underset{\text{hydrogen}}{3(2)} \;+\; \underset{\text{oxgyen}}{8} \;+\; \underset{\text{hydrogen}}{2} \;=\; 24$$

The number of valence electrons is $4 + 3(1) + 6 + 1 = 14$.
The number of bonding electrons is $24 - 14 = 10$.
Using these bonding electrons for single bonds in the predicted arrangement gives:

$$\begin{array}{c} H \\ | \\ H-C-O-H \\ | \\ H \end{array}$$

The number of unshared electrons is $14 - 10 = 4$.
The only atom still missing an octet of electrons is oxygen, which needs four more electrons. Putting the unshared electrons in the structure around oxygen, we get a complete Lewis structure of methyl alcohol.

$$\begin{array}{c} H \\ | \\ H-C-\overset{\cdot\cdot}{\underset{\cdot\cdot}{O}}-H \\ | \\ H \end{array}$$

Problem 7.3 Draw the complete Lewis structure of:

    a. $CHCl_3$ **b.** CH_3NH_2 **c.** C_2H_5CHO

C. Lewis Structures of Ions

To draw the Lewis structure of an ion, we follow the same steps as for drawing the Lewis structure of a molecule with one exception: In calculating the number of valence electrons available, one additional electron is added for each negative charge on the ion or one electron is subtracted for each positive charge on the ion. The entire structure is enclosed in brackets, and the charge is shown as a superscript outside the brackets.

Example 7.4   Draw Lewis structures for the following ions.

   a. hydronium ion, H_3O^+ **b.** ammonium ion, NH_4^+
   c. chlorate ion, ClO_3^-

Solution   **a.** The atomic arrangement for the hydronium ion, H_3O^+, is:

$$H$$
$$H \quad O \quad H$$

The number of spaces is $3(2) + 8 = 14$.

The number of valence electrons is $\underset{\text{hydrogen}}{3(1)} + \underset{\text{oxygen}}{6} - \underset{\text{charge}}{1} = 8$.

The number of shared electrons is $14 - 8 = 6$.
The number of unshared electrons is $8 - 6 = 2$.
The Lewis structure is:

$$\left[\begin{array}{c} H \\ | \\ H-O-H \end{array} \right]^+$$

b. The atomic arrangement for the ammonium ion, NH_4^+, is:

$$H$$
$$H \quad N \quad H$$
$$H$$

The number of spaces is $\underset{\text{hydrogen}}{4(2)} + \underset{\text{nitrogen}}{8} = 16$.

The number of valence electrons is $\underset{\text{hydrogen}}{4(1)} + \underset{\text{nitrogen}}{5} - \underset{\text{charge}}{1} = 8$.

The number of shared electrons is $16 - 8 = 8$.
The number of unshared electrons is $8 - 8 = 0$.

The Lewis structure is:

$$\left[\begin{array}{c} H \\ | \\ H-N-H \\ | \\ H \end{array}\right]^{+}$$

c. The most symmetrical arrangement of atoms for the chlorate ion, ClO_3^-, is:

O   Cl   O

O

The number of spaces is $3(8) + 8 = 32$.
oxygen   chlorine

The number of valence electrons is $3(6) + 7 + 1 = 26$.
oxygen   chlorine   charge

The number of shared electrons is $32 - 26 = 6$.
The number of nonbonding electrons is $26 - 6 = 20$.
The Lewis structure of this ion is:

$$\left[\begin{array}{ccc} :\ddot{O}-\ddot{Cl}-\ddot{O}: \\ | \\ :\ddot{O}: \end{array}\right]^{-}$$

Problem 7.4   Draw Lewis structures for the following ions.

    a. hydroxide ion, OH^- **b.** bromate ion, BrO_3^-

    c. cyanide ion, CN^-

D. Resonance (Optional)

As chemists began to work with Lewis structures, it became more and more obvious that, for a great many molecules and ions, no single Lewis structure provided a truly accurate representation. For example, a Lewis structure for the carbonate ion, CO_3^{2-}, shows carbon bonded to three oxygen atoms by a combination of one double bond and two single bonds. Three possible Lewis structures for CO_3^{2-} are shown in Figure 7.4. Each implies that one carbon oxygen

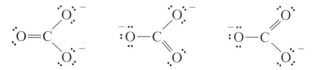

FIGURE 7.4   Three possible Lewis structures for the carbonate ion, CO_3^{2-}.

bond is different from the other two. However, this difference in bonding is not the case; rather, it has been shown that all three bonds are identical.

To describe molecules and ions, like the carbonate ion, for which no single Lewis structure is adequate, the theory of **resonance** was developed by Linus Pauling in the 1930s. According to resonance theory, many molecules and ions are best described by drawing two or more Lewis structures and considering the real molecule or ion as a hybrid (composite) of these structures. The individual Lewis structures are called contributing structures. We show that the real structure is a hybrid of the various contributing structures by connecting them with double-headed arrows as in Figure 7.5.

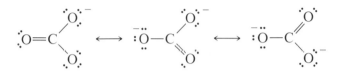

FIGURE 7.5 The carbonate ion can be represented as its three possible contributing structures connected with double-headed arrows to imply resonance.

Remember that the carbonate ion, or any other compound we describe in this way, has one and only one real structure. The problem is that our systems of representation are not adequate to describe the real structures of molecules and ions. The resonance method is a particularly useful way to describe the structure of these compounds, for it retains the use of Lewis structures with electron-pair bonds. We fully realize that the carbonate ion is not accurately represented by any single contributing structure (Figure 7.4).

Example 7.5

Show that sulfur trioxide can be represented by a resonance hybrid of three contributing structures.

Solution

The most symmetrical arrangement for sulfur trioxide is:

O S O

O

The number of electron spaces is $8 + 3(8) = 32$.
The number of valence electrons is $6 + 3(6) = 24$.
The number of shared electrons is $32 - 24 = 8$.
The number of unshared electrons is $24 - 8 = 16$.
The Lewis structure of sulfur trioxide is either

$$\ddot{\text{O}}\!-\!\text{S}\!-\!\ddot{\text{O}} \quad \text{or} \quad \ddot{\text{O}}\!-\!\text{S}\!=\!\ddot{\text{O}} \quad \text{or} \quad \ddot{\text{O}}\!=\!\text{S}\!-\!\ddot{\text{O}}$$

All of these structures are equivalent, therefore the molecule exhibits resonance and might be represented by:

$$:\ddot{O}\!-\!S\!-\!\ddot{O}: \longleftrightarrow :\ddot{O}\!-\!S\!=\!\ddot{O}: \longleftrightarrow :\ddot{O}\!=\!S\!-\!\ddot{O}:$$

Problem 7.5   Show that ozone, O_3, is a resonance hybrid.

<hr>

7.3   ## Bond Angles and the Shapes of Molecules

In Section 7.2, a shared pair of electrons was presented as the fundamental unit of the covalent bond, and Lewis structures were drawn for several small molecules and ions containing various combinations of single, double, and triple bonds. In this section, we use the **valence-shell electron-pair repulsion (VSEPR) model** to predict the geometry of these and other covalently bonded molecules and ions.

The VSEPR model can be explained in the following way. We know that an atom has an outer shell of valence electrons. These valence electrons may be involved in the formation of single, double, or triple bonds, or they may be unshared. Each set of electrons, whether unshared or in a bond, creates a negatively charged region of space. We have already learned that like charges repel each other. The VSEPR model states that the various regions containing electrons or electron clouds around an atom spread out so that each region is as far from the others as possible.

A. Linear Molecules

If a molecule contains only two atoms, those two atoms are in a straight line and thus form a **linear molecule.** Some three-atom molecules also have straight-line geometry. For example:

$$\ddot{O}\!=\!C\!=\!\ddot{O} \qquad H\!-\!C\!\equiv\!N: \qquad H\!-\!C\!\equiv\!C\!-\!H$$

carbon dioxide   hydrogen cyanide   acetylene

Notice that, in the Lewis structure of these molecules, the central atom(s) bonds with only two other atoms and has no unshared electrons. Only two electron clouds emerge from that central atom. For these two clouds to be as far away from each other as possible, they must be on opposite sides of the central atom, forming a **bond angle** of 180° with each other. An angle of 180° gives a straight line. The VSEPR theory says, then, that the geometry around an atom that has only two bonds and no unshared electrons is a straight line. Figure 7.6 shows the linear nature of these molecules.

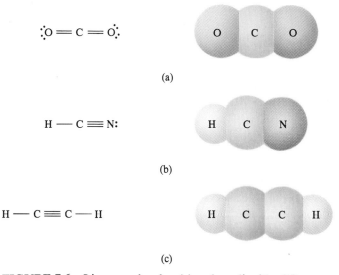

(a)

(b)

(c)

FIGURE 7.6 Linear molecules: (a) carbon dioxide, CO_2; (b) hydrogen cyanide, HCN; and (c) acetylene, C_2H_2.

B. Structures with Three Regions of High Electron Density around the Central Atom

Look at the following Lewis structures:

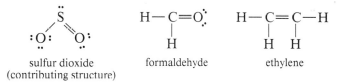

In these molecules, each central atom has three electron clouds emanating from it. In sulfur dioxide, the sulfur atom is bonded to two oxygen atoms and has one unshared pair of electrons. In formaldehyde and ethylene, each carbon atom has two single bonds to hydrogen, a double bond to another atom, and no unshared pair. The sulfur atom in sulfur dioxide and the carbon atom in ethylene and formaldehyde is surrounded by three clouds of high electron density. For these clouds to be as far as possible from one another, they will form a plane containing the central atom and will emanate from the central atom at angles of 120° to each other. The structure will be **trigonal planar.** The central atom will be in the center of the triangle, and the ends of the electron clouds at the corners of the triangle. If you experiment with a marshmallow as the central atom and three toothpicks as electron clouds, you can prove to yourself that the toothpicks are farthest apart when using a trigonal planar structure. Figure 7.7 illustrates these structures. Note that the angles are not exactly 120° but are remarkably close to that predicted value.

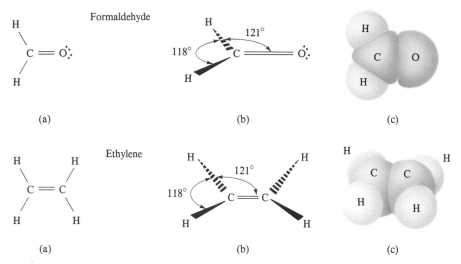

FIGURE 7.7 Shapes of the formaldehyde and ethylene molecules: (a) Lewis structures; (b) trigonal planar arrangement of the three regions of high electron density around the carbon atom; (c) space-filling models. In these figures, ◣ represents a bond projecting in front of the plane of the paper, and ⠒⠒⠒ represents a bond projecting behind the plane of the paper.

Although the electron clouds of these molecules give a trigonal planar shape around each carbon atom, one describes the geometry of a molecule only on the basis of the relationships between its atoms. A formaldehyde molecule is trigonal planar because it has an atom at the end of each electron cloud. The ethylene molecule has trigonal planar geometry around each of its carbon atoms. The whole molecule is planar, and its shape resembles two triangles joined point to point. In sulfur dioxide, there are three electron clouds around the sulfur. Only two of these connect two atoms. In the molecule, the oxygen–sulfur–oxygen atoms make a 120° angle. The molecule is **bent.**

A central atom surrounded by three clouds of high electron density will have trigonal planar geometry if it is bonded to three atoms. Its geometry will be called bent if it is bonded to two atoms and also has an unshared pair of electrons.

C. Structures with Four Regions of High Electron Density around the Central Atom

The following Lewis structures show three molecules whose central atom is surrounded by four clouds of high electron density:

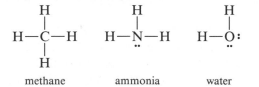

These molecules are alike in that each central atom is surrounded by four pairs of electrons, but they differ in the number of unshared electron pairs on the central atom. Remember that, although we have drawn them in a plane, the molecules are three-dimensional and atoms may be in front of or behind the plane of the paper. What geometry does the VSEPR theory predict for these molecules?

Let us predict the shape of methane, CH_4. The Lewis structure of methane shows a central atom surrounded by four separate regions of high electron density. Each region consists of a pair of electrons bonding the carbon atom to a hydrogen atom. According to the VSEPR model, these regions of high electron density spread out from the central carbon atom in such a way that they are as far from one another as possible.

You can predict the resulting shape using a styrofoam ball or marshmallow and four toothpicks. Poke the toothpicks into the ball, making sure that the free ends of the toothpicks are as far from one another as possible. If you have positioned them correctly, the angle between any two toothpicks will be 109.5°. If you now cover this model with four triangular pieces of paper, you will have built a four-sided figure called a regular tetrahedron. Figure 7.8 shows (a) the Lewis structure for methane, (b) the tetrahedral arrangement of the four regions of high electron density around the central carbon atom, and (c) a space-filling model of methane.

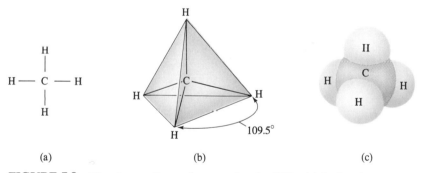

(a)   (b)   (c)

FIGURE 7.8   The shape of a methane molecule, CH_4: (a) its Lewis structure; (b) its tetrahedral shape; (c) a space-filling model.

According to the VSEPR model, the H—C—H bond angle in methane should be 109.5°. This angle has been measured experimentally and found to be 109.5°. Thus, the bond angle predicted by the VSEPR model is identical to that observed. We say that methane is a **tetrahedral molecule.** The carbon atom is at the center of a tetrahedron. Each hydrogen is at one of the corners of the tetrahedron.

We can predict the shape of the ammonia molecule in exactly the same manner. The Lewis structure of NH_3 (see Figure 7.9) shows a central nitrogen atom surrounded by four separate regions of high electron density. Three of these regions consist of a single pair of electrons forming a covalent bond with a

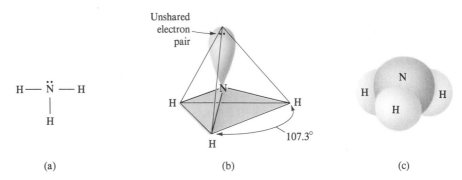

FIGURE 7.9   The shape of an ammonia molecule, NH_3: (a) its Lewis structure; (b) its geometry; (c) a space-filling model. Notice how the unshared electrons serve to create its shape.

hydrogen atom; the fourth region contains an unshared pair of electrons. According to the VSEPR model, the four regions of high electron density around the nitrogen are arranged in a tetrahedral manner, so we predict that each H—N—H bond angle should be 109.5°. The observed bond angle is 107.3°. This small difference between the predicted angle and the observed angle can be explained by proposing that the unshared pair of electrons on nitrogen repels the adjacent bonding pairs more strongly than the bonding pairs repel each other.

Ammonia is not a tetrahedral molecule. The atoms of ammonia form a **pyramidal molecule** with nitrogen at the peak and the hydrogen atoms at the corners of a triangular base. Just as the unshared pair of electrons in sulfur dioxide contribute to the geometry of the molecule but are not included in the description of its geometry, the unshared pair of electrons in ammonia gives it a tetrahedral shape but its geometry is based only on the arrangement of atoms, which is pyramidal.

Figure 7.10 shows the Lewis structure of the water molecule. In H_2O, a central oxygen atom is surrounded by four separate regions of high electron density. Two of these regions contain a pair of electrons forming a covalent bond between oxygen and hydrogen; the other two regions contain an unshared electron pair. The four regions of high electron density in water are arranged in a tetrahedral manner around oxygen. Based on the VSEPR model, we predict an H—O—H bond angle of 109.5°. Experimental measurements show that the actual bond angle is 104.5°. The difference between the predicted and observed bond angles can be explained by proposing, as we did for NH_3, that unshared pairs of electrons repel adjacent bonding pairs more strongly than the bonding pairs repel each other. Note that the variation from 109.5° is greatest in H_2O, which has two unshared pairs of electrons; it is smaller in NH_3, which has one unshared pair; and there is no variation in CH_4.

To describe the geometry of the water molecule, remember that the geometry of a molecule describes only the geometric relationships between its atoms.

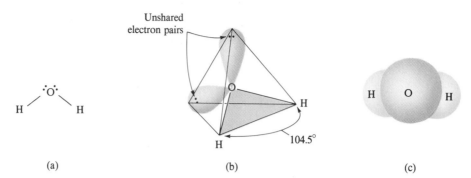

FIGURE 7.10 The shape of a water molecule, H_2O: (a) its Lewis structure; (b) its geometry; (c) a space-filling model. Notice how the unshared pairs of electrons affect the tetrahedral geometry.

The three atoms of a water molecule are in a bent line like those of sulfur dioxide. We say the water molecule is bent.

A general prediction emerges from our discussions of the shapes of methane, ammonia, and water: Whenever four separate regions of high electron density surround a central atom, we can accurately predict a tetrahedral distribution of electron clouds and bond angles of approximately 109.5°.

The geometry of molecules can be predicted. A molecule whose central atom is bonded to four other atoms is tetrahedral. One in which the central atom has one unshared pair of electrons and bonds to three other atoms will be pyramidal, and one in which the central atom has two unshared pairs of electrons and bonds to two other atoms will be bent. Table 7.2 summarizes this geometry.

TABLE 7.2   Molecular shapes and bond angles

Number of regions of high electron density around central atom	Arrangement of regions of high electron density in space	Predicted bond angles	Example	Geometry of molecule
4	tetrahedral	109.5°	CH_4, methane NH_3, ammonia H_2O, water	tetrahedral pyramidal bent
3	trigonal planar	120°	H_2CO, formaldehyde C_2H_4, ethylene SO_2, sulfur dioxide	trigonal planar planar bent
2	linear	180°	CO_2, carbon dioxide C_2H_2, acetylene	linear linear

Example 7.6   Predict all bond angles in the following molecules.

                  a CH_3Cl **b.** CH_3CN **c.** CH_3COOH

Solution   **a.** The Lewis structure of methyl chloride is:

$$H-\overset{\overset{\displaystyle H}{|}}{\underset{\underset{\displaystyle H}{|}}{C}}-\ddot{\underset{..}{Cl}}:$$

In the Lewis structure of CH_3Cl, carbon is surrounded by four regions of high electron density, each of which forms a single bond. Based on the VSEPR model, we predict a tetrahedral distribution of electron clouds around carbon, H—C—H and H—C—Cl bond angles of 109.5°, and a tetrahedral shape for the molecule. Note the use of ⸺ to represent a bond projecting behind the plane of the paper and ► to represent a bond projecting forward from the plane of the paper.

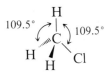

b. The Lewis structure of acetonitrile, CH_3CN, is:

$$H-\overset{\overset{\displaystyle H}{|}}{\underset{\underset{\displaystyle H}{|}}{C}}-C\equiv N:$$

The methyl group, CH_3—, is tetrahedral. The carbon of the —CN group is in the middle of a straight line stretching from the carbon of the methyl group through the nitrogen.

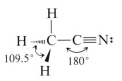

c. The Lewis structure of acetic acid is:

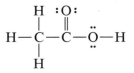

Both the carbon bonded to three hydrogens and the oxygen bonded to carbon and hydrogen are centers of tetrahedral structures. The central carbon will have 120° bond angles:

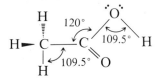

The geometry around the first carbon is tetrahedral, around the second carbon atom is trigonal planar, and around the oxygen is bent.

Problem 7.6   Predict the bond angles and the resulting geometry of the following molecules and ions.

      a. CH_3CHO **b.** CO_3^{2-} **c.** CH_2Cl_2 **d.** NH_4^+

7.4   The Polarity of Molecules

Some molecules contain only nonpolar bonds—for example, methane, CH_4. Such molecules are **nonpolar molecules.** Other molecules contain polar bonds —that is, bonds between atoms whose electronegativities differ by more than 0.4 units. Whether these latter molecules are polar or nonpolar depends on the arrangement in space of these bonds and the resulting geometry of the molecules. If we picture the geometry of a molecule and show its polar bonds with an arrow ($\longmapsto$) aimed at the more electronegative atom, we can usually obtain a picture of the molecule that indicates whether or not it is polar. Figure 7.11 illustrates this method for several molecules.

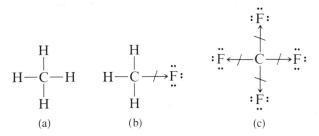

FIGURE 7.11   Polarity in molecules: (a) methane has no polar bonds and is a nonpolar molecule; (b) methyl fluoride has one polar bond (denoted by $\longmapsto$) and is therefore a polar molecule; (c) carbon tetrafluoride has four counterbalanced polar bonds and thus is a nonpolar molecule.

Which of the molecules in Figure 7.11 is polar? Methane, which contains no polar bonds, is clearly nonpolar. Methyl fluoride contains one polar bond between carbon (EN = 2.5) and fluorine (EN = 4.0). Methyl fluoride is a polar molecule; the negative end of the dipole is at the fluoride atom. Now look at carbon tetrafluoride. It contains four polar carbon–fluorine bonds but they counteract one another, so the molecule itself is nonpolar.

The principle is like that of erecting an antenna tower (Figure 7.12). If the guy wires are correctly balanced against one another, the tower stays erect.

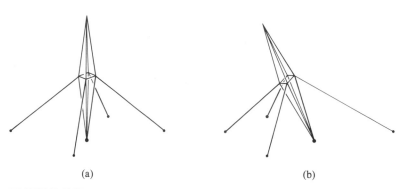

(a)   (b)

FIGURE 7.12   Erecting an antenna tower: (a) antenna tower with balanced guy wires (compare with carbon tetrafluoride); (b) antenna tower with unbalanced guy wires (compare with methyl fluoride).

(Compare the balanced pull of the carbon–fluorine bonds in carbon tetrafluoride.) If the guy wires are not balanced, the tower topples. (Compare the unbalanced pull of the carbon–fluorine bond in methyl fluoride.)

Figure 7.13 shows three other molecules and their polarities. Notice that both ammonia and water have unshared electrons on the atom at the negative end of the dipole. These unshared electrons enhance the polarity of the bond to make these molecules very polar.

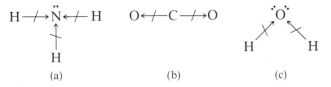

(a)   (b)   (c)

FIGURE 7.13   Predicting the polarity of molecules: (a) ammonia, NH_3, has unbalanced polar bonds and is a polar molecule; (b) carbon dioxide, CO_2, has balanced polar bonds and is a nonpolar molecule; (c) water, H_2O, has unbalanced polar bonds and is a polar molecule.

Example 7.7   Predict whether the following molecules are polar or nonpolar.

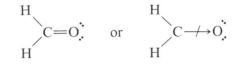

a. CH_2O **b.** $SiBr_4$ **c.** SO_2

Solution   **a.** The Lewis structure of CH_2O is:

$$\begin{array}{c} H \\ \diagdown \\ \diagup \\ H \end{array} C = \ddot{\underset{..}{O}} \colon \qquad or \qquad \begin{array}{c} H \\ \diagdown \\ \diagup \\ H \end{array} C \longrightarrow\!\!\!|\ \ddot{\underset{..}{O}} \colon$$

The electronegativity difference (ΔEN) between carbon and oxygen is 1.0. Thus, the C=O bond will be polar, with a partial positive charge on the carbon and a partial negative charge on the oxygen. The C—H bond is not polar. By showing the polar bond as an arrow, the diagram shows clearly that the molecule is polar (the bond is not balanced). The unshared electrons on the oxygen should enhance this polarity.

b. The Lewis structure of silicon tetrabromide is:

$$\begin{array}{c} \overset{..}{\underset{}{:Br:}} \\ | \\ :\!\overset{..}{\underset{..}{Br}}\!-\!\overset{..}{\underset{}{Si}}\!-\!\overset{..}{\underset{..}{Br}}\!: \\ | \\ \overset{}{\underset{..}{:Br:}} \end{array}$$

The electronegativity of silicon is 1.8, that of bromine is 2.8. The silicon–bromine bond is polar. Showing these bonds as arrows in a tetrahedral structure (Figure 7.14) clarifies that silicon tetrabromide is a nonpolar molecule. The polar bonds balance each other as in carbon tetrafluoride (Figure 7.11), and the molecule is nonpolar.

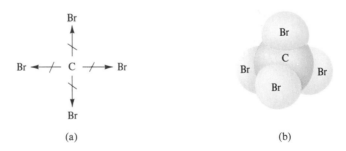

(a)   (b)

FIGURE 7.14   Silicon tetrabromide: (a) its geometry showing polar bonds; (b) a space-filling model.

c. The Lewis structure of sulfur dioxide is:

The electronegativity of sulfur is 2.5 and that of oxygen is 3.5; thus the sulfur–oxygen bonds are polar. By drawing these polar bonds as arrows in the bent molecule of sulfur dioxide, we show its polar nature:

$$S^{\delta+}$$

$$\delta^-O \qquad O^{\delta-}$$

The molecule is polar. This molecule is a resonance hybrid, but this fact does not affect its polarity (see Section 7.2D).

Problem 7.7   Predict the polarity of the following molecules.

      a. HBr   **b.** CH_3CHO **c.** PCl_3

7.5 Interactions of Water with Molecules; Electrolytes and Nonelectrolytes

A. Interaction between Polar Molecules

Polar molecules interact with one another. This interaction is governed by the fact that like charges repel and opposite charges attract. Figure 7.15 shows a collection of polar molecules aligned so that the partially positive end of one molecule is near the partially negative end of another.

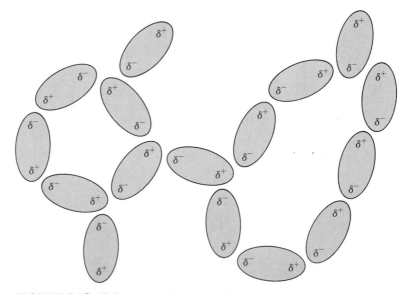

FIGURE 7.15 Polar molecules, when free to move around in the liquid state, will align themselves, as in this diagram, with the positive end nearest the negative end of another molecule.

Polar molecules aligned by charge need not be all of the same compound. Figure 7.13 showed that water is a polar molecule. When another polar molecule such as ammonia (also shown in Figure 7.13) is added to water, it dissolves because of the interaction between the two types of molecules. Figure 7.16 shows the polarity of water molecules and ammonia molecules and the interactions between them.

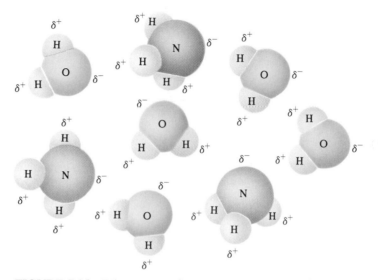

FIGURE 7.16   When ammonia (a polar molecule) is dissolved in water (another polar molecule) the two kinds of molecules orient themselves, negative end to positive end, as in this diagram.

B. Water Solutions of Ionic Compounds

When an ionic compound dissolves in water, the solution contains ions rather than neutral particles. For example, when sodium chloride, NaCl, dissolves in water, the solution contains sodium ions (Na^+) and chloride ions (Cl^-), rather than neutral units of NaCl. A polyatomic ion does not break up into separate atoms in solution. Thus, a solution of sodium nitrate, $NaNO_3$, contains sodium ions and nitrate ions; no nitrogen ions or oxygen ions are present. A solution of ammonium sulfate, $(NH_4)_2SO_4$, contains ammonium ions and sulfate ions.

This dissolution of ionic compounds takes place because of interactions like that between ammonia and water. Figure 7.17 shows at the molecular level what happens when sodium chloride, an ionic compound, dissolves in water. Each ion of the solid crystal becomes surrounded by water molecules, with the negative end of the water molecules approaching closest to the positive sodium ions, and the positive end of the water molecules surrounding the negative chloride ions. The water molecules pull these ions, one by one, away from the

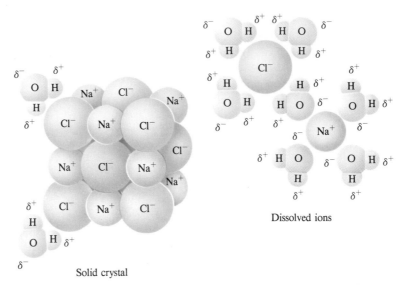

Solid crystal

Dissolved ions

FIGURE 7.17   The dissolution of sodium chloride in water. Notice how the polar water molecules are oriented in one way around the positive sodium ions and in another way around the oppositely charged chloride ions.

rest of the crystal. Other ionic compounds that are virtually insoluble in water have such strong interactions between their ions that the pull of the polar water molecules is not strong enough to break the ions apart.

Frequently, when a very polar molecule such as hydrogen chloride dissolves in water, there is sufficient interaction between the two molecules to cause one of them to break up into ions. With hydrogen chloride, the equation for this reaction is

$$HCl + H_2O \longrightarrow H_3O^+ + Cl^-$$

To some extent the same process occurs when ammonia is dissolved in water. The equation for that reaction is

$$NH_3 + H_2O \longrightarrow NH_4^+ + OH^-$$

C. Electrolytes, Nonelectrolytes, and Weak Electrolytes

The presence of ions in a solution of an ionic compound can be demonstrated with an apparatus like the one shown in Figure 7.18. The apparatus consists of two electrodes—one connected to one pole of a power source and the other connected to a light bulb that is, in turn, connected to the other pole of the power source. If the two electrodes touch, an electric current flows through the completed circuit and the bulb lights (Figure 7.18b). If the two electrodes are separated and then immersed in water, no current flows (Figure 7.18c). Pure

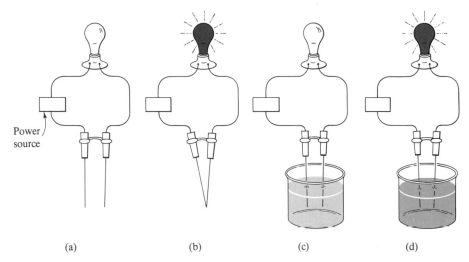

(a)   (b)   (c)   (d)

FIGURE 7.18 Conductivity apparatus for showing the presence of ions in an aqueous solution: (a) electrodes apart, circuit broken; (b) electrodes touching, circuit complete; (c) electrodes in pure water, circuit broken; (d) electrodes in an ionic solution, circuit complete.

water cannot carry an electric current. However, if the pure water is replaced by an aqueous (water) solution of an ionic compound, the bulb lights, indicating a flow of electricity through the solution (Figure 7.18d). The electric current is carried through the solution by the dissolved ions.

Compounds whose aqueous solutions conduct electricity are called **electrolytes.** This group of compounds includes hydroxides, salts (metal–nonmetal and metal–polyatomic ion compounds), and some acids (nitric, sulfuric, hydrochloric). Compounds that dissolve in water without forming a solution that carries an electric current are called **nonelectrolytes.** These compounds dissolve as molecules. Many compounds containing only nonmetals are nonelectrolytes. If a compound containing only nonmetals ionizes, its molecule will usually contain hydrogen bonded to a very electronegative atom like oxygen (as in acetic acid) or chlorine (as in hydrochloric acid).

Compounds that react somewhat but not completely with water to form ions are said to be partially ionized; their solutions are poor conductors of electricity. They are called **weak electrolytes.** Acetic acid is a weak electrolyte. In the apparatus in Figure 7.18, a solution of acetic acid causes the bulb to glow only faintly, showing that not many ions are present in solution.

D. Other Differences between Ionic and Covalent Compounds

Ionic compounds are usually solids at room temperature. They have high melting points. They dissolve only in very polar liquids like water. Some are only sparingly soluble in water.

Covalent compounds are found in all three states—solids, liquids, and gases—at room temperature. Those of low molecular weight and those that are polar may dissolve in water, but a covalent solid is much more apt to be soluble in a nonpolar (carbon tetrachloride) or slightly polar (ethyl alcohol) liquid.

Example 7.8   Predict whether water solutions of the following will conduct an electric current.

   a. potassium chloride, KCl   **b.** methyl chloride, CH_3Cl

   c. silane, SiH_4

Solution   **a.** Potassium chloride is an ionic compound. It dissolves in water to form ions that can conduct an electric current.

b. Methyl chloride contains only nonmetals. It does *not* contain hydrogen bonded to a very electronegative atom. The difference in electronegativity between carbon and chlorine is 0.5, therefore the molecule is only very slightly polar. It would not be expected to ionize, and its aqueous solution (if formed) would not be expected to carry an electric current.

c. Silane is a nonpolar molecule. (Its shape is tetrahedral like methane. See Figure 7.8.) Its water solution will not conduct a current.

Problem 7.8   Predict whether water solutions of the following will conduct an electric current.

a. hydrogen bromide, HBr

b. calcium nitrate, $Ca(NO_3)_2$

c. chloroform, $CHCl_3$

7.6   **Summary**

Atoms are joined together in compounds by chemical bonds. A covalent bond is formed by two atoms sharing electrons, either equally to yield a nonpolar bond or unequally to form a polar bond. An ionic bond is formed by an almost complete transfer of electrons from one atom or group to another atom or group. Electronegativity measures the attraction an atom has for the electrons in a chemical bond between that atom and another. If the difference in electronegativity is slight, the bond will be nonpolar covalent. If the difference is greater, the bond will be polar covalent. A large difference in electronegativity results in an ionic bond. Atoms may share two, four, or six electrons to form single, double, or triple bonds. One atom may donate both electrons of a single bond.

The Lewis structure of a molecule shows the arrangement of its atoms and the distribution of their electrons. The arrangement of atoms is usually the most symmetrical. If a molecule (or ion) has several equally likely arrangements of its electrons, that molecule (or ion) is said to exhibit resonance.

The number of electron clouds, or regions of high electron density around the central atom of a molecule, determines the geometry around that atom. Two clouds result in linear geometry, three result in trigonal planar or bent geometry, and four may yield either bent, or pyramidal, or tetrahedral geometry.

A molecule is polar if its atoms are so arranged that the polar bonds it contains are not balanced. Polar molecules and ionic compounds are most apt to dissolve in polar solvents like water. Nonpolar molecules dissolve most readily in nonpolar solvents. Ionic compounds and some polar molecules dissolve in water to yield ions; such compounds are called electrolytes. Compounds that are partially ionized in water solution are weak electrolytes. Compounds that do not dissociate into ions in water are called nonelectrolytes.

Key Terms

bent molecules (7.3B)
bond angles (7.3A)
chemical bond (7.1)
coordinate covalent bond (7.1C)
covalent bond (7.1)
covalent compound (7.5D)
dipole (7.1A)
double bond (7.1C)
electrolyte (7.5C)
electronegativity (7.1B)
ionic bond (7.1A)
ionic compounds (7.5D)
Lewis structure (7.2)
linear molecule (7.3A)
lone pair (7.1C)
nonelectrolyte (7.5C)
nonpolar covalent bond (7.1A)

nonpolar molecule (7.4)
octet rule (7.1)
partial charge (7.1B)
percent ionic character (7.1B)
polar covalent bond (7.1A)
polarity (7.1A)
pyramidal molecule (7.3C)
resonance (7.2D)
single bond (7.1C)
symmetry (of a Lewis structure)
(7.2A)
tetrahedral molecule (7.3C)
trigonal planar molecule (7.3B)
triple bond (7.1C)
valence-shell electron-pair repulsion
(VSEPR) model (7.3)
weak electrolyte (7.5C)

Multiple-Choice Questions

For Questions 1–4 consider the following compounds. You may want to draw the Lewis structure of each.

a. CCl_4 **b.** HCl **c.** CO_2 **d.** H_2S **e.** CO

MC1. Which of these molecules is tetrahedral?

MC2. Which is a nonpolar linear molecule?

MC3. Which contains a triple bond?

MC4. Which would you expect to be like water in geometry?

For Questions 5–8, consider allylamine, which has the structure

$$\overset{1}{C}H_2 = \overset{2}{C}H\overset{3}{C}H_2NH_2$$

MC5. How many electrons must be shown in the Lewis structure of this molecule?
   a. 18   **b.** 12   **c.** 28   **d.** 24   **e.** 42

MC6. Of the carbon atoms in allylamine, which would be surrounded by a tetrahedral structure? (The carbon atoms are numbered above the formula.)
   a. none   **b.** #1   **c.** #2   **d.** #3   **e.** all

MC7. How many unshared electrons are in this structure?
   a. 0   **b.** 1   **c.** 2   **d.** 3   **e.** 4

MC8. Allylamine is most apt to be:
   a. an ionic compound.   **b.** a polar compound.
   c. a nonpolar compound.   **d.** an acid.   **e.** a salt.

MC9. With which of the following elements will chlorine form the least polar bond?
   a. Na   **b.** Al   **c.** C   **d.** I   **e.** Cl

MC10. The preferred atomic arrangement for sulfur trioxide would be:

   a. S—O—O   **b.** O—O—O—S   **c.** O—O—S—O
         |
         O

   d. O—S   **e.** O—S—O
       | | |
       O—O   O

Problems

7.1 The Chemical Bond

*7.9. Chlorine bonds with each of the elements of period 3. Using electronegativity values (Figure 7.3), predict which of these bonds would be ionic, which polar covalent, and which nonpolar covalent.

7.10. In each of the following polar bonds, indicate which atom has a slightly negative charge (δ^-) and which atom has a slightly positive charge (δ^+).
   a. BrF   **b.** NI   **c.** PBr   **d.** CS

7.11. Predict the bond type (nonpolar, polar covalent, or ionic) in each of the following compounds:
   a. HI   **b.** LiF   **c.** Br_2 **d.** H_2S

*7.12. What is the difference between a single, a double, and a triple bond?

7.13. Phosphorus forms the compound phosphine, PH_3. Draw the Lewis structure of this compound. The compound reacts with hydrogen ion to form the phosphonium ion, PH_4^+. Draw the Lewis structure of the ion. Using these structures, tell the difference between covalent and coordinate covalent bonds. Is the difference significant once the phosphonium ion is formed?

7.14. Look back in the chapter to find an example of a compound that contains a single covalent bond. Draw the Lewis structure of this compound. Find an example of a compound containing a double bond and draw its Lewis structure. Do the same for a compound containing a triple bond.

7.2 Lewis Structures

*7.15. Draw Lewis structures of these molecules:
   a. silicon tetrachloride, $SiCl_4$
   b. sulfur dichloride, SCl_2
   c. arsine, AsH_3
   d. iodoform, CHI_3

7.16. Draw Lewis structures of the following molecules:
   a. C_2H_5OH **b.** $CH_3CH_2CCl_3$
   c. $CH_2=CHCH_3$ **d.** $C_2H_5NH_2$

*7.17. Draw Lewis structures of the following ions:
   a. $CH_3NH_3^+$ **b.** OH^- **c.** CN^-
   d. NH_2^-

7.18. Draw the Lewis structure of sulfur trioxide, SO_3, and show that this molecule exhibits resonance.

7.19. The following molecules are exceptions to the octet rule. All contain only single bonds. Draw a Lewis structure for each and state how it is an exception.
   a. BF_3 **b.** $BeCl_3$ **c.** PCl_5
   d. SF_6

7.3 Bond Angles and the Shapes of Molecules

*7.20. Predict the geometry of each molecule in Problem 7.15.

7.21. Predict the bond angles in each compound in Problem 7.16.

7.22. Draw the Lewis structure and predict the bond angles and geometry of:
   a. methyl acetylene, $CH_3C=CH$
   b. carbon disulfide, CS_2

*7.23. Draw the Lewis structure of acetaldehyde, CH_3CHO. Predict the geometry around each carbon atom.

7.4 The Polarity of Molecules

7.24. Ethyl alcohol, C_2H_5OH, and methyl ether, CH_3OCH_3, have the same empirical formula but one is more polar than the other. Draw the Lewis structures of these molecules. Predict the geometry of each and state which is more polar.

7.25. Why is ammonia polar but the ammonium ion (NH_4^+) nonpolar?

*7.26. Silicon and chlorine have different electronegativities. Why, then, is silicon tetrachloride not a polar molecule?

7.5 Interactions of Water with Molecules; Electrolytes and Nonelectrolytes

*7.27. Predict which of the following molecules are electrolytes:
   a. LiF   **b.** CO   **c.** CaI_2
   d. N_2O

7.28. Sulfur dioxide is a polar molecule. Sketch how the molecules of sulfur dioxide would line up in liquid sulfur dioxide.

7.29. Explain why bromine is only slightly soluble in water but hydrogen bromide is very soluble.

7.30. Why does nitrogen of the air not dissolve to an appreciable extent in the ocean?

Review Problems

7.31. What is the empirical formula of a compound that contains 39.7% potassium, 27.8% manganese, and the rest oxygen? Will this compound be ionic or covalent? soluble or insoluble in water?

7.32. Consider the element selenium. Give its electron configuration. Draw its Lewis structure. Predict the formula of its compound with hydrogen. Selenium forms a chloride that contains 35.8% selenium. What is the formula of this compound?

7.33. A compound melts at $-9.6°C$ and boils at $216°C$. The combustion of a 0.529 g sample of this compound yielded only 1.640 g carbon dioxide and 0.727 g water. What is the empirical formula of this compound? Is it ionic or covalent?

soluble or insoluble in water? Explain your answers.

7.34. A sample of a metal weighing 56.3 g displaced a volume of 2.9 mL water. The same sample required 546 J to increase its temperature from $23°C$ to $98°C$. The sample contained 0.286 mol of metal. What are the atomic weight, density, and specific heat of this metal? Use the tables in Chapter 2 to decide whether this metal is more in demand in the jewelry or the hardware business.

7.35. An oxide of nitrogen contains 25.9% nitrogen. What is the formula of this compound? What is the oxidation number of nitrogen in this formula? Predict the formula of the acid that would form if this oxide were dissolved in water.

8

Chemical Reactions and Stoichiometry

Many people think of a chemist as a person dressed in a white coat standing in front of a laboratory bench crowded with mysteriously shaped glass containers. The contents of these containers bubble, change color, and give off strange vapors. One feels certain that the bubbling and fuming substances are about to combine and form a new miracle drug or fiber.

In spite of this picture, our study of chemistry thus far has included little about what happens when chemicals combine and react to form new substances. In this chapter, we will consider chemical reactions. The concepts that we will cover include:

1. Two ways of classifying chemical reactions:
   a. As combination, decomposition, displacement, or double-displacement reactions.
   b. As reactions that do or do not show a change in oxidation numbers of the elements involved.
2. The quantitative aspects of chemical reactions: how to calculate the theoretical yield of a reaction, how to compare the theoretical yield with the actual yield to learn the percent yield of a reaction, and how to calculate the yield based on a limiting reactant.
3. The energy associated with a reaction: whether this energy is absorbed or released and how to calculate the energy change that would be associated with the reaction of a given amount of reactant.

8.1   Physical versus Chemical Changes

In Section 1.3 we divided the variety of observable properties of matter into two classes: physical and chemical. **Physical properties** are those properties whose observation does not involve a change in the composition of the sample. A sample of water does not change its composition if we pour it from a tall pitcher into a flat bowl. It can be frozen to a solid or vaporized to steam. Yet it remains water, with the formula H_2O. Through all of these **physical changes,** the composition of water is unchanged. Another physical change, the crushing of limestone, is illustrated in Figure 8.1. Even though the limestone is crushed to smaller particles, the composition of the limestone does not change.

The observation of a **chemical property** involves a change in the composition of the sample—that is, a **chemical change.** When an electric current is passed through water that contains a few drops of sulfuric acid, the water decomposes to hydrogen and oxygen. Water molecules are no longer present; instead, we have two new substances, hydrogen and oxygen. A chemical change is also illustrated in Figure 8.1. When heated, limestone (known chemically as calcium carbonate) is converted into two new substances, lime (calcium oxide) and carbon dioxide, that have very different properties from those of limestone.

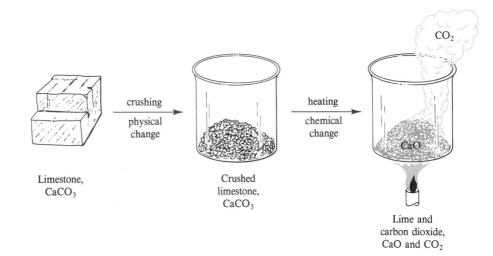

FIGURE 8.1   The crushing of limestone is a physical change; it does not alter the chemical composition of the limestone. The heating of limestone is a chemical change; the limestone decomposes into two other substances, lime and carbon dioxide.

As shown in Section 3.4, a chemical reaction can be described with an equation that shows the formula of each reactant and each product. An equation must be balanced to show that no mass is lost or gained during the reaction. An equation may also show the physical state of each component and the energy change associated with the reaction. For example, a complete equation for the burning of propane, C_3H_8, is:

$$C_3H_8(g) + 5\ O_2(g) \longrightarrow 3\ CO_2(g) + 4\ H_2O(l) \qquad \Delta H = -2220\ kJ$$

The symbols after each formula show the physical state of that substance. The equation can be read as: One molecule of gaseous propane reacts with five molecules of gaseous oxygen to yield three molecules of gaseous carbon dioxide and four molecules of liquid water.

An equation can also be read in terms of moles. This equation states: One mole of gaseous propane reacts with five moles of gaseous oxygen to yield three moles of gaseous carbon dioxide and four moles of liquid water.

The ΔH term following the equation describes the **enthalpy** change ΔH that accompanies the reaction. For this equation, the enthalpy term states that the combustion of one mole of gaseous propane to gaseous carbon dioxide and liquid water is accompanied by the release of 2220 kJ of energy. We will discuss enthalpy changes accompanying a reaction more thoroughly in Section 8.5.

The equation given for the burning of propane is a complete equation. When a discussion centers on the reactants and products of a reaction, rather than on the associated energy change, the physical states and the enthalpy change (ΔH) are often omitted.

Table 8.1 repeats the information given in Table 3.7 about how to write a chemical equation.

Example 8.1   State the complete meaning of the equation:

$$2\ H_2O(l) \xrightarrow{\text{elec}} 2\ H_2(g) + O_2(g) \qquad \Delta H = 591\ kJ$$

Solution   The equation shows that, when an electric current is passed through water, two molecules (or two moles) of liquid water will decompose to yield two molecules (or two moles) of gaseous hydrogen and one molecule (or one mole) of gaseous oxygen. Energy of 591 kJ is absorbed for every two moles of water decomposed. The equation also shows that a molecule of water contains two atoms of hydrogen and one atom of oxygen and that hydrogen and oxygen are diatomic gases.

Problem 8.1   State the complete meaning of the equation:

$$2\ HgO(s) \longrightarrow 2\ Hg(l) + O_2(g) \qquad \Delta H = 90.7\ kJ$$

8.2   **Kinds of Chemical Changes**

One common way of classifying chemical reactions is to separate them into four categories: combination, decomposition, displacement, and double displacement. We give several examples of each type for two reasons: (1) to

TABLE 8.1   Parts of a chemical equation

Reactants	The substances that combine in the reaction. Formulas must be correct.
Products	The substances that are formed by the reaction. Formulas must be correct.
ΔH	The enthalpy (heat energy) change accompanying the reaction. Energy is released if $\Delta H < 0$; energy is absorbed if $\Delta H > 0$.
Arrows $\rightarrow$	Found between reactants and products, means "reacts to form."
$\nrightarrow$	Means equation is not balanced.
$(\uparrow)$	Placed after the formula of a product that is a gas.
$(\downarrow)$	Placed after the formula of a product that is an insoluble solid—that is, a precipitate.
Physical state	Indicates the physical state of the substance whose formula it follows.
(g)	Indicates that the substance is a gas.
(l)	Indicates that the substance is a liquid.
(s)	Indicates that the substance is a solid.
(aq)	Means that the substance is in aqueous (water) solution.
Coefficients	The numbers placed in front of the formulas to balance the equation.
Conditions	Words or symbols placed over the arrow ($\rightarrow$) to indicate conditions used to make the reaction occur. For example:
Δ	Heat is added.
hv	Light is added.
elec	Electrical energy is added.

ensure that you understand the scope of each category and (2) to help you gain experience in interpreting and balancing equations. You may want to refer back to Section 3.4A for a list of the rules and steps to follow in balancing equations.

A. Combination Reactions

In a **combination reaction,** two substances combine to form a single compound. Two examples are the reaction of solid magnesium with gaseous oxygen to form magnesium oxide, a solid:

$$2\ Mg(s) + O_2(g) \longrightarrow 2\ MgO(s)$$

and the reaction of hydrogen gas with chlorine gas to form gaseous hydrogen chloride:

$$H_2(g) + Cl_2(g) \longrightarrow 2\ HCl(g)$$

FIGURE 8.2   The brilliant white light associated with some fireworks is due to the release of energy when magnesium reacts with oxygen.

Figure 8.2 illustrates an example of a combination reaction.

Other combination reactions have compounds as reactants. The reaction of gaseous carbon dioxide with solid calcium oxide to form solid calcium carbonate is an example of such a reaction.

$$CaO(s) + CO_2(g) \longrightarrow CaCO_3(s)$$

Example 8.2   Write balanced equations for the following combination reactions:

a. When solid phosphorus, P_4, is burned in chlorine gas, solid phosphorus trichloride is formed.

b. When gaseous dinitrogen pentoxide is bubbled through a solution of water, nitric acid is formed.

Solution   **a.** Write the reactants, an arrow, and then the products, with the physical state of each reactant or product shown after its formula. Write conditions for the reaction over the arrow.

$$P_4(s) + Cl_2(g) \longrightarrow\!\!\!/ \ PCl_3(s)$$

Four atoms of phosphorus will yield four molecules of phosphorus trichloride, which will require 12 atoms (six molecules) of chlorine. Putting in these coefficients gives the balanced equation

$$P_4(s) + 6\ Cl_2(g) \longrightarrow 4\ PCl_3(s)$$

b. Write the reactants, an arrow, and the products.

$$N_2O_5(g) + H_2O(l) \longrightarrow\!\!\!/ \ HNO_3(aq)$$

Include the physical states and conditions of the reaction as given in its statement. Starting with nitrogen (it is always wise to leave hydrogen and oxygen to the last), two atoms of nitrogen in one molecule of N_2O_5 will form two molecules of nitric acid. Each molecule of nitric acid contains one hydrogen atom, thus two molecules will require two hydrogen atoms or one molecule of water. We then have six atoms of oxygen in the reactants — the same number of oxygen atoms required by two molecules of nitric acid. The equation is now balanced:

$$N_2O_5(g) + H_2O(l) \longrightarrow 2\ HNO_3(aq)$$

Problem 8.2   Write balanced equations for the following combination reactions:

a. When solid carbon burns in a limited supply of oxygen gas, the gas carbon monoxide, CO, is formed. This gas is deadly to humans because it combines with hemoglobin in the blood, making it impossible for the blood to transport oxygen. Write the equation for the formation of carbon monoxide.

b. Lithium oxide, Li_2O, dissolves in water to form a solution of lithium hydroxide.

B. Decomposition Reactions

In a **decomposition reaction,** a compound is decomposed to its component elements or to other compounds. Although some compounds decompose spontaneously, usually light or heat is needed to initiate the decomposition. Following are three examples of decomposition — one induced by light, one induced chemically by a catalyst, and the third caused by heat.

The antiseptic hydrogen peroxide is sold in opaque brown bottles because hydrogen peroxide decomposes in light (Figure 8.3). The equation for this decomposition is:

$$2\ H_2O_2(aq) \xrightarrow{\ h\nu\ } 2\ H_2O(l) + O_2(g)$$

Oxygen can be prepared by heating solid potassium chlorate in the presence of manganese dioxide, a catalyst. A **catalyst** is a chemical that, when added to a reaction mixture, hastens the reaction but can be recovered unchanged after the reaction is complete.

$$2\ KClO_3(s) \xrightarrow{\ MnO_2\ } 2\ KCl(s) + 3\ O_2(g)$$

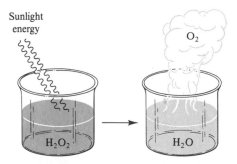

FIGURE 8.3 Light hastens the decomposition of hydrogen peroxide. The dark glass in which hydrogen peroxide is usually stored keeps out the light, thus protecting the hydrogen peroxide from decomposition.

When slaked lime, $Ca(OH)_2(s)$, is heated, lime (CaO) and water vapor are produced:

$$Ca(OH)_2(s) \longrightarrow CaO(s) + H_2O(g)$$

Example 8.3

Write balanced equations for the following decomposition reactions:

a. Solid ammonium carbonate decomposes at room temperature to ammonia, carbon dioxide, and water. (Because of the ease of decomposition and the penetrating odor of ammonia, ammonium carbonate can be used as smelling salts.)

b. On heating, lead(II) nitrate crystals decompose to yield a solid lead(II) oxide and the gases oxygen and nitrogen dioxide.

Solution

a. The unbalanced equation for the decomposition of ammonium carbonate is:

$$(NH_4)_2CO_3(s) \xrightarrow{\hspace{0.3cm}/\hspace{0.3cm}} NH_3(g) + CO_2(g) + H_2O(l)$$

Inspection of this equation indicates that two nitrogen atoms, therefore two ammonia molecules, are needed on the right. With this change, all other atoms are balanced:

$$(NH_4)_2CO_3(s) \xrightarrow{\hspace{0.3cm}/\hspace{0.3cm}} 2\,NH_3(g) + CO_2(g) + H_2O(l)$$

b. Writing the formulas for the reactant and the products in the form of an equation gives:

$$Pb(NO_3)_2(s) \xrightarrow{\hspace{0.3cm}/\hspace{0.3cm}} PbO(s) + O_2(g) + NO_2(g)$$

Balance the elements in the order Pb, N, O, to leave oxygen for the last, which is in general a good practice. The lead is balanced as it stands, one atom on each side. There are two nitrogen atoms on the left, therefore we need 2 NO_2 as a product. The oxygen is unbalanced, with six atoms in the reactants and five in the products. Try 2 $Pb(NO_3)_2$, which will give 2 PbO, 4 NO_2, and two atoms of oxygen, making one molecule of O_2. Now the equation is balanced.

$$2\ Pb(NO_3)_2(s) \longrightarrow 2\ PbO(s) + 4\ NO_2(g) + O_2(g)$$

Problem 8.3 Write balanced equations for the following decomposition reactions:

  a. Solid silver(I) oxide decomposes on heating to yield silver and gaseous oxygen.

  b. In the chemical test for arsenic, the gas arsine, AsH_3, is prepared. When arsine is decomposed by heating, arsenic deposits as a mirror-like coating on the surface of the glass container and hydrogen comes off as a gas. Write the balanced equation for the decomposition of arsine.

C. Displacement Reactions

In **displacement reactions,** an uncombined element reacts with a compound and displaces an element from that compound. For example, bromine is found in seawater as sodium bromide. When chlorine is bubbled through seawater, bromine gas is released and a solution of sodium chloride is formed:

$$2\ NaBr(aq) + Cl_2(g) \longrightarrow 2\ NaCl(aq) + Br_2(g)$$

As another example, when an iron nail is dropped into a solution of copper(II) sulfate, iron(II) sulfate is formed in solution and metallic copper is deposited:

$$CuSO_4(aq) + Fe(s) \longrightarrow FeSO_4(aq) + Cu(s)$$

Another displacement reaction, the reaction of metallic copper with silver nitrate, is shown in Figure 8.4.

Example 8.4 Write balanced equations for the following displacement reactions:

  a. When a piece of aluminum is dropped into hydrochloric acid, hydrogen is released as a gas and a solution of aluminum chloride is formed.

  b. When chlorine is bubbled through a solution of sodium iodide, crystals of iodine appear in a solution of sodium chloride.

Solution   **a.** The unbalanced equation is:

$$Al(s) + HCl(aq) \longrightarrow\!\!\!/ \ AlCl_3(aq) + H_2(g)$$

FIGURE 8.4   A displacement reaction. In the first beaker a copper wire has just been placed in a solution of silver nitrate. Crystals of silver are beginning to appear on the wire. In the second beaker the reaction is almost complete, and a great deal of silver has been deposited. The copper has displaced silver. The equation for this reaction is $2 \text{ AgNo}_3 + \text{Cu} \longrightarrow \text{Cu(NO}_3)_2 + 2 \text{ Ag}$.

Aluminum is balanced. To balance the chlorine, we need 3 HCl on the reactant side, which will give $1\frac{1}{2}$ molecules of hydrogen.

$$\text{Al}(s) + 3 \text{ HCl}(aq) \longrightarrow\!\!\!| \text{ AlCl}_3(aq) + 1\tfrac{1}{2} \text{ H}_2(g)$$

We need whole molecules of hydrogen. To get a whole number coefficient without unbalancing the other elements, we can multiply the whole equation by 2 to get:

$$2 \text{ Al}(s) + 6 \text{ HCl}(aq) \longrightarrow 2 \text{ AlCl}_3(aq) + 3 \text{ H}_2(g)$$

b. The unbalanced equation is:

$$\text{Cl}_2(g) + \text{NaI}(aq) \longrightarrow\!\!\!| \text{ NaCl}(aq) + \text{I}_2(s)$$

To balance the chlorine, we must form 2 NaCl. Two NaCl require two sodium atoms, or 2 NaI, which gives two iodine atoms, as needed. The balanced equation then is:

$$\text{Cl}_2(g) + 2 \text{ NaI}(aq) \longrightarrow 2 \text{ NaCl}(aq) + \text{I}_2(s)$$

Problem 8.4   Write the balanced equations for the following displacement reactions:

   a. When a piece of zinc is dropped into sulfuric acid, bubbles of hydrogen appear. Zinc(II) sulfate, ZnSO_4, is also formed in the solution.

   b. If a cloth bag containing mercury is suspended in a solution of silver(I) nitrate, silver crystals form on the surface of the bag. The second product, mercury(I) nitrate, is found in the surrounding solution.

Be aware that the displacement reactions we have discussed will occur. Much more information is needed before you can predict whether or not a proposed displacement will take place. That information is given in Chapter 14 (Oxidation–Reduction).

D. Double-Displacement Reactions

In **double displacement,** sometimes called metathesis or ion exchange, two ionic compounds react to form two different compounds. These reactions fall into a pattern that can be expressed as:

$$AB + CD \longrightarrow CB + AD$$

in which A and C are cations, B and D are anions. These reactions are often called "exchanging-partner" reactions because the cations A and C exchange the anions with which they are associated.

Double-displacement reactions fall into two categories: (1) those in which an acid reacts with a base to form a salt and water, which are known as neutralization reactions, and (2) those in which one of the products is insoluble, which are usually precipitation reactions, although occasionally the insoluble product is a gas.

1. Reaction of an acid with a base: Neutralization reactions

In **neutralization reactions,** an acid reacts with a base to form a salt and water. Recall from Section 5.7D that an acid is a compound that liberates hydrogen ions in solution and a base (we will center here on hydroxides, a subgroup of bases) is a compound that liberates hydroxide ions in solution. A salt is defined as an ionic compound in which the cation is not hydrogen and the anion is not hydroxide. These reactions are called neutralization reactions because the base neutralizes the acid. Some examples are:

a. The reaction of sodium hydroxide with hydrochloric acid to form sodium chloride and water:

$$NaOH(aq) + HCl(aq) \longrightarrow NaCl(aq) + H_2O(l)$$

Note that the salt formed, sodium chloride, combines the cation of the base, Na^+, with the anion of the acid, Cl^-. The formula of the salt is the neutral combination of these ions, here a $1:1$ combination in sodium chloride, NaCl.

b. The reaction of magnesium hydroxide with phosphoric acid to form magnesium phosphate and water:

$$3\ Mg(OH)_2(aq) + 2\ H_3PO_4(aq) \longrightarrow Mg_3(PO_4)_2(aq) + 6\ H_2O(l)$$

Here again the salt formed, magnesium phosphate, $Mg_3(PO_4)_2$, is a neutral combination of the cation of the base, Mg^{2+}, with the anion of the acid, PO_4^{3-}.

A **polyprotic acid** is one whose molecules ionize to yield more than one hydrogen ion. Sulfuric acid, H_2SO_4, phosphoric acid, H_3PO_4, and carbonic acid, H_2CO_3, are examples of polyprotic acids. When a polyprotic acid is one of the reactants, neutralization may be incomplete and an **acid salt** may form. An example of this reaction is:

$$NaOH(aq) + H_2SO_4(aq) \longrightarrow NaHSO_4(aq) + H_2O(l)$$

In this reaction, only one of the hydrogens of the diprotic acid reacted; the product is an acid salt, sodium hydrogen sulfate. The addition of more sodium hydroxide neutralizes the second hydrogen of this diprotic acid:

$$NaHSO_4(aq) + NaOH(aq) \longrightarrow Na_2SO_4(aq) + H_2O(l)$$

Example 8.5

Write the balanced equations for the following neutralization reactions:

   a. the complete reaction of sulfuric acid with calcium hydroxide

   b. the complete reaction of magnesium hydroxide with hydrochloric acid

   c. the reaction of sodium hydroxide with carbonic acid to form an acid salt

Solution

a. The reactants are H_2SO_4 and $Ca(OH)_2$. The products will show Ca^{2+} with SO_4^{2-} instead of with OH^- and H^+ with OH^- (HOH is the same as H_2O). Write these facts in the form of an equation:

$$Ca(OH)_2(aq) + H_2SO_4(aq) \longrightarrow\!\!\!\!/ \;\; CaSO_4(aq) + H_2O(l)$$

To balance the equation, note that there are two H^+ and two OH^-. They will combine to give 2 H_2O and the balanced equation:

$$Ca(OH)_2(aq) + H_2SO_4(aq) \longrightarrow CaSO_4(aq) + 2\ H_2O(l)$$

b. The formulas of the reactants are $Mg(OH)_2$ and HCl. The products will show Mg^{2+} with Cl^- instead of OH^- and H^+ with OH^- instead of Cl^-. We write the equation as:

$$Mg(OH)_2(aq) + HCl(aq) \longrightarrow\!\!\!\!/ \;\; MgCl_2(aq) + H_2O(l)$$

We need two chloride ions to combine with the Mg^{2+}, so we write 2 HCl. We now have 2 H^+ and 2 OH^- to combine with them to form 2 H_2O:

$$Mg(OH_2(aq) + 2\ HCl(aq) \longrightarrow MgCl_2(aq) + 2\ H_2O(l)$$

c. The reactants are NaOH and H_2CO_3. If an acid salt is to be formed, only one of the H^+ in carbonic acid will be replaced. The cations changing partners are Na^+ and H^+. The anions are HCO_3^- and OH^-. Writing the equation, we get

$$NaOH(aq) + H_2CO_3(aq) \longrightarrow NaHCO_3(aq) + H_2O(l)$$

Problem 8.5   Write the balanced equations for the following acid-base reactions:

  a. the complete reaction of aluminum hydroxide and acetic acid

  b. the complete reaction of potassium hydroxide and sulfuric acid

  c. ammonium hydroxide with phosphoric acid to form the acid salt $NH_4H_2PO_4$

---

2. Double-displacement reactions that form insoluble ionic products

Precipitation reactions, the second group of double-displacement reactions, result in the formation of **insoluble** ionic compounds. Ionic compounds differ enormously in the extent to which they dissolve in water, or their **solubility**. Table 8.2 illustrates this point by listing the solubilities of several ionic

TABLE 8.2   Solubilities of ionic solids in cold water

Name	Formula	Solubility (g/0.1 L)
barium iodide	BaI	170
silver(I) nitrate	$AgNO_3$	122
sodium nitrate	$NaNO_3$	92.1
ammonium chloride	NH_4Cl	29.7
lead(II) chloride	$PbCl_2$	0.99
calcium carbonate	$CaCO_3$	1.4×10^{-3}
barium sulfate	$BaSO_4$	2.22×10^{-4}
silver(I) chloride	$AgCl$	8.9×10^{-5}

compounds in cold water. Notice that several, such as barium iodide and silver(I) nitrate, are very soluble in water, whereas others, such as lead(II) chloride, are only slightly soluble. Others, such as silver(I) chloride, are virtually insoluble. Generally, if more than 0.1 g of an ionic solid dissolves in 100 mL (0.1 L) of water, the compound is said to be **soluble.** Less than 0.1 g calcium carbonate, barium sulfate, and silver(I) chloride dissolve in 100 mL water. Therefore, they are classified as insoluble compounds.

Table 8.3 lists some **solubility rules** by which the solubility of an ionic compound in water can be predicted.

Example 8.6   Write the formulas of the following salts and predict whether each is soluble in water.

  a. lead(II) nitrate   **b.** iron(II) chloride

  c. ammonium sulfide   **d.** barium sulfate

TABLE 8.3   Solubility rules for ionic compounds

NH_4^+	All common salts of ammonium ion are soluble.
Na^+ K^+	All common salts of sodium and potassium are soluble.
NO_3^-	All nitrates are soluble.
$C_2H_3O_2^-$	All acetates are soluble except iron(III) acetate, $Fe(C_2H_3O_2)_3$.
Cl^- Br^- I^-	All chlorides, bromides, and iodides are soluble except those of Ag^+, Hg^+, and Pb^{2+}. $PbCl_2$ and $PbBr_2$ are slightly soluble in hot water.
SO_4^{2-}	All sulfates are soluble except $CaSO_4$, $BaSO_4$, $PbSO_4$, and Ag_2SO_4.
PO_4^{3-} CO_3^{2-}	Only alkali metal and NH_4^+ phosphates and carbonates are soluble.
S^{2-}	Only alkali metal and NH_4^+ sulfides are soluble.
OH^-	Only alkali metal and NH_4^+ hydroxides are soluble. Ca^{2+}, Ba^{2+}, and Sr^{2+} hydroxides are slightly soluble.

Solution

	Formula	*Solubility*	*Reason*
a. lead(II) nitrate	$Pb(NO_3)_2$	soluble	It is a nitrate.
b. iron(II) chloride	$FeCl_2$	soluble	It is a chloride, but not one of the listed exceptions.
c. ammonium sulfide	$(NH_4)_2S$	soluble	It is an ammonium salt.
d. barium sulfate	$BaSO_4$	insoluble	It is listed as an insoluble sulfate.

Problem 8.6   Write formulas for the following salts and predict whether each is soluble in water.

a. barium acetate   **b.** silver(I) sulfide

c. ammonium phosphate   **d.** calcium carbonate

In precipitation reactions, solutions of two ionic compounds are combined. If two of the ions in the resulting mixture combine to form an insoluble compound or precipitate, a reaction occurs. (Figure 8.5 shows the formation of a precipitate.) If no insoluble product is produced, no reaction occurs. For

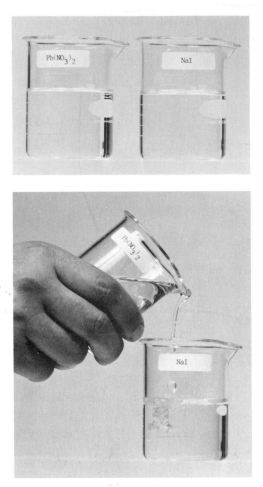

FIGURE 8.5   The formation of a precipitate.
When a clear colorless solution of lead nitrate
($Pb(NO_3)_2$) is added to a clear colorless solution
of sodium iodide (NaI), a yellow precipitate of
lead iodide (PbI_2) appears. The equation for
this reaction is given in Example 8.7a.

example, if a solution of barium iodide is added to a solution of ammonium
nitrate, no reaction takes place because the predicted products barium nitrate
and ammonium iodide are both soluble.

$$BaI_2(aq) + 2\ NH_4NO_3(aq) \longrightarrow Ba(NO_3)_2(aq) + NH_4I(aq)$$

A reaction would occur if a solution of barium iodide were added to a solution
of silver nitrate, because one of the products, silver iodide, is insoluble.

$$BaI_2(aq) + 2\ AgNO_3(aq) \longrightarrow Ba(NO_3)_2(aq) + 2\ AgI(\downarrow)$$

In the equations for these reactions, the physical state of the insoluble product, the precipitate, is indicated either by (s) or by a downward-pointing arrow after its formula; the soluble components of the reaction are shown as (aq).

Example 8.7

Write the balanced equation for the following reactions. Indicate with a downward-pointing arrow any precipitate formed; name the precipitate.

  a. Solutions of lead(II) nitrate and sodium iodide react to form a yellow precipitate.

  b. The reaction between a solution of copper(II) nitrate and a solution of potassium sulfide yields a heavy black precipitate.

Solution

a. The formulas for the reactants are $Pb(NO_3)_2$ and NaI. The formulas of the products of a reaction between these two compounds would have an interchange of anions, yielding PbI_2 and $NaNO_3$. Arranging these formulas in an unbalanced equation, we get:

$$Pb(NO_3)_2(aq) + NaI(aq) \longrightarrow\!\!\!\!| \; PbI_2 + NaNO_3$$

Balancing this equation requires two iodide ions and therefore 2 NaI. Two sodium nitrate are formed:

$$Pb(NO_3)_2(aq) + 2\,NaI(aq) \longrightarrow PbI_2(\downarrow) + 2\,NaNO_3(aq)$$

Because all sodium salts are soluble, the precipitate must be lead(II) iodide; we place an arrow after that formula.

b. The formulas of the reactants are $Cu(NO_3)_2$ and K_2S. The formulas of the products are CuS and KNO_3. From Table 8.3 we know that potassium nitrate is soluble, so the precipitate must be CuS, copper(II) sulfide. The unbalanced equation is:

$$Cu(NO_3)_2(aq) + K_2S(aq) \longrightarrow\!\!\!\!| \; CuS(\downarrow) + KNO_3(aq)$$

Balancing this equation requires two potassium nitrate. The balanced equation is:

$$Cu(NO_3)_2(aq) + K_2S(aq) \longrightarrow CuS(\downarrow) + 2\,KNO_3$$

All nitrates are soluble, so the precipitate is copper(II) sulfide. A downward-pointing arrow is put after its formula.

Problem 8.7

Write balanced equations for the following reactions, each of which produces a precipitate. Indicate the precipitate by a downward-pointing arrow after its formula. Name the products.

  a. When chromium(III) chloride is added to a solution of sodium hydroxide, a green precipitate forms.

  b. When sulfuric acid is added to a solution of barium chloride, a white precipitate forms.

As mentioned earlier, occasionally the insoluble product is a gas, as in the following examples:

1. Hydrogen chloride is released as a gas when concentrated sulfuric acid is added to solid sodium chloride. Although hydrogen chloride is very soluble in water, it is quite insoluble in concentrated sulfuric acid. The acid salt sodium hydrogen sulfate is the second product.

$$NaCl(s) + H_2SO_4 \longrightarrow HCl(g) + NaHSO_4(s)$$

2. Acetic acid may be released as a gas in a reaction similar to that in the first example. The equation for the reaction of concentrated hydrochloric acid with sodium acetate is:

$$NaC_2H_3O_2(s) + HCl \longrightarrow HC_2H_3O_2(g) + NaCl(s)$$

3. A carbonate reacts with an acid to form carbonic acid, which immediately decomposes to gaseous carbon dioxide and water.

$$Na_2CO_3(s) + 2\ HCl(aq) \longrightarrow 2\ NaCl(aq) + CO_2(g) + H_2O(l)$$

8.3   Oxidation–Reduction: A Second Way to Classify Reactions

Another way of classifying reactions separates them into only two groups: (1) those that do not involve a change in oxidation number but do result in a decrease in the number of ions in solution and (2) those that involve a transfer of electrons and changes in oxidation number.

Those that result in a decrease in the number of ions in solution are usually double-displacement reactions (Section 8.2D). In a neutralization reaction, hydrogen ion combines with hydroxide ion to form the covalent, un-ionized compound water, thus decreasing the number of ions in solution. In a precipitation reaction, the insoluble product removes ions from the solution. Other reactions in this category are those that form weak electrolytes (Section 7.5C), such as acetic acid or aqueous ammonia (ammonium hydroxide). Examples are the reaction between sodium acetate and hydrochloric acid

$$NaC_2H_3O_2(aq) + HCl(aq) \longrightarrow NaCl(aq) + HC_2H_3O_2(aq)$$

and the reaction between sodium hydroxide and ammonium chloride

$$NaOH(aq) + NH_4Cl(aq) \longrightarrow NaCl(aq) + NH_3(aq) + H_2O(l)$$

Those reactions that involve a transfer of electrons include combination, displacement, and decomposition reactions. Reactions that involve a transfer of electrons are known as **oxidation–reduction** or **redox** reactions. The reaction of sodium with chlorine is a redox reaction:

$$2\ Na + Cl_2 \longrightarrow 2\ NaCl$$

During the reaction, each sodium loses an electron to form a sodium ion:

$$Na^{\bullet} \longrightarrow Na^+ + e^-$$

Each chlorine atom gains an electron to form a chlorine ion:

$$2 \text{ e}^- + \; :\!\ddot{\text{Cl}}\!\!-\!\!\ddot{\text{Cl}}\!: \; \longrightarrow 2 \; :\!\ddot{\text{Cl}}\!:^-$$

The element that loses electrons is **oxidized.** In the reaction of sodium with chlorine, sodium is oxidized. The element that gains electrons is **reduced.** In this reaction, chlorine is reduced.

Displacement reactions are usually oxidation–reduction reactions. A typical displacement reaction is that of copper with silver nitrate:

$$\text{Cu}(s) + 2 \text{ AgNO}_3(aq) \longrightarrow 2 \text{ Ag} + \text{Cu(NO}_3)_2(aq)$$

In this reaction, the copper loses electrons (is oxidized):

$$\text{Cu}^0 \longrightarrow \text{Cu}^{2+} + 2 \text{ e}^-$$

and the silver ion gains electrons (is reduced):

$$\text{Ag}^+ + \text{e}^- \longrightarrow \text{Ag}^0$$

Oxidation without reduction is impossible. If an element in a reaction loses electrons, another element in the reaction must gain electrons.

A. Identifying Oxidation–Reduction Reactions

Oxidation numbers (Section 6.2A) have many uses, but the one that concerns us here is their role in determining whether or not a particular reaction involves oxidation–reduction. In an oxidation–reduction reaction, at least two oxidation numbers change. The element that is oxidized increases its oxidation number, and the element that is reduced decreases its oxidation number. In the reaction of sodium with chlorine, sodium atoms are oxidized to sodium ions; the oxidation number of sodium increases from 0 to $+1$. At the same time, chlorine is reduced to chloride ions; the oxidation number of chlorine decreases from 0 to -1.

By assigning oxidation numbers to all the elements in the reactants and the products of a reaction, we can determine whether the reaction results in a change in oxidation number. If a change does occur, the reaction is an oxidation–reduction reaction. For example, consider the reaction between magnesium and oxygen:

$$\underset{0}{2 \text{ Mg}} + \underset{0}{\text{O}_2} \longrightarrow \underset{+2, -2}{2 \text{ MgO}}$$

Below each element in each substance in the equation, we have written its oxidation number. The oxidation number of magnesium increases from 0 to $+2$; magnesium is oxidized. The oxidation number of oxygen decreases from 0 to -2; oxygen is reduced. Thus, the reaction of magnesium with oxygen is an oxidation–reduction reaction.

A reaction that is not an oxidation–reduction reaction will cause no changes in oxidation numbers. Consider the reaction of sodium hydroxide with hydrochloric acid:

$$\underset{+1,\,-2,\,+1}{\text{NaOH}} + \underset{+1,\,-1}{\text{HCl}} \longrightarrow \underset{+1,\,-1}{\text{NaCl}} + \underset{+1,\,-2}{\text{H}_2\text{O}}$$

Below each element in the equation, we have written its oxidation number. Because each element has the same oxidation number in the products as it does in the reactants, we know that this neutralization reaction is not an oxidation–reduction reaction.

Example 8.8   For the following reactions decide: (1) Is it an oxidation–reduction reaction? (2) If so, which element is oxidized and which element is reduced?

   a. Bromine can be prepared by bubbling chlorine gas through a solution of sodium bromide. The equation for this reaction is:

$$\text{2 NaBr}(aq) + \text{Cl}_2(g) \longrightarrow \text{2 NaCl}(aq) + \text{Br}_2(g)$$

   b. If you blow through a straw into limewater, the solution becomes milky. In chemical terms, if carbon dioxide is bubbled through a solution of calcium hydroxide in water, a milky white precipitate of calcium carbonate forms:

$$\text{CO}_2(g) + \text{Ca(OH)}_2(aq) \longrightarrow \text{CaCO}_3(\downarrow) + \text{H}_2\text{O}(l)$$

Solution   **Reaction a**

   1. Write the oxidation number under each element in the equation.

$$\underset{+1,\,-1}{\text{2 NaBr}(aq)} + \underset{0}{\text{Cl}_2(g)} \longrightarrow \underset{+1,\,-1}{\text{2 NaCl}(aq)} + \underset{0}{\text{Br}_2(g)}$$

   2. Do any elements change oxidation number? Yes, both chlorine and bromine do. Therefore, this reaction is oxidation–reduction.

   3. The oxidation number of chlorine changes from 0 to -1; chlorine is reduced. The oxidation number of bromine changes from -1 to 0; bromine is oxidized.

   Reaction b

$$\underset{+4,\,-2}{\text{CO}_2(g)} + \underset{+2,\,-2,\,+1}{\text{Ca(OH)}_2(aq)} \longrightarrow \underset{+2,\,+4,\,-2}{\text{CaCO}_3(\downarrow)} + \underset{+1,\,-2}{\text{H}_2\text{O}(l)}$$

Below each element we have written its oxidation number. None of these numbers changed during the reaction; the reaction is not an oxidation–reduction reaction.

Problem 8.8   Determine which of the following are oxidation–reduction reactions. For those reactions that are oxidation–reduction reactions, identify the element oxidized and the element reduced.

   a. $NH_3(g) + HCl(g) \longrightarrow NH_4Cl(s)$
   b. $Zn(s) + 2\ HCl(aq) \longrightarrow ZnCl_2(aq) + H_2(g)$

In an oxidation–reduction reaction, the substance that gains electrons is the **oxidizing agent.** The substance that loses electrons is the **reducing agent.** In the reaction of magnesium with oxygen,

$$2\ Mg(s) \quad + \quad O_2(g) \quad \longrightarrow 2\ MgO(s)$$
   0   0   +2, −2

loses electrons	gains electrons
is oxidized to Mg^{2+}	is reduced to 2 O^{2-}
is the reducing agent	is the oxidizing agent

Example 8.9   In the reaction of sodium with chlorine to form sodium chloride, which substance is the oxidizing agent? Which is the reducing agent?

Solution   Write the equation for the reaction and assign oxidation numbers:

$$2\ Na(s) + Cl_2(g) \longrightarrow 2\ NaCl(s)$$
     0   0   +1, −1

Because chlorine changes oxidation number from 0 to −1, it is reduced; it is the oxidizing agent. Because sodium changes oxidation number from 0 to +1, it is oxidized; it is the reducing agent.

Problem 8.9   When a piece of copper is put in a solution of mercury(II) nitrate, droplets of mercury appear on the copper coil and the solution becomes blue due to the presence of copper(II) nitrate. Show that this reaction is oxidation–reduction, and identify the oxidizing and reducing agents.

The characteristics of oxidation and reduction are summarized in Table 8.4. Chapter 14 covers oxidation–reduction in greater depth. Our discussion here is merely an introduction.

B. Combustion Reactions

Combustion reactions are a special type of oxidation–reduction reaction. We correctly associate combustion reactions with burning. In the usual combustion reaction, the elements in the reacting compound combine with oxygen to

TABLE 8.4   Characteristics of oxidation–reduction reactions

Substance oxidized	Substance reduced
loses electrons	gains electrons
attains a more positive oxidation number	attains a more negative oxidation number
is the reducing agent	is the oxidizing agent

form oxides as in the combustion of propane to form carbon dioxide and water:

$$2\ C_3H_8(l) + 10\ O_2(g) \longrightarrow 6\ CO_2(g) + 8\ H_2O(l)$$

Figure 8.6 shows the combustion of gasoline in oxygen.

The above reactions take place when an adequate supply of oxygen is present. In the absence of an adequate supply of oxygen, carbon monoxide may be formed instead of carbon dioxide.

$$2\ C_3H_8(l) + 7\ O_2(g) \longrightarrow 6\ CO(g) + 8\ H_2O(l)$$

Example 8.10   Write the balanced equations for the complete combustion in oxygen of:

a. sulfur to form sulfur dioxide

b. butane, C_4H_{10}

c. ethyl alcohol, C_2H_5OH

Solution   Note that the physical states of these substances are not given. They are therefore omitted from the equations.

a. Write the formulas of the reactants, sulfur and oxygen, and of the product, SO_2.

$$S_8 + 8\ O_2 \longrightarrow 8\ SO_2$$

The equation is balanced.

b. Butane contains only carbon and hydrogen and the combustion is complete, so the products are carbon dioxide and water. The unbalanced equation is:

$$C_4H_{10} + O_2 \longrightarrow\!\!\!\!/\ \ CO_2 + H_2O$$

Four atoms of carbon on the left give four molecules of carbon dioxide on the right. Ten atoms of hydrogen on the left give five molecules of

FIGURE 8.6 Combustion as illustrated by burning gasoline storage tanks. The combustion of gasoline or other petroleum products is usually accompanied by yellow flames and dense black smoke.

water on the right, which requires thirteen atoms of oxygen ($6\frac{1}{2}$ molecules) on the left.

$$C_4H_{10} + 6\tfrac{1}{2}\,O_2 \longrightarrow 4\,CO_2 + 5\,H_2O$$

To write the equation using only whole numbers of molecules, we must multiply through by 2 to get:

$$2\,C_4H_{10} + 13\,O_2 \longrightarrow 8\,CO_2 + 10\,H_2O$$

c. Ethyl alcohol contains carbon, hydrogen, and oxygen. The products of complete combustion will be carbon dioxide and water.

$$C_2H_5OH + O_2 \longrightarrow\!\!\!/\ CO_2 + H_2O$$

Initial balancing gives two atoms of carbon on the left, two molecules of carbon dioxide on the right; six atoms of hydrogen on the left, three molecules of water on the right; seven atoms of oxygen on the right, $3\frac{1}{2}$ molecules of oxygen on the left:

$$C_2H_5OH + 3\tfrac{1}{2}\,O_2 \longrightarrow 2\,CO_2 + 3\,H_2O$$

We multiply by 2 to clear the fraction and give the balanced equation:

$$2\,C_2H_5OH + 7\,O_2 \longrightarrow 4\,CO_2 + 6\,H_2O$$

Problem 8.10   Write balanced equations for the combustion of the following compounds:

    a. methyl alcohol, CH_3OH

    b. heptane, C_7H_{16}

    c. dioxane, $C_4H_8O_2$

8.4   Mass Relationships in an Equation

A. Simple Problems

A balanced equation is a quantitative statement of a reaction. It relates amounts of reactants to amounts of products. Let us see what this statement means in terms of a particular reaction.

Pentane, C_5H_{12}, burns in oxygen to form carbon dioxide and water. The balanced equation for the combustion of pentane is:

$$C_5H_{12} + 8\,O_2 \longrightarrow 5\,CO_2 + 6\,H_2O$$

In qualitative terms, this equation shows that pentane reacts with oxygen to form carbon dioxide and water. In quantitative terms, the equation states that 1 molecule of pentane reacts with 8 molecules of oxygen to form 5 molecules of carbon dioxide and 6 molecules of water. For 15 molecules of pentane, (8×15) or 120 molecules of oxygen would be needed for complete reaction; (5×15) molecules of carbon dioxide and (6×15) molecules of water would be formed. Whatever number (n) of molecules of pentane were used in the reaction, $5n$ molecules of carbon dioxide and $6n$ molecules of water would be formed.

If 6.02×10^{23} molecules (1 mole) of pentane are burned, $[8(6.02 \times 10^{23})]$ molecules (8 moles) of oxygen are needed. The reaction would form $[5(6.02 \times 10^{23})]$ molecules (5 moles) of carbon dioxide and $[6(6.02 \times 10^{23})]$ molecules (6 moles) of water. These quantitative relationships are summarized in Table 8.5.

Any balanced equation gives the ratio between moles of reactants and moles of products. Given the number of moles of one component used or produced in a reaction, the number of moles or grams of any other component used or produced can be calculated. Such calculations are called **stoichiometry.**

A stoichiometric problem can be stated in many ways, but it always contains the following information:

1. The reaction involved.

2. A stated amount of one component of the reaction.

3. A question asking "how much" of another substance is needed or formed in the reaction.

The quantitative problems in previous chapters were solved by answering a series of questions:

1. What is wanted?

2. What is given?

3. What conversion factors are needed to go from "given" to "wanted"?

4. How should the arithmetic equation be set up so that the units of the "given" are converted to the units of the "wanted"?

TABLE 8.5   Quantitative relationships in an equation

C_5H_{12}	$+ 8\ O_2$	$\longrightarrow 5\ CO_2$	$+ 6\ H_2O$
1 molecule	+ 8 molecules	$\longrightarrow$ 5 molecules	+ 6 molecules
1 mole	+ 8 moles	$\longrightarrow$ 5 moles	+ 6 moles
72 g	+ 256 g	$\longrightarrow$ 220 g	+ 108 g

Stoichiometric problems can be solved by answering the same set of questions. The only difference is that some of the conversion factors are derived from the balanced chemical equation for the reaction involved. The steps to follow in solving a stoichiometric problem are the following:

■ **1.** Write the balanced equation for the reaction.

■ **2.** Decide which substance is wanted and in what units.

■ **3.** Decide which substance is given and in what units and what amount.

■ **4.** Determine the conversion factors required to convert:
   a. the amount of given substance into moles
   b. the moles of the given substance into moles of the wanted substance
   c. the moles of wanted substance into the units wanted in the problem.

■ **5.** Combine the amount of given substance and its units along with the appropriate conversion factors into an arithmetic equation in such a way that all factors except the wanted substance in the proper units will cancel.

We will now do some stoichiometry problems. Before we begin, several points should be emphasized. First, to prevent confusion and errors, the name and units of each item in the equation are always shown. Second, no arithmetic is done until the whole equation is written out. Third, all units in the final equation must cancel except for those required in the answer.

Example 8.11

How many grams of carbon dioxide are formed when 61.5 g pentane are burned in oxygen?

Solution

Equation

$$C_5H_{12} + 8\ O_2 \longrightarrow 5\ CO_2 + 6\ H_2O$$

Wanted
? g CO_2

Given
61.5 g C_5H_{12}

Conversion factors

C_5H_{12}, mass to moles: 72.2 g C_5H_{12} = 1 mol C_5H_{12}
mol C_5H_{12} to mol CO_2: 1 mol C_5H_{12} yields 5 mol CO_2
CO_2, moles to mass: 1 mol CO_2 = 44.0 g CO_2

Arithmetic equation

$$? \text{ g } CO_2 = 61.5 \text{ g } C_5H_{12} \times \frac{1 \text{ mol } C_5H_{12}}{72.2 \text{ g } C_5H_{12}} \times \frac{5 \text{ mol } CO_2}{1 \text{ mol } C_5H_{12}} \times \frac{44.0 \text{ g } CO_2}{1 \text{ mol } CO_2}$$

Answer
187 g CO_2

Problem 8.11

Hydrogen burns in oxygen to form water. What mass of oxygen is necessary for the complete combustion of 1.74 g of hydrogen?

Example 8.12

What mass of iron(III) oxide (rust) is formed when 6.23 g iron react completely with oxygen in the air to form this product?

Solution

Equation
The name iron(III) oxide shows that the iron is present in the product as Fe^{3+}, the oxide ion is O^{2-}. Iron(III) oxide then is Fe_2O_3. The equation for the reaction is:

$$4\ Fe + 3\ O_2 \longrightarrow 2\ Fe_2O_3$$

Wanted

$? \text{ g Fe}_2\text{O}_3$

Given

6.23 g Fe

Conversion factors

Fe, mass to moles: 55.85 g Fe = 1 mol Fe

mol Fe to mol Fe_2O_3: 4 mol Fe yield 2 mol Fe_2O_3

Fe_2O_3, moles to mass: 1 mol Fe_2O_3 = 2(55.85) + 3(16.0)
$$= 159.7 \text{ g Fe}_2\text{O}_3$$

Arithmetic equation

$$? \text{ g Fe}_2\text{O}_3 = 6.23 \text{ g Fe} \times \frac{1 \text{ mol Fe}}{55.85 \text{ g Fe}} \times \frac{2 \text{ mol Fe}_2\text{O}_3}{4 \text{ mol Fe}} \times \frac{159.7 \text{ g Fe}_2\text{O}_3}{1 \text{ mol Fe}_2\text{O}_3}$$

Answer

8.91 g Fe_2O_3

Problem 8.12   Bromine is prepared by the reaction of chlorine with sodium bromide. How many grams of chlorine are necessary for the preparation of 2.12 g of bromine?

Example 8.13   How many molecules of hydrogen will be formed by the reaction of 2.65×10^{-3} g zinc with hydrochloric acid?

Solution   **Equation**

$$\text{Zn} + 2 \text{ HCl} \longrightarrow \text{ZnCl}_2 + \text{H}_2$$

Wanted

? molecules H_2

Given

2.65×10^{-3} g Zn

Conversion factors

Zn, mass to moles: 65.4 g Zn = 1 mol Zn

mol Zn to mol H_2: 1 mol Zn yields 1 mol H_2

H_2, moles to number of molecules: 1 mol H_2
$$= 6.02 \times 10^{23} \text{ molecules H}_2$$

Arithmetic equation

$$? \text{ molecules H}_2 = 2.65 \times 10^{-3} \text{ g Zn} \times \frac{1 \text{ mol Zn}}{65.4 \text{ g Zn}} \times \frac{1 \text{ mol H}_2}{1 \text{ mol Zn}}$$

$$\times \frac{6.02 \times 10^{23} \text{ molecules H}_2}{1 \text{ mol H}_2}$$

Answer

2.44×10^{19} molecules H_2

Problem 8.13   How many molecules of oxygen are required for the complete combustion of 1.67 g methane, CH_4?

B. Percent Yield

In Examples 8.11, 8.12, and 8.13 we have calculated the **theoretical yield** of the reaction—that is, the amount of product that would be obtained if the reaction proceeded completely and only as stated in the equation. Very often—in fact, in most cases—the actual yield of a reaction does not equal the theoretical yield but is a lesser amount. The reasons for such discrepancies are varied: The reactant may have been impure, some of the product may have been lost, or small amounts of substances other than those shown in the equation may have been formed. In these cases the actual yield is less than the calculated theoretical yield. The **percent yield** is calculated by comparing the actual yield with the theoretical yield.

The percent yield of a reaction depends on results obtained in the laboratory. If we know the amount of reactants used, we can calculate the theoretical yield. We then measure the amount of obtained product, the actual yield. The ratio between the two gives the percent yield.

$$\text{Percent yield} = \frac{\text{actual yield}}{\text{theoretical yield}} \times 100\%$$

As an example of this process, suppose that, for the reaction in Example 8.11, the actual yield is only 165 g CO_2. The theoretical yield, the one calculated, is 187 g CO_2. The percent yield of the reaction would be:

$$\frac{165 \text{ g}}{187 \text{ g}} \times 100\% = 88.2\% \text{ yield}$$

Example 8.14 illustrates how to calculate percent yield.

Example 8.14   The reaction of ethane with chlorine yields ethyl chloride and hydrogen chloride:

$$\underset{\text{ethane}}{C_2H_6} + Cl_2 \longrightarrow \underset{\text{ethyl chloride}}{C_2H_5Cl} + HCl$$

When 5.6 g ethane react with chlorine, 8.2 g ethyl chloride are obtained. Calculate the percent yield of ethyl chloride.

Solution   Because this problem involves percent yield, the first step is to calculate the theoretical yield. We use the steps outlined in Section 8.4A.

Equation

$$C_2H_6 + Cl_2 \longrightarrow C_2H_5Cl + HCl$$

Wanted

? g C_2H_5Cl (theoretical yield)

Given

5.6 g C_2H_6, 8.2 g C_2H_5Cl (actual yield)

Conversion factors

C_2H_6, mass to moles: 30.1 g C_2H_6 = 1 mol C_2H_6

mol C_2H_6 to mol C_2H_5Cl: 1 mol C_2H_6 yields 1 mol C_2H_5Cl

C_2H_5Cl, moles to mass: 1 mol C_2H_5Cl = 64.5 g C_2H_5Cl

Arithmetic equation

$$? \text{ g } C_2H_5Cl = 5.6 \text{ g } C_2H_6 \times \frac{1 \text{ mol } C_2H_6}{30.1 \text{ g } C_2H_6} \times \frac{1 \text{ mol } C_2H_5Cl}{1 \text{ mol } C_2H_6} \times \frac{64.5 \text{ g } C_2H_5Cl}{1 \text{ mol } C_2H_5Cl}$$

$$= 12 \text{ g } C_2H_5Cl$$

The theoretical yield is 12 g C_2H_5Cl. The actual yield is 8.2 g ethyl chloride.

Answer

$$\text{Percent yield} = \frac{8.2 \text{ g } C_2H_5Cl}{12 \text{ g } C_2H_5Cl} \times 100\% = 68\%$$

Problem 8.14   When 1.6 g oxygen reacts with an excess of nitrogen, 1.3 g nitrogen oxide is formed. Calculate the percent yield of nitrogen oxide. The balanced equation is:

$$N_2 + O_2 \longrightarrow 2 \text{ NO}$$

The ratio of actual yield to theoretical yield can be used to measure the purity of a sample. Such a determination uses a reaction in which the active ingredient of the sample reacts quantitatively — that is, the reaction gives 100% yield. The percent yield of the reaction measures the percent of active ingredient in the sample. Example 8.15 illustrates this process. The problem investigates the purity of a zinc sample and utilizes the reaction of zinc with hydrochloric acid, according to the equation:

$$Zn + 2 \text{ HCl} \longrightarrow ZnCl_2 + H_2$$

This reaction is quantitative. A weighed sample of the impure zinc is immersed in excess hydrochloric acid. An excess of the acid assures that more than enough is present to react with all the zinc in the sample.

Example 8.15   When a sample of impure zinc weighing 7.45 g reacts with an excess of hydrochloric acid, 0.214 g hydrogen is released. Calculate the percent zinc in the sample.

Analysis of the problem

1. The zinc sample is impure. The part that is zinc reacts to form hydrogen; the rest does not.

2. We can calculate the amount of zinc needed to form 0.214 g hydrogen. This amount is the actual amount of zinc in the sample; it must be less than the actual weight of the sample.

3. We can calculate the ratio between the amount of zinc needed to form 0.214 g H_2 and the actual weight of the sample. This ratio multiplied by 100% is the percent purity of the sample.

Solution   1. Calculate the amount of zinc that would yield 0.214 g hydrogen using the steps in Section 8.4A.

Equation

$$Zn + 2\ HCl \longrightarrow ZnCl_2 + H_2$$

Wanted
? g Zn

Given
0.214 g H_2, 7.45 g sample

Conversion factors

H_2, mass to moles: 2.016 g H_2 = 1 mol H_2

mol H_2 to mol Zn: 1 mol H_2 requires 1 mol Zn (from the equation)

Zn, moles to mass: 1 mol Zn = 65.4 g Zn

Arithmetic equation

$$? \text{ g Zn} = 0.214 \text{ g } H_2 \times \frac{1 \text{ mol } H_2}{2.016 \text{ g } H_2} \times \frac{1 \text{ mol Zn}}{1 \text{ mol } H_2} \times \frac{65.4 \text{ g Zn}}{1 \text{ mol Zn}}$$

$$= 6.94 \text{ g Zn}$$

This amount is the actual amount of zinc in the sample.

2. Calculate the percent ratio, which will equal the percentage of zinc in the sample.

$$\% \text{ Zn} = \frac{6.94 \text{ g Zn}}{7.45 \text{ g sample}} \times 100\% = 93.2\%$$

Problem 8.15   When a piece of impure copper weighing 0.54 g is added to a solution of silver nitrate, 1.5 g silver precipitates. What is the percent purity of the copper sample?

C. Problems Involving a Limiting Reactant

All the stoichiometry problems encountered thus far have had two reactants, but the amount was given for only one of them. In calculating the solutions, we assumed that enough of the second reactant was present to allow complete reaction of the first. This is not always the case. We will encounter many problems in which we know the amounts of all reactants and must calculate which reactant is present in excess before we can calculate a theoretical yield.

A problem of this type might be encountered if you owned a bicycle shop. Among other parts, each bicycle requires two wheels and one set of handlebars. Suppose that, after a busy season, your stockroom contains only 14 wheels and 6 sets of handlebars, although it contains huge quantities of all the other necessary parts. How many bicycles can you build before a new shipment of wheels and handlebars arrives? You have enough wheels for seven bicycles (14 ÷ 2). You have enough handlebars for six bicycles (6 ÷ 1). Clearly, you can put together only six bicycles and will have two unused wheels. No matter how you juggle the parts, you have only enough handlebars for six bicycles. Even with 100 wheels, you could make only six bicycles. The number of handlebars is the limiting factor in the number of bicycles you can make.

Similarly, in a chemical reaction in which the amount of each reactant is known, one reactant is usually present in excess; the amount of the other limits the amount of product that can be formed. The problem becomes how to decide which is the **limiting reactant.**

Consider the reaction of hydrogen with chlorine to form hydrogen chloride. The balanced equation for this reaction is:

$$H_2 + Cl_2 \longrightarrow 2\ HCl$$

According to the equation, each mole of hydrogen that reacts requires one mole of chlorine to form two moles of hydrogen chloride (Figure 8.7a). If only 0.5 mol hydrogen is present, only 0.5 mol chlorine is needed, and 1.0 mol hydrogen chloride is formed (Figure 8.7b). If 2.0 mol hydrogen are to react, then 2.0 mol chlorine are necessary, and 4.0 mol hydrogen chloride are formed. Suppose you have 2.0 mol hydrogen but only 1.0 mol chlorine (Figure 8.7d). Only 1.0 mol hydrogen can react, because you have only 1.0 mol chlorine. In this instance, chlorine is the limiting reactant; only 2.0 mol hydrogen chloride are formed, and the second mole of hydrogen remains unreacted.

A limiting reactant problem, then, consists of two simpler problems. Suppose we have the reaction

$$A + B \longrightarrow C$$

and we know how much A and B are available. To determine how much C will be formed, we must calculate (1) how much C can be prepared from the given amount of A and (2) how much C can be prepared from the given amount of B. The smaller of these two amounts is the theoretical yield of C.

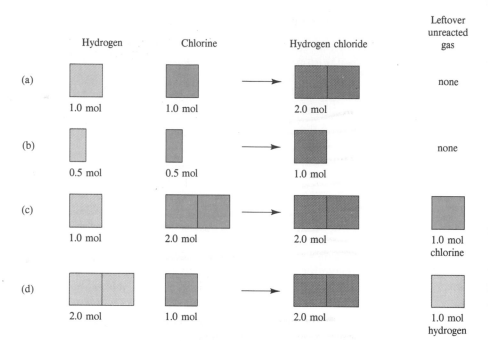

FIGURE 8.7   The amount of product formed by a chemical reaction is limited by the amounts of reactants. In (a) and (b), the reactants are present in the ratio of the balanced equation, $H_2 + Cl_2 \longrightarrow 2\ HCl$, so they react completely. In (c) and (d), one of the reactants is present in excess; some of this reactant is left unreacted after the reaction is complete.

Example 8.16

What mass of lithium chloride can be formed by the reaction of 5.00 g lithium with 5.00 g chlorine?

Analysis of the problem
We know how much of each reactant is present. The theoretical yield of product might be calculated from either. The true or limiting theoretical yield is the smaller of the two.

Solution

Equation

$$2\ Li + Cl_2 \longrightarrow 2\ LiCl$$

Wanted
? g LiCl

Given
Amounts of the two reactants: 5.00 g Li and 5.00 g Cl

Conversion factors

For lithium: 6.94 g Li = 1 mol Li

1 mol Li yields 1 mol LiCl

For chlorine: 70.9 g Cl_2 = 1 mol Cl_2

1 mol Cl_2 yields 2 mol LiCl

For both: 1 mol LiCl = 42.4 g LiCl

Arithmetic equation

There are two equations: Equation A is based on the amount of lithium present. Equation B is based on the amount of chlorine present.

Equation A is:

$$? \text{ g LiCl} = 5.00 \text{ g Li} \times \frac{1 \text{ mol Li}}{6.94 \text{ g Li}} \times \frac{1 \text{ mol LiCl}}{1 \text{ mol Li}} \times \frac{42.4 \text{ g LiCl}}{1 \text{ mol LiCl}}$$

$$= 30.6 \text{ g LiCl}$$

Equation B is:

$$? \text{ g LiCl} = 5.00 \text{ g Cl}_2 \times \frac{1 \text{ mol Cl}_2}{70.9 \text{ g Cl}_2} \times \frac{2 \text{ mol LiCl}}{1 \text{ mol Cl}_2} \times \frac{42.4 \text{ g LiCl}}{1 \text{ mol LiCl}}$$

$$= 5.98 \text{ g LiCl}$$

We now ask ourselves "Will this reaction yield 30.6 g or 5.98 g of lithium chloride?" We must choose the smaller amount, 5.98 g, because only enough chlorine is available to prepare 5.98 g lithium chloride; we cannot get 30.6 g. Thus, chlorine is the limiting reagent, and lithium is present in excess.

Problem 8.16 Hydrogen reduces copper(II) oxide to pure copper according to the equation:

$$CuO + H_2 \longrightarrow Cu + H_2O$$

What mass of pure copper would be formed by the reaction of 2.96 g hydrogen with 105 g copper(II) oxide?

8.5 Energy Changes Accompanying Chemical Reactions

All changes, whether chemical or physical, are accompanied by a change in energy. Each reacting molecule possesses a certain amount of energy due to the nature of its chemical bonds. So does each product molecule. As the bonds of the reacting molecules break and the new bonds of the products form, energy is released or absorbed, depending on whether the reactants have higher or lower energy than the products.

We can measure energy changes in several ways. The two kinds of energy change of most interest to us are: (1) the change in **free energy** (ΔG), which is the energy available to do useful work (discussed in Chapter 13), and (2) the change in **enthalpy** (ΔH), which is the heat energy absorbed or released by the reaction and measured at constant pressure. Most chemical reactions take place under the constant pressure of the atmosphere. The energy released or absorbed by such reactions is the change in enthalpy, ΔH, which can be shown as

$$\Delta H_{\text{reaction}} = H_{\text{products}} - H_{\text{reactants}}$$

In reporting values of ΔH, a superscript is used to show the temperature at which the measurements were made. For example, the symbol $\Delta H^{0°C}$ shows that the change in enthalpy was measured at $0°C$. If no temperature is shown, the enthalpy change was measured at $25°C$. All changes are measured at one atmosphere pressure.

The value of ΔH given with an equation refers to that particular equation. When the enthalpy change was measured, the physical states of the components were those stated in the equation. If the physical states are different, there will be a different enthalpy change. This difference is illustrated by the next two equations for the formation of water. They differ in enthalpy change. In the first, gaseous water is formed, and in the second, liquid water is formed; the difference between their enthalpy changes reflects the difference in energy content between a gas and a liquid. (See Chapter 9 for more discussion of this point.)

$$H_2(g) + \tfrac{1}{2}O_2(g) \longrightarrow H_2O(g) \qquad \Delta H = -241 \text{ kJ}$$
$$H_2(g) + \tfrac{1}{2}O_2(g) \longrightarrow H_2O(l) \qquad \Delta H = -286 \text{ kJ}$$

The enthalpy change given for a reaction also depends on the coefficients used in the equation for the reaction. Thus, if the equation for the formation of water is written

$$2\,H_2(g) + O_2(g) \longrightarrow 2\,H_2O(g) \qquad \Delta H = -482 \text{ kJ}$$

the enthalpy change is twice what is was in the previous equation for the formation of gaseous water when the coefficient of water was 1. This last problem can be resolved by doing as we do in several equations where we report the enthalpy change per mole of one component of the reaction, thus removing any ambiguity in interpretation.

A. Endothermic and Exothermic Reactions

A reaction that absorbs energy is an **endothermic** reaction; its enthalpy change (ΔH) is positive. The enthalpy of the products of the reaction is greater than that of the reactants. Energy is absorbed from the surroundings. The following reactions are endothermic.

1. The formation of hydrogen iodide:

$$\tfrac{1}{2}H_2(g) + \tfrac{1}{2}I_2(s) \longrightarrow HI(g) \qquad \Delta H = +25.9 \text{ kJ/mol HI}$$
$$\text{(equation a)}$$

2. The decomposition of water:

$$H_2O(l) \longrightarrow H_2(g) + \tfrac{1}{2}O_2(g) \qquad \Delta H = +285.8 \text{ kJ/mol } H_2O$$
$$\text{(equation b)}$$

A reaction that releases energy is an **exothermic** reaction; its enthalpy change is negative. The enthalpy of the products is less than that of the reactants. Energy is released to the surroundings. The following reactions are exothermic.

1. The combustion of methane:

$$CH_4(g) + 2 O_2(g) \longrightarrow CO_2(g) + 2 H_2O(l)$$
$$\Delta H = -891 \text{ kJ/mol } CH_4 \qquad \text{(equation c)}$$

2. The formation of water:

$$H_2(g) + \tfrac{1}{2}O_2(g) \longrightarrow H_2O(l) \qquad \Delta H = -285.8 \text{ kJ/mol } H_2O$$
$$\text{(equation d)}$$

Notice that the decomposition of water (equation b) is endothermic and requires the input of 285.8 kJ energy per mole of water decomposed. The reverse reaction, the formation of one mole of water from hydrogen and oxygen (equation d), is exothermic and releases 285.8 kJ energy. The amount of energy is the same, but the sign of the energy change is different.

Another example of the relationship between energy change and the direction of a reaction is the formation and decomposition of glucose. Glucose $(C_6H_{12}O_6)$ is formed from carbon dioxide and oxygen in the cells of green plants in the process called photosynthesis. Photosynthesis is an endothermic reaction. The source of the energy for the formation of glucose is light (radiant energy), usually from the sun.

$$6 CO_2(g) + 6 H_2O(l) \xrightarrow{\text{photosynthesis}} C_6H_{12}O_6(s) + 6 O_2(g)$$
$$\Delta H = +2.80 \times 10^3 \text{ kJ/mol glucose}$$

In the reverse of this reaction, the glucose formed is metabolized (broken down) in plant and animal cells to form carbon dioxide, water, and energy:

$$C_6H_{12}O_6(s) + 6 O_2(g) \xrightarrow{\text{metabolism}} 6 CO_2(g) + 6 H_2O(l)$$
$$\Delta H - -2.80 \times 10^3 \text{ kJ/mol glucose}$$

Thus, green plants have the remarkable ability to trap the energy of sunlight and use that energy to produce glucose from carbon dioxide and water. The energy is stored in the glucose. Animal and plant cells have the equally remarkable ability to metabolize glucose and use the energy released to maintain body temperature or do biological work, such as contracting muscles or thinking.

Example 8.17   For each of the following reactions: (1) Decide whether the reaction is exothermic or endothermic. (2) Write the equation for the reverse reaction, and state the accompanying enthalpy change, ΔH.

a. $N_2(g) + O_2(g) \longrightarrow 2 NO(g) \qquad \Delta H = +181 \text{ kJ}$

b. $2 NO_2(g) \longrightarrow N_2O_4(g) \qquad \Delta H = -57.6 \text{ kJ}$

c. $PCl_3(g) + Cl_2(g) \longrightarrow PCl_5(g) \qquad \Delta H = -92.5 \text{ kJ}$

Solution   **a.** The enthalpy change is positive; the reaction is endothermic. The reverse reaction is:

$$2 \text{ NO}(g) \longrightarrow \text{N}_2(g) + \text{O}_2(g) \qquad \Delta H = -181 \text{ kJ}$$

b. The enthalpy change is negative; the reaction is exothermic. The reverse reaction is:

$$\text{N}_2\text{O}_4(g) \longrightarrow 2 \text{ NO}_2(g) \qquad \Delta H = +57.6 \text{ kJ}$$

c. The enthalpy change is negative; the reaction is exothermic. The reverse reaction is:

$$\text{PCl}_5(g) \longrightarrow \text{PCl}_3(g) + \text{Cl}_2(g) \qquad \Delta H = +92.5 \text{ kJ}$$

Problem 8.17   For each of the following reactions: (1) State whether the reaction is exothermic or endothermic. (2) Write the equation for the reverse reaction, and state its enthalpy change.

a. $\text{N}_2(g) + 3 \text{ H}_2(g) \longrightarrow 2 \text{ NH}_3(g) \qquad \Delta H = -92.0 \text{ kJ}$
b. $2 \text{ H}_2\text{O}(g) + 2 \text{ Cl}_2(g) \longrightarrow 4 \text{ HCl}(g) + \text{O}_2(g) \qquad \Delta H = -120 \text{ kJ}$
c. $\text{C}_2\text{H}_5\text{OH}(l) + 3 \text{ O}_2(g) \longrightarrow 2 \text{ CO}_2(g) + 3 \text{ H}_2\text{O}(l)$
$$\Delta H = -1.37 \times 10^3 \text{ kJ}$$

B. The Stoichiometry of Energy Changes

The energy change associated with a reaction is a stoichiometric quantity and can be treated arithmetically, as were mass changes in Section 8.4. For many reactions, enthalpy changes have been determined and tabulated in the chemical literature. The changes listed in such sources apply only to the form of the equation they accompany, as explained previously.

Example 8.18   Calculate the enthalpy change for the combustion of 35.5 g gaseous propane (C_3H_8).

$$\text{C}_3\text{H}_8(g) + 5 \text{ O}_2(g) \longrightarrow 3 \text{ CO}_2(g) + 4 \text{ H}_2\text{O}(l)$$
$$\Delta H = -2.22 \times 10^3 \text{ kJ}$$

Solution   **Equation**
Given above.

Wanted
? kJ released

Given
35.5 g C_3H_8

Conversion factors

Propane, C_3H_8, mass to moles: 44.1 g C_3H_8 = 1 mol C_3H_8

The combustion of 1 mol propane releases 2.22×10^3 kJ energy.

Arithmetic equation

$$? \text{ kJ} = 35.5 \text{ g } C_3H_8 \times \frac{1 \text{ mol } C_3H_8}{44.1 \text{ g } C_3H_8} \times \frac{-2.22 \times 10^3 \text{ kJ}}{1 \text{ mol } C_3H_8}$$

Answer
-1.79×10^3 kJ

Problem 8.18 Calculate the enthalpy change when 45.6 g liquid water are formed by the reaction of gaseous hydrogen with gaseous oxygen according to the equation:

$$H_2(g) + \tfrac{1}{2}O_2(g) \longrightarrow H_2O(l) \qquad \Delta H = -286 \text{ kJ}$$

Example 8.19 Calculate the enthalpy change when 15.0 g glucose are metabolized at 25°C to gaseous carbon dioxide and liquid water. The equation for the reaction is given in Section 8.5A.

Solution   **Equation**

$$C_6H_{12}O_6(s) + 6 \text{ } O_2(g) \longrightarrow 6 \text{ } CO_2(g) + 6 \text{ } H_2O(l)$$
$$\Delta H = -2.80 \times 10^3 \text{ kJ}$$

Wanted
? kJ

Given
15.0 g glucose

Conversion factors

Glucose, mass to moles: 180 g glucose = 1 mol glucose

The metabolism of 1 mol glucose releases 2.80×10^3 kJ energy.

Arithmetic equation

$$? \text{ kJ} = 15.0 \text{ g glucose} \times \frac{1 \text{ mol glucose}}{180 \text{ g glucose}} \times \frac{-2.80 \times 10^3 \text{ kJ}}{1 \text{ mol glucose}}$$

Answer
-2.33×10^2 kJ

Problem 8.19 Calculate the enthalpy change when 16.5 g methane are burned to gaseous carbon dioxide and liquid water at 25°C.

8.6   **Summary**

A chemical reaction can be described by an equation that gives the correct formula of each reactant and product and the relative numbers of formula units that react or are formed. Symbols in parentheses after the formula of the substance may be used to show the physical state of the substances involved in the reaction. An equation must be balanced to be a quantitative description of a reaction.

Reactions can be classified as to type. One such classification divides reactions into four categories: combination, decomposition, displacement, and double displacement. A second classification separates reactions into those that involve a transfer of electrons and those that do not. If electrons are transferred, the reaction is an oxidation–reduction reaction. In such a reaction, one substance is oxidized and loses electrons, and another substance is reduced and gains electrons. Oxidation numbers are useful in identifying oxidation–reduction reactions.

Stoichiometric calculations are based on balanced chemical equations. They allow us to calculate the amount of product that would be formed from a given amount of reactant in the equation. In some instances, where amounts of two reactants are given, one reactant may be limiting. In some instances, only a percent of the theoretical yield is obtained.

Chemical reactions obey the Law of Conservation of Mass/Energy. The change in enthalpy (ΔH) for a reaction measures the heat released or absorbed by a reaction carried out at constant pressure and determines whether the reaction is exothermic or endothermic. The amount of energy absorbed or released by a particular reaction can be calculated.

Key Terms

acid salt (8.2D1)

catalyst (8.2B)

chemical change (8.1)

chemical property (8.1)

combination reaction (8.2A)

combustion reaction (8.3B)

decomposition reaction (8.2B)

displacement reactions (8.2C)

double displacement (8.2D)

endothermic (8.5A)

enthalpy (8.1, 8.5)

exothermic (8.5A)

free energy (8.5)

insoluble compounds (8.2D2)

limiting reactant (8.4C)

neutralization reactions (8.2D1)

oxidation (8.3)

oxidation number (8.3A)

oxidation–reduction reactions (8.3)

oxidized (8.3)

oxidizing agent (8.3A)

percent yield (8.4B)

physical change (8.1)

physical properties (8.1)

polyprotic acid (8.2D1)

precipitation reactions (8.2D2)

redox reactions (8.3)

reduced (8.3)

reducing agent (8.3A)
reduction (8.3)
solubility (8.2D2)
solubility rules (Table 8.3)

soluble (8.2D2)
stoichiometry (8.4A)
theoretical yield (8.4B)

Multiple-Choice Questions

MC1. Which of the following equations would not yield a magnesium salt as a product?

    a. $Mg + H_2SO_4 \longrightarrow$

    b. $MgO + HCl \longrightarrow$

    c. $Mg(OH)_2 + H_2SO_4 \longrightarrow$

    d. $MgSO_4 + BaCl_2 \longrightarrow$

    e. All of these reactions would yield a magnesium salt.

MC2. Which of the following equations shows correctly the complete neutralization of phosphoric acid with sodium hydroxide?

    a. $NaOH + H_3PO_4 \longrightarrow Na_3PO_4 + H_2O$

    b. $2\ NaOH + H_2PO_4 \longrightarrow Na_2PO_4 + 2\ H_2O$

    c. $3\ Na(OH)_2 + 2\ H_3PO_4 \longrightarrow Na_3(PO_4)_2 + 6\ H_2O$

    d. $3\ NaOH + H_3PO_4 \longrightarrow Na_3PO_4 + 3\ H_2O$

    e. $3\ NaOH + H_3PO_3 \longrightarrow Na_3PO_3 + 3\ H_2O$

MC3. How many molecules of water are formed when the following equation is balanced?

$$Al_2O_3 + HCl \not\longrightarrow AlCl_3 + H_2O$$

    a. 1   **b.** 2   **c.** 3   **d.** 4   **e.** 6

MC4. When a solution of nickel(II) sulfate is added to a solution of potassium sulfide, a precipitate is formed. The precipitate is:

    a. NiS **b.** K_2SO_4 **c.** Ni_2SO_3 **d.** Ni_2S **e.** K_2S

MC5. Which of the following equations show oxidation–reduction?

    1. $NaOH + HCl \longrightarrow NaCl + H_2O$

    2. $H_2 + Cl_2 \longrightarrow 2\ HCl$

    3. $MnO_2 + 4\ HCl \longrightarrow Cl_2 + MnCl_2 + 2\ H_2O$

    4. $Na_2SO_3 + 2\ HCl \longrightarrow 2\ NaCl + SO_2 + H_2O$

    a. none   **b.** all   **c.** 2 and 3   **d.** 2, 3, and 4   **e.** 1, 2, and 4

For questions 6–10 consider the following equation:

$$C_3H_8(g) + 5\ O_2(g) \longrightarrow 3\ CO_2(g) + 4\ H_2O(l) \qquad \Delta H = -2220\ kJ$$

Useful formula weights are: propane, C_3H_8, 44.1; carbon dioxide, CO_2, 44.0

MC6. How many moles of carbon dioxide will be formed by the combustion of 36.1 g propane?
a. 0.820 mol   **b.** 2.46 mol   **c.** 108 mol   **d.** 3.00 mol
e. None of these answers is correct.

MC7. What energy change accompanies the combustion of 36.1 g propane?
a. 1.82×10^3 kJ are released
b. 1.82×10^3 kJ are absorbed
c. 5.46×10^2 kJ are released
d. 6.00×10^2 kJ are absorbed
e. 3.64×10^3 kJ are released

MC8. If the equation is rewritten as

$$C_3H_8(l) + 5\ O_2(g) \longrightarrow 3\ CO_2(g) + 4\ H_2O(g)$$

will the enthalpy change be the same?
a. yes   **b.** no

MC9. When 36.1 g propane was burned, 77.0 g carbon dioxide was formed. What was the percent yield for this reaction?
a. 21.3%   **b.** 43.7%   **c.** 71.3%   **d.** 84.1%   **e.** 100%

MC10. When 6.25 mol propane is reacted with 6.25 mol oxygen, how much carbon dioxide should be formed?
a. 18.75 mol CO_2 **b.** 3.75 mol CO_2 **c.** 8.53×10^2 mol CO_2
d. 0.629 mol CO_2 **e.** 165 mol CO_2

Problems

8.1 Physical versus Chemical Changes

8.20. Explain each term in the following equations:
a. $C(s) + O_2(g) \longrightarrow CO_2(g)$
$$\Delta H = -3.93 \times 10^2 \text{ kJ}$$
b. $2\ H_2O(g) + 2\ Cl_2(g) \longrightarrow$
$$4\ HCl(g) + O_2(g)$$
$$\Delta H = +1.20 \times 10^2 \text{ kJ}$$
c. $N_2(g) + O_2(g) \longrightarrow 2\ NO(g)$
$$\Delta H = 1.81 \times 10^2 \text{ kJ}$$

8.2 Kinds of Chemical Changes

8.21. Write balanced equations for the following reactions:
a. $K + O_2 \xrightarrow{\ \ } K_2O$
b. $Se + O_2 \xrightarrow{\ \ } SeO_2$
c. $Cr + O_2 \xrightarrow{\ \ } Cr_2O_3$
d. $S_8 + O_2 \xrightarrow{\ \ } SO_3$

8.22. Write balanced equations for the following combination reactions:
a. $Cu + S_8 \xrightarrow{\ \ } Cu_2S$
b. $Al + N_2 \xrightarrow{\ \ } Al_2N_3$
c. $N_2 + I_2 \xrightarrow{\ \ } NI_3$
d. $Ag + S_8 \xrightarrow{\ \ } Ag_2S$

8.23. Write balanced equations for the following displacement reactions:
a. $Fe + Cu(NO_3)_2 \xrightarrow{\ \ }$
$$Fe(NO_3)_2 + Cu$$
b. $Mg + HCl \xrightarrow{\ \ } MgCl_2 + H_2$
c. $Li + H_2O \xrightarrow{\ \ } LiOH + H_2$
d. $Al + H_2SO_4 \xrightarrow{\ \ } Al_2(SO_4)_3 + H_2$

8.24. Write balanced equations for the following displacement reactions:
a. $Hg + AgNO_3 \xrightarrow{\ \ } Hg(NO_3)_2 + Ag$
b. $Cl_2 + KI \xrightarrow{\ \ } KCl + I_2$

c. $Ca + H_2O \xrightarrow{\quad} Ca(OH)_2 + H_2$

d. $Ni + HCl \xrightarrow{\quad} NiCl_2 + H_2$

8.25. An important industrial chemical process is the electrolysis of a water solution of sodium chloride to give chlorine gas. Sodium hydroxide is also a product. Write a balanced equation for this important process.

*8.26. Write complete and balanced equations for the following reactions. Name the products.

a. $HCl + Mg(OH)_2 \longrightarrow$

b. $Cu(OH)_2 + H_2SO_4 \longrightarrow$

c. $H_3PO_4 + Al(OH)_3 \longrightarrow$

d. $H_2SO_4 + KOH \longrightarrow$

e. $H_2CO_3 + NaOH \longrightarrow$

8.27. Write equations that will show the formation of acid salts from the following reactants:

a. $H_2CO_3 + KOH \longrightarrow$

b. $H_2SO_4 + Ca(OH)_2 \longrightarrow$

c. $H_3PO_4 + Mg(OH)_2 \longrightarrow$

*8.28. Write balanced equations for the following reactions, all of which form precipitates. Identify and name the precipitate.

a. $Ba(NO_3)_2 + K_2SO_4 \longrightarrow$

b. $ZnCl_2 + H_2S \longrightarrow$

c. $Pb(NO_3)_2 + K_2CrO_4 \longrightarrow$

d. $Ni(NO_3)_2 + NaOH \longrightarrow$

e. $AgNO_3 + HBr \longrightarrow$

8.3 Oxidation–Reduction: A Second Way to Classify Reactions

8.29. In each of the following, an element has changed oxidation number. Name the element and state its change in oxidation number.

a. KCl to Cl_2 b. Na to NaOH

c. $MnSO_4$ to MnO_2 d. H_2S to S

e. Fe to $FeCl_2$ f. $C_2O_4^{2-}$ to CO_3^{2-}

*8.30. Identify the following changes as oxidation or reduction:

a. CrO_4^{2-} to Cr^{3+} b. MnO_4^- to Mn^{2+}

c. SO_4^{2-} to SO_2 d. K to K^+

e. Ag^+ to Ag   f. N_2 to NO

8.31. What characterizes an oxidation–reduction reaction?

*8.32. Write balanced equations for the following reactions and identify those that are oxidation–reduction.

a. $Ca(OH)_2 + HCl \xrightarrow{\quad} CaCl_2 + H_2O$

b. $SO_3 + BaO \xrightarrow{\quad} BaSO_4$

c. $AgNO_3 + Fe \xrightarrow{\quad} Fe(NO_3)_2 + Ag$

d. $Na_2SO_4 + BaCl_2 \xrightarrow{\quad}$
$$BaSO_4 + NaCl$$

e. $NaI + Cl_2 \xrightarrow{\quad} NaCl + I_2$

8.33. In the oxidation–reduction reactions in Problem 8.32, which element is oxidized and which is reduced? What changes in oxidation number have occurred?

8.4 Mass Relationships in an Equation

*8.34. Chlorine dioxide is used for bleaching paper. It is prepared by the reaction:

$$2 \, NaClO_2 + Cl_2 \longrightarrow 2 \, ClO_2 + 2 \, NaCl$$

a. Name the oxidizing and reducing agents in this reaction.

b. Calculate the mass of chlorine dioxide that would be prepared by the reaction of 5.50 kg sodium chlorite ($NaClO_2$).

8.35. Write the balanced equation for the reaction of chlorine with sodium to form sodium chloride. Calculate the mass of sodium that will react completely with 5.00 g chlorine.

8.36. Glucose, $C_6H_{12}O_6$, burns to form carbon dioxide and water.

a. Write the balanced equation for this reaction.

b. Calculate the moles of glucose that would be needed to form 1.55 mol carbon dioxide.

c. Calculate the mass of water formed by the reaction in **b.**

8.37. Pure aluminum is prepared by electrolysis of aluminum oxide according to the balanced equation:

$$2\ Al_2O_3 \xrightarrow{\text{elec}} 4\ Al + 3\ O_2$$

What mass of aluminum would be prepared from 6.06 g aluminum oxide?

8.38. a. Write the balanced equation for the reaction of sodium hydroxide with sulfuric acid to form sodium sulfate.
   b. Calculate how many moles of sulfuric acid would be neutralized by 8.00 g sodium hydroxide.

8.39. Each of the reactions in Problem 8.23 produces a free element. For each equation, calculate the number of grams of this element that would be formed by the reaction of 5.15 g of the first reactant with an excess of the second reactant.

8.40. For Problem 8.24, calculate the mass of free element formed by the reaction of 23.6 g of the first reactant with an excess of the second.

*8.41. a. Write the balanced equation for the complete combustion of octane, C_8H_{18}, with oxygen to form carbon dioxide and water.
   b. Calculate the mass of water formed by the complete combustion of 1.8 L octane (density = 0.775 g/mL).

8.42. When silver(I) carbonate is heated, it decomposes to silver, oxygen, and carbon dioxide according to the unbalanced equation:

$$Ag_2CO_3 \not\longrightarrow Ag + O_2 + CO_2$$

   a. Balance the equation.
   b. Calculate the weight of silver that would be isolated by heating 0.565 g of silver carbonate.

8.43. Aspirin is formed by the reaction of salicylic acid with acetyl chloride according to the balanced equation:

$$\underset{\text{salicylic acid}}{C_7H_6O_3} + \underset{\text{acetyl chloride}}{C_2H_3OCl} \longrightarrow$$

$$\underset{\text{aspirin}}{C_9H_8O_4} + HCl$$

What weight of aspirin is formed by the reaction of 5.00 g salicylic acid with excess acetyl chloride?

8.4B Percent Yield

*8.44. Chlorine can be prepared by the reaction of manganese(IV) oxide with excess hydrogen chloride according to the balanced equation:

$$MnO_2 + 4\ HCl \longrightarrow$$

$$MnCl_2 + Cl_2 + 2\ H_2O$$

If 0.85 g chlorine is obtained by the reaction of 1.35 g manganese(IV) oxide, what is the percent yield of the reaction?

8.45. A 1.126-g sample of a mixture of potassium chloride and potassium chlorate ($KClO_3$) was heated, yielding 0.38 g oxygen. The balanced equation for the reaction is:

$$2\ KClO_3 \longrightarrow 2\ KCl + 3\ O_2$$

Remembering that only the potassium chlorate will decompose to yield oxygen, calculate what percent of the original sample is potassium chlorate.

8.46. When bromine reacts with benzene, bromobenzene is the product:

$$\underset{\text{benzene}}{C_6H_6} + \underset{\text{bromine}}{Br_2} \longrightarrow$$

$$\underset{\text{bromobenzene}}{C_6H_5Br} + \underset{\text{hydrogen bromide}}{HBr}$$

If 5.05 g benzene yields 7.81 g bromobenzene, what is the percent yield of the reaction?

8.4C Limiting Reactant

8.47. **a.** What is a limiting reactant?
   b. What weight of which reactant remains after 6.34 g hydrogen are reacted with 6.24 g oxygen to form water?

8.48. Carbon dioxide will react with calcium oxide and water to form calcium bicarbonate.
   a. Write the balanced equation for this reaction.
   b. Calculate the yield of calcium bicarbonate if 4.65 g carbon dioxide reacts with 5.10 g calcium oxide in water.

*8.49. Calculate the weight of magnesium oxide formed by the reaction of 1.56 g magnesium with 2.63 g oxygen.

8.50. Calculate the mass of aspirin formed by the reaction of 4.67 g salicylic acid with 7.64 g acetyl chloride. (See Problem 8.43 for the balanced equation for this reaction.)

8.5 Energy Changes Accompanying Chemical Reactions

8.51. For the following reactions, calculate the energy change when 5.0 g of the boldfaced reactant is used. Express your answer in kilojoules. State whether the energy is absorbed or released.

$$H_2(g) + CO_2(g) \longrightarrow CO(g) + H_2O(l)$$
$$\Delta H = 1883 \text{ kJ/mol}$$

$$C_{12}H_{22}O_{11}(s) + 12 \text{ O}_2(g) \longrightarrow$$
$$12 \text{ CO}_2(g) + 11 \text{ H}_2O(l)$$
$$\Delta H = -5641 \text{ kJ/mol}$$

$$C_3H_6O_3(l) + 3 \text{ O}_2(g) \longrightarrow$$
$$3 \text{ CO}(g) + 3 \text{ H}_2O(l)$$
$$\Delta H = -1367 \text{ kJ/mol}$$

8.52. **a.** Calculate the energy required for the formation of 0.5 mol glucose from carbon dioxide and water. (The equation is shown in Section 8.5A.) Express your answer in kilojoules.
   b. Calculate the energy change when 0.50 mol glucose is metabolized to carbon dioxide and water. Express your answer in kilojoules.

*8.53. Calculate the energy released by the metabolism of 6.5 g ethyl alcohol according to the equation:

$$C_2H_5OH(l) + 3 \text{ O}_2(g) \longrightarrow$$
$$2 \text{ CO}_2(g) + 3 \text{ H}_2O(l)$$
$$\Delta H = -367 \text{ kJ/mol ethanol}$$

8.54. A plant requires 4178.8 kcal for the formation of 1.00 kg starch from carbon dioxide and water. Calculate the energy required for the formation of 6.32 g starch. Express your answer in kilojoules. Is this reaction endothermic or exothermic?

Review Problems

8.55. What mass of lead(II) iodide can be prepared by adding an excess of sodium iodide solution to 167 mL of a solution that contains 5.00 g lead(II) nitrate in each liter of solution?

8.56. A 5.39-g sample of barium is ignited. The oxide formed is dissolved in water. What weight of barium sulfate can be prepared by adding sulfuric acid to this solution?

8.57. The heat of combustion of a compound is the enthalpy associated with the reaction of one mole of the compound with oxygen.
   a. Given the following data, predict the heat of combustion of butane.

Name	Formula	Heat of Combustion
propane	C_3H_8; $CH_3CH_2CH_3$	-2.220×10^3 kJ/mol
butane	C_4H_{10}; $CH_3CH_2CH_2CH_3$	?
pentane	C_5H_{12}; $CH_3CH_2CH_2CH_2CH_3$	-3.510×10^3 kJ/mol
hexane	C_6H_{14}; $CH_3CH_2CH_2CH_2CH_2CH_3$	-4.163×10^3 kJ/mol
heptane	C_7H_{16}; $CH_3CH_2CH_2CH_2CH_2CH_2CH_3$	-4.811×10^3 kJ/mol

b. Calculate the amount of pentane needed to release 5.000×10^3 kJ.

8.58. Given the following data, show by calculation whether ethanol (ethyl alcohol, an ingredient of gasohol) or octane (an ingredient of regular gasoline) yields the most energy per liter.

Name	Formula	Density	Heat of combustion
ethanol	C_2H_5OH	0.7025 g/mL	-1.337×10^3 kJ/mol
octane	C_8H_{18}	0.7893 g/mL	-5.445×10^3 kJ/mol

8.59. An impure sample of sodium weighing 0.945 g is added to water. The hydrogen released weighs 0.0163 g. What is the percent purity of the sample?

■ 9 ■

The Properties of Gases

Although atoms and molecules are too small to be seen by the naked eye, chemists have learned a great deal about their properties by experimental observations. Chemists have also learned a great deal about collections of molecules as they exist as samples of matter in the three physical states: gas, liquid, and solid. In the next two chapters, we will consider the properties of matter in the three states. Among the things we will consider in this chapter are:

1. The characteristics of the three states of matter.
2. The kinetic energy of molecules.
3. The kinetic molecular theory, which describes how the properties of a gas are related to the behavior of its molecules.
4. How to measure the volume, temperature, and pressure of a gas.
5. The gas laws showing the relationships that exist between the volume, temperature, and pressure of a gas sample.
6. The ideal gas law, which relates these parameters to the number of moles of sample.
7. Partial pressures in a mixture of gases.
8. The differences between the ideal gas described by the kinetic molecular theory and real gases.

Characteristics of the Solid, Liquid, and Gaseous States

In Sections 1.3 and 2.5A3, we noted that the physical properties of a particular substance determine its state at room temperature. If both its **normal melting point** and its **normal boiling point** are below room temperature (20°C), the substance is a gas under normal conditions. The normal melting point of oxygen is −218°C; its normal boiling point is −189°C. Oxygen is a gas at room temperature. If the normal melting point of a substance is below room temperature and its normal boiling point is above room temperature, the substance is a liquid at room temperature. Benzene melts at 6°C and boils at 80°C; it is a liquid at room temperature. If both the normal melting point and the normal boiling point are above room temperature, the substance is a solid. Sodium chloride melts at 801°C and boils at 1413°C. Sodium chloride is a solid under normal conditions. Figure 9.1 illustrates the relationship between physical state and normal melting and boiling points.

| | Room temperature (20°C) | | | | |
| | Low temperature | | | High temperature | |
	Melting point	Boiling point		Melting point	Boiling point
Solids					
sodium chloride				801°C	1413°C
naphthalene				81°C	218°C
Liquids					
water	0°C				100°C
benzene	6°C				80°C
Gases					
oxygen	−218°C	−189°C			
methane	−182°C	−164°C			

FIGURE 9.1   The physical state as related to normal melting and boiling points. Notice that the solids melt and boil above room temperature, the liquids melt below room temperature and boil above room temperature, and the gases melt and boil below room temperature.

A. Shape and Volume

A solid has a fixed shape and volume that do not change with the shape of its container. Consider a rock and how its size and shape stay the same, regardless of where you put it. A liquid has a constant volume, but its shape conforms to the shape of its container. Consider a sample of milk. Its volume stays the same, whether you put it in a saucer for the cat to drink or in a glass for yourself; clearly its shape changes to match the shape of the container. A gas changes both its shape and volume to conform to the shape and volume of its container.

Consider a sample of air. It will fill an empty room, a balloon, a tire, or a rubber raft. Its shape and volume conform to the shape and volume of the container in which it is placed. Figure 9.2 illustrates these points.

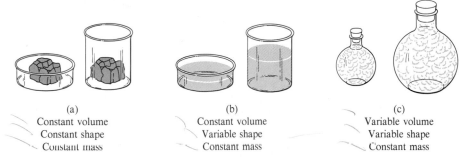

(a)	(b)	(c)
Constant volume	Constant volume	Variable volume
Constant shape	Variable shape	Variable shape
Constant mass	Constant mass	Constant mass

FIGURE 9.2 Constancy of volume, shape, and mass in the three states of matter: (a) solid, (b) liquid, (c) gas.

B. Density

The densities of liquids and solids are measured in grams per milliliter and grams per cubic centimeter, respectively, and change very little as the temperature of the sample changes. Gases have much lower densities, so much lower that gas densities are measured in grams per liter instead of grams per milliliter. The density of a gas varies considerably as the temperature of the gas changes. Table 9.1 shows the densities of three common substances, one in each of the three physical states, at two different temperatures.

TABLE 9.1 Densities of three common substances

	Density at 20°C	*Density at 100°C*
solid: sodium chloride	2.16 g/cm³	2.16 g/cm³
liquid: water	0.998 g/mL	0.958 g/mL
gas: oxygen	1.33 g/L	1.05 g/L

C. Compressibility

The volume of a solid or a liquid does not change very much with pressure. You cannot change the volume of a brick by squeezing it, nor can you squeeze one liter of liquid into a 0.5-L bottle. The volume of a gas does change a great deal with pressure; you can squeeze a 1.0-L balloon into a 0.5-L space.

D. Inferences about Intermolecular Structure

The constant shape and volume of a solid suggest that its particles (atoms, ions, or molecules) are held together by fairly rigid bonds. The variable shape and constant volume of a liquid suggest that there is some bonding between its particles but that these bonds are not rigid and probably are less strong than those in a solid. The fact that a gas has neither constant shape nor constant volume suggests that there are no bonds and only very slight interactive forces between the particles of a gas.

The variety in **compressibility** suggests other hypotheses. If solids and liquids cannot be compressed, the particles of which they are composed must be very close together. The high compressibility of a gas implies that the particles of a gas are very far apart with a great deal of space between them. This last hypothesis is supported by the difference between the densities of solids and liquids and the densities of gases. One mL of a solid or liquid always has much more mass than does one mL of a gas.

9.2   Kinetic Energy

Any consideration of the properties of a collection of particles such as molecules requires knowledge of their energy. Part of this energy is **kinetic energy,** the energy of motion. The kinetic energy (KE) of an object is determined by the equation

$$KE = \tfrac{1}{2}mv^2 \qquad \text{where } m = \text{mass}, v = \text{velocity}$$

This equation states that the kinetic energy of an object is dependent on both its mass and its velocity. A semitrailer truck and a subcompact car traveling at the same velocity have different kinetic energies. You would be aware of this difference if they crashed, for the truck would demolish the subcompact. For the two vehicles to have the same kinetic energy, the subcompact would have to be traveling at a much higher velocity than the truck.

A. The Distribution of Kinetic Energy

In a collection of molecules, each molecule has a kinetic energy that can be calculated by the equation given. Even if the molecules have a constant mass, they differ in velocity, so that a collection of molecules will have a wide range of kinetic energies, from very low to very high. Each molecule may change its kinetic energy often, but the overall distribution will remain the same.

Figure 9.3 shows a typical distribution of kinetic energies in a collection of molecules. In the graph, kinetic energy is plotted along the horizontal axis, and the percent of molecules having a particular kinetic energy is shown by the height of the curve at that point. Several observations can be made by studying the graph:

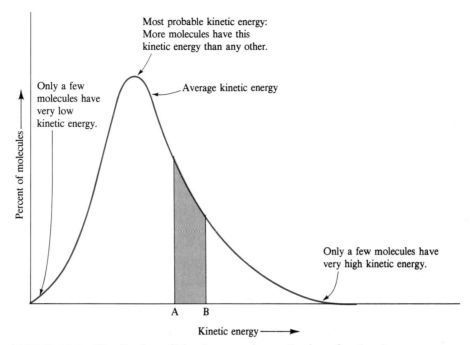

FIGURE 9.3   Distribution of kinetic energy in a collection of molecules.

1. The area under the curve represents the total number of molecules in the sample. Between any two points on the horizontal axis, the area under the curve represents the number of molecules that have kinetic energies in that range. For example, the shaded area between A and B represents the number of molecules that have kinetic energies between A and B.

2. The peak of the curve shows the **most probable kinetic energy.** More molecules have this energy than any other.

3. The average kinetic energy is slightly greater than the most probable kinetic energy.

4. Some molecules have a kinetic energy much higher than the average value.

5. Some molecules have a kinetic energy much lower than the average value.

Notice that the distribution of energies is much like the distribution of grades on a test. Figure 9.4 shows the distribution of grades on a standardized test. There is a most probable score. Most of the grades fall close to the most probable score; some grades are higher and others are lower.

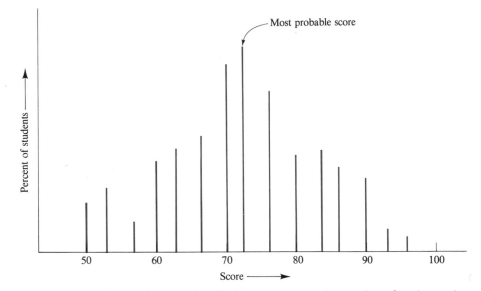

FIGURE 9.4   Graph of test grades. Each bar represents the number of students who received a particular grade.

B. Kinetic Energy and Temperature

The average kinetic energy of a collection of molecules is directly proportional to its temperature. At absolute zero ($-273°C$), the molecules have a minimum kinetic energy. As the temperature of the sample increases, so does its average kinetic energy. As the temperature rises, the distribution of kinetic energies among the molecules in the sample also changes. Figure 9.5 shows the distribution of kinetic energies in a sample at two different temperatures. Curve A is at the lower temperature; curve B is at the higher temperature. Notice the following differences between the two curves:

1. The peak of curve B is lower and broader than the peak of curve A. This difference in the curves means that, at the higher temperature, fewer molecules have the average kinetic energy and the distribution of energies is more spread out.

2. The peak of curve B is at a higher kinetic energy than the peak of curve A. This difference means that, at the higher temperature, the average kinetic energy of the molecules is higher.

We can conclude that, as the temperature of a sample increases, not only does the average kinetic energy increase but also fewer molecules have the average energy and the distribution of energies among the molecules is more uniform.

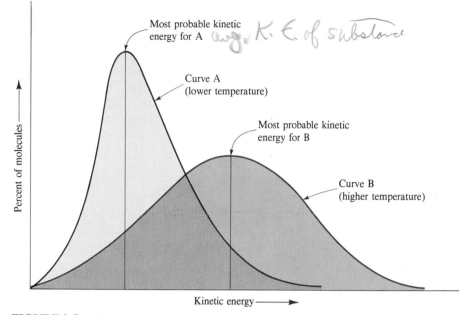

FIGURE 9.5 Distribution of kinetic energy in the same collection of molecules at two different temperatures.

9.3 The Kinetic Molecular Theory

The **kinetic molecular theory** describes the properties of molecules in terms of motion (kinetic energy) and of temperature. The theory is most often applied to gases but is helpful in explaining molecular behavior in all states of matter. As applied to gases, the kinetic molecular theory has the following postulates:

1. Gases are composed of very tiny particles (molecules). The actual volume of these molecules is so small as to be negligible compared with the total volume of the gas sample. A gas sample is, then, mostly empty space. This fact explains the compressibility of gases.

2. There are no attractive forces between the molecules of a gas. This postulate explains why, over a period of time, the molecules of a gas do not cluster together at the bottom of its container.

3. The molecules of a gas are in constant, rapid, random, straight-line motion. This postulate explains why a gas spreads so rapidly through the available space — for example, why the smell of hot coffee can spread quickly from the kitchen throughout the house.

4. During their motion, the gas molecules constantly collide with one another and with the walls of the container. (The collision with the walls

provides the pressure exerted by a gas.) None of these collisions is accompanied by any loss of energy; instead, they are what is known as **elastic collisions.** A "new" tennis ball collides more elastically than a "dead" tennis ball.

5. The average kinetic energy of the molecules in a gas sample is proportional to its temperature (Kelvin) and is independent of the composition of the gas. In other words, at the same temperature, all gases have the same average kinetic energy. It also follows from this postulate that at zero Kelvin all molecular motion has ceased.

These postulates and the experimental evidence for them are summarized in Table 9.2.

TABLE 9.2   The kinetic molecular theory

Postulate	Evidence
1. Gases are tiny molecules in mostly empty space.	The compressibility of gases.
2. There are no attractive forces between molecules.	Gases do not clump.
3. The molecules move in constant, rapid, random, straight-line motion.	Gases mix rapidly.
4. The molecules collide elastically with container walls and one another.	Gases exert pressure that does not diminish over time.
5. The average kinetic energy of the molecules is proportional to the Kelvin temperature of the sample.	Charles' Law (Section 9.5B)

Clearly, the actual properties of individual gases vary somewhat from these postulates, for their molecules do have a real volume and there is some attraction between the molecules. However, our discussion will ignore these variations and concentrate on an **ideal gas,** one that behaves according to this model.

9.4   Measuring Gas Samples

A gas sample obeys a number of laws that relate its volume to its pressure, temperature, and mass. How are these properties measured? Mass and volume are familiar concepts and can be measured with familiar apparatus. Temperature can be measured on any scale — Celsius, Fahrenheit, or Kelvin; however, if the temperature is to be used in a calculation involving gases, the Kelvin scale must be used. **Standard temperature** for gases, the temperature at which the properties of different gases are compared, is 273 K (0°C).

Pressure is defined as force per unit area and is measured in units that have dimensions of force per unit area. For example, the air pressure in tires is measured in pounds per square inch (psi). The pressure of the atmosphere is frequently measured with a mercury barometer.

Pressure can be more easily understood if we consider how a **barometer** measures pressure. The basic features of a mercury barometer are shown in Figure 9.6. In preparing a barometer, a glass tube at least 760 mm long and

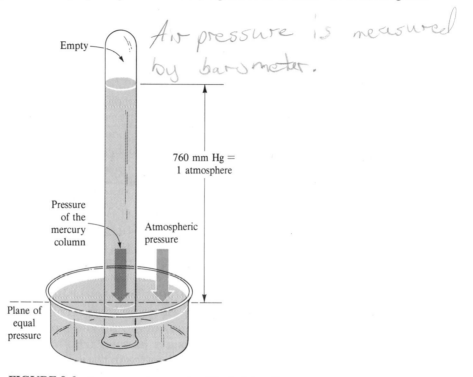

Air pressure is measured by barometer.

Empty

760 mm Hg = 1 atmosphere

Pressure of the mercury column

Atmospheric pressure

Plane of equal pressure

FIGURE 9.6 A mercury barometer. The height of mercury in the column is proportional to the pressure of the atmosphere.

closed at one end is filled with mercury, then carefully inverted into a pool of mercury. The level of the mercury in the column will fall slightly and then become steady.

The height of the column of mercury measures the pressure of the atmosphere. To understand this concept, consider the pressure on the surface of the mercury pool at the base of the column. Above this surface rises the "sea" of air (the atmosphere) that surrounds the Earth. On each square centimeter of the surface, we can visualize a 20-km column of air pressing down. On the surface under the mercury column, the mercury is pressing down. The two pressures must be equal. If they were not, mercury would be flowing into or out of the column, and the height of the column would not be steady. The atmosphere must be exerting a pressure equal to that exerted by the mercury column. Remember that pressure is force per unit area. The total area under the atmosphere or under the column of mercury is not critical, because the force that is measured is the force on each unit of area under the column or the atmosphere, not the total force.

When this experiment is performed in dry air at sea level and at $0°C$, the column of mercury is 760 mm high; therefore, we say that the **atmosphere** is exerting a pressure equal to that of 760 mm of mercury. This amount of pressure has been defined as one atmosphere (1 atm) of pressure and designated as **standard pressure.** Thus **STP** is used to mean **standard temperature and pressure** or standard conditions ($0°C$, 1 atm). The values of standard pressure measured in units other than atmospheres are shown:

$$1 \text{ atmosphere} = 1.01325 \times 10^5 \text{ Pascals (the } \textbf{Pascal, } Pa, \text{ is the SI unit)}$$

$$= 76 \text{ cm, or } 760 \text{ mm, mercury}$$

$$= 760 \text{ torr (1 } \textbf{torr} = \text{the pressure exerted by } 1 \text{ mm mercury)}$$

$$= 29.92 \text{ in. mercury (used to report atmospheric pressure in weather reports)}$$

$$= 1.013 \text{ bar (used in meteorology)}$$
$$(1 \text{ cm mercury} = 13.3 \text{ millibars})$$

Each of these relationships can be used as a conversion factor, as shown in the following problems.

Example 9.1

a. How many atmospheres pressure is exerted by a column of mercury 654 mm high?

b. What is this pressure in Pascals?

Solution

a. Wanted
? atm (pressure in atmospheres)

Given
A column of mercury 654 mm high

Conversion factor

1 atm = 760 mm Hg

Equation

$$? \text{ atm} = 654 \text{ mm Hg} \times \frac{1 \text{ atm}}{760 \text{ mm Hg}}$$

Answer
0.861 atm

b. **Wanted**
? Pa (pressure in Pascals)

Given
A pressure of 0.861 atm

Conversion factor

1 atm = 1.01325×10^5 Pa

Equation

$$? \text{ Pa} = 0.861 \text{ atm} \times \frac{1.01325 \times 10^5 \text{ Pa}}{1 \text{ atm}}$$

Answer
8.72×10^4 Pa

Problem 9.1   What pressure in Pascals is equal to a pressure of 1.65 atm? to 369 torr?

In dry air at sea level, the average air pressure is 1 atm. Atmospheric pressure decreases as altitude increases, because the sea of air above becomes less dense. Our bodies become adjusted to the normal pressure of the altitude at which we live. Minor problems of adjustment can occur when we move from sea level to the mountains, and vice versa. Major problems develop at higher altitudes. Commercial jet-aircraft cabins must therefore be pressurized, because humans cannot survive the low pressure of the atmosphere at the altitudes at which jet aircraft fly. For the same reason, travelers in space must wear pressurized suits.

Barometers measure the pressure of the atmosphere. **Manometers** measure the pressure of isolated gas samples. Some manometers measure pressure with a column of mercury, like a mercury barometer. This type of manometer has a U-shaped tube partially filled with mercury (Figure 9.7). One end of the tube is open to a chamber holding a gas sample, and the other end is open to the atmosphere. If the mercury level on the side of the tube open to the gas sample is lower than that on the side open to the atmosphere, the pressure of the gas is greater than that of the atmosphere by an amount equal to the difference in height between the two mercury columns. If the mercury level on the side of the gas sample is higher than that on the side open to the atmosphere, the pressure of the gas is less than the atmospheric pressure by the difference in the heights of the two columns.

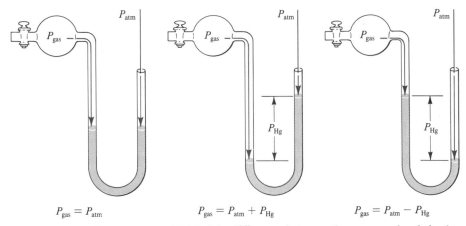

$$P_{gas} = P_{atm} \qquad P_{gas} = P_{atm} + P_{Hg} \qquad P_{gas} = P_{atm} - P_{Hg}$$

FIGURE 9.7   A manometer. The height difference between the mercury levels in the two sides of the tube measures the pressure difference between the gas sample and the atmosphere.

9.5   The Gas Laws

A. Boyle's Law

Boyle's Law states: If the temperature of a gas sample is kept constant, the volume of the sample will vary inversely as the pressure varies. This statement means that, if the pressure increases, the volume will decrease. If the pressure decreases, the volume will increase. This law can be expressed as an equation that relates the initial volume (V_1) and the initial pressure (P_1) to the final volume (V_2) and the final pressure (P_2). At constant temperature,

$$\frac{V_1}{V_2} = \frac{P_2}{P_1}$$

Rearranging this equation gives:

$$V_1 P_1 = V_2 P_2 \quad \text{or} \quad V_2 = V_1 \times \frac{P_1}{P_2}$$

Boyle's Law is illustrated in Figure 9.8, which shows a sample of gas enclosed in a container with a movable piston. The container is kept at a constant temperature and subjected to a regularly increasing amount of pressure. When the piston is stationary, the pressure it exerts on the gas sample is equal to the pressure the gas exerts on it. When the pressure on the piston is doubled, it moves downward until the pressure exerted by the gas equals the pressure exerted by the piston. At this point the volume of the gas is halved. If the pressure on the piston is again doubled, the volume of gas decreases to one-fourth its original volume.

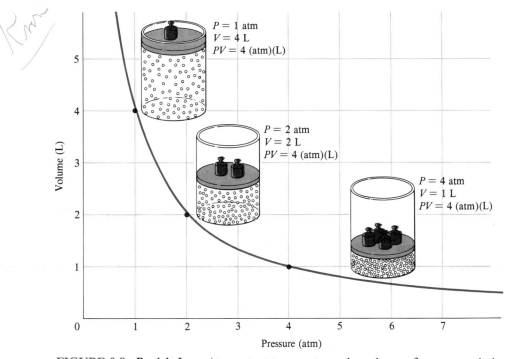

FIGURE 9.8 Boyle's Law: At constant temperature, the volume of a gas sample is inversely proportional to the pressure. The curve is a graph based on the data listed in the figure.

At the molecular level, the pressure of a gas depends on the number of collisions its molecules have with the walls of the container. If the pressure on the piston is doubled, the volume of the gas decreases by one-half. The gas molecules, now confined in a smaller volume, collide with the walls of the container twice as often and their pressure once again equals that of the piston.

How does Boyle's Law relate to the kinetic molecular theory? The first postulate of the theory states that a gas sample occupies a relatively enormous empty space containing molecules of negligible volume. Changing the pressure on the sample changes only the volume of that empty space — not the volume of the molecules.

Example 9.2   A sample of gas has a volume of 6.20 L at 20°C and 0.980 atm pressure. What is its volume at the same temperature and at a pressure of 1.11 atm?

Solution   **1.** Tabulate the data.

	Initial conditions	*Final conditions*
volume	$V_1 = 6.20$ L	$V_2 = ?$
pressure	$P_1 = 0.980$ atm	$P_2 = 1.11$ atm

2. Check the pressure units. If they are different, use a conversion factor to make them the same. (Pressure conversion factors are found in Section 9.4).

3. Substitute in the Boyle's Law equation:

$$V_2 = V_1 \times \frac{P_1}{P_2} = 6.20 \text{ L} \times \frac{0.980 \text{ atm}}{1.11 \text{ atm}} = 5.47 \text{ L}$$

4. Check that your answer is reasonable. The pressure has increased; the volume should decrease. The calculated final volume is less than the initial volume, as predicted.

Problem 9.2   A sample of gas has a volume of 253 mL at 0.50 atm. What is its volume at the same temperature and at a pressure of 1.0 atm?

■

B. Charles' Law

Charles' Law states: If the pressure of a gas sample is kept constant, the volume of the sample will vary directly with the temperature in Kelvin (Figure 9.9). As the temperature increases, so will the volume; if the temperature decreases, the volume will decrease. This relationship can be expressed by an equation relating the initial volume (V_1) and initial temperature (T_1 measured in K) to the final volume (V_2) and final temperature (T_2 measured in K). At constant pressure,

$$\frac{V_1}{V_2} = \frac{T_1}{T_2}$$

Rearranging this equation gives:

$$V_2 = V_1 \times \frac{T_2}{T_1} \quad \text{or} \quad \frac{V_2}{T_2} = \frac{V_1}{T_1}$$

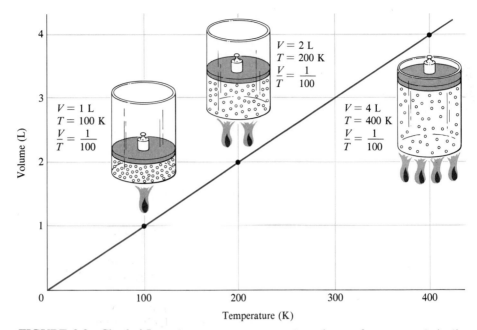

FIGURE 9.9 Charles' Law: At constant pressure, the volume of a gas sample is directly proportional to the temperature in degrees Kelvin.

How does Charles' Law relate to the postulates of the kinetic molecular theory? The theory states that the molecules in a gas sample are in constant, rapid, random motion. This motion allows the tiny molecules to effectively occupy the relatively large volume filled by the entire gas sample.

What is meant by "effectively occupy"? Consider a basketball game, with thirteen persons on the court during a game (ten players and three officials). Standing still, they occupy only a small fraction of the floor. During play they are in constant, rapid motion effectively occupying the entire court. You could not cross the floor without danger of collision. The behavior of the molecules in a gas sample is similar. Although the actual volume of the molecules is only a tiny fraction of the volume of the sample, the constant motion of the molecules allows them to effectively fill that space. As the temperature increases, so does the kinetic energy of the molecules. As they are all of the same mass, an increased kinetic energy must mean an increased velocity. This increased velocity allows the molecules to occupy or fill an increased volume, as do the basketball players in fast action. Similarly, with decreased temperature, the molecules move less rapidly and fill a smaller space.

The next example shows how Charles' Law can be used in calculations.

Example 9.3   The volume of a gas sample is 746 mL at 20°C. What is its volume at body temperature (37°C)? Assume the pressure remains constant.

Solution   **1.** Tabulate the data.

	Initial conditions	*Final conditions*
volume	$V_1 = 746$ mL	$V_2 = ?$
temperature	$T_1 = 20°C$	$T_2 = 37°C$

2. Do the units match? Charles' Law requires that the temperature be measured in Kelvin in order to give the correct numerical ratio. Therefore, change the given temperature to Kelvin:

$$T_1 = 20 + 273 = 293 \text{ K} \qquad T_2 = 37 + 273 = 310 \text{ K}$$

3. Calculate the new volume:

$$V_2 = V_1 \times \frac{T_2}{T_1} = 746 \text{ mL} \times \frac{310 \text{ K}}{293 \text{ K}} = 789 \text{ mL}$$

4. Is the answer reasonable? This volume is larger than the original volume, as was predicted from the increase in temperature. The answer is thus reasonable.

Problem 9.3   A balloon has a volume of 1.56 L at 25°C. If the balloon is cooled to −10°C, what will be its new volume? Assume the pressure remains constant.

C. The Combined Gas Law

Frequently, a gas sample is subjected to changes in both temperature and pressure. In such cases, the Boyle's Law and Charles' Law equations can be combined into a single equation, representing the **Combined Gas Law**, which states: The volume of a gas sample changes inversely with its pressure and directly with its Kelvin temperature.

$$V_2 = V_1 \times \frac{T_2}{T_1} \times \frac{P_1}{P_2}$$

As before, V_1, P_1, and T_1 are the initial conditions, and V_2, P_2, and T_2 are the final conditions. The Combined Gas Law equation can be rearranged to another frequently used form:

$$\frac{P_1V_1}{T_1} = \frac{P_2V_2}{T_2}$$

Example 9.4   A gas sample occupies a volume of 2.5 L at 10°C and 0.95 atm. What is its volume at 25°C and 0.75 atm?

Solution

Initial conditions	Final conditions
$V_1 = 2.5$ L	$V_2 = ?$
$P_1 = 0.95$ atm	$P_2 = 0.75$ atm
$T_1 = 10°C = 283$ K	$T_2 = 25°C = 298$ K

Check that P_1 and P_2 are measured in the same units and that both temperatures have been changed to Kelvin. Substitute in the equation:

$$\frac{0.95 \text{ atm} \times 2.5 \text{ L}}{283 \text{ K}} = \frac{0.75 \text{ atm} \times V_2}{298 \text{ K}}$$

Solving this equation, we get:

$$V_2 = \frac{0.95 \text{ atm} \times 2.5 \text{ L} \times 298 \text{ K}}{283 \text{ K} \times 0.75 \text{ atm}} = 3.3 \text{ L}$$

This answer is reasonable. Both the pressure change (lower) and the temperature change (higher) would cause an increased volume.

Problem 9.4   A sample of gas occupies a volume of 5.7 L at 37°C and 9.76×10^4 Pa. What is its volume at standard temperature and pressure?

Example 9.5   A gas sample originally occupies a volume of 0.546 L at 745 mm Hg and 95°C. What pressure will be needed to contain the sample in 155 mL at 25°C?

Solution

Initial conditions	Final conditions
$V_1 = 0.546$ L	$V_2 = 155$ mL $= 0.155$ L
$P_1 = 745$ mm Hg	$P_2 = ?$
$T_1 = 95°C = 368$ K	$T_2 = 25°C = 298$ K

Notice that the units of each property are now the same in the initial and final state. Substituting into the equation:

$$\frac{(0.546 \text{ L})(745 \text{ mm Hg})}{368 \text{ K}} = \frac{(0.155 \text{ L})(P_2)}{298 \text{ K}}$$

$$P_2 = \frac{(0.546 \text{ L})(745 \text{ mm Hg})(298 \text{ K})}{(368 \text{ K})(0.155 \text{ L})}$$

$$P_2 = 2.13 \times 10^3 \text{ mm Hg}$$

$$\text{or} \quad P_2 = 2.13 \times 10^3 \text{ mm Hg} \times \frac{1 \text{ atm}}{760 \text{ mm Hg}} = 2.80 \text{ atm}$$

Problem 9.5   A gas sample expands from a volume of 4.19 mL at 0°C and a pressure of 987 torr to a volume of 398 mL at a temperature of 39°C. What must be the pressure on the gas after its expansion?

D. Avogadro's Hypothesis and Molar Volume

Avogadro's Hypothesis states: At the same temperature and pressure, equal volumes of gases contain equal numbers of molecules (Figure 9.10). This statement means that, if one liter of nitrogen at a particular temperature and pres-

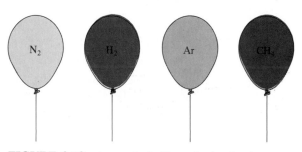

FIGURE 9.10   Avogadro's Hypothesis: At the same temperature and pressure, equal volumes of different gases contain the same number of molecules. Each balloon holds 1.0 L of gas at 20°C and 1 atm pressure. Each contains 0.045 mol or 2.69×10^{22} molecules of gas.

sure contains 1.0×10^{22} molecules, then one liter of any other gas at the same temperature and pressure also contains 1.0×10^{22} molecules. The reasoning behind Avogadro's Hypothesis is not always immediately apparent. But consider that the properties of a gas that relate its volume to its temperature and pressure have been described using the postulates of the kinetic molecular theory without mentioning the composition of the gas. One of the conclusions we drew from those postulates was that, at any pressure, the volume a gas sample occupies depends on the kinetic energy of its molecules and the average of those kinetic energies is dependent only on the temperature of the sample. Stated slightly differently, at a given temperature, all gas molecules, regardless of their chemical composition, have the same average kinetic energy and therefore occupy the same effective volume.

One corollary of Avogadro's Hypothesis is the concept of **molar volume.** The molar volume (the volume occupied by one mole) of a gas under 1.0 atm pressure and at $0°C$ (273.15 K) (STP or standard conditions) is, to three significant figures, 22.4 L. Molar volume can be used to calculate **gas densities,** d_{gas}, under standard conditions. The equation for this calculation is:

$$\text{At STP, } d_{gas} = \frac{\text{mole weight in grams}}{22.4 \text{ liters per mole}}$$

Example 9.6

Calculate the density of nitrogen under standard conditions (STP).

Solution

The mole weight of nitrogen is (2×14.0) or 28.0 g/mol. The molar volume is 22.4 L. Density is the ratio of mass to volume (mass/volume). Therefore:

$$\text{Density of nitrogen at STP} = \frac{28.0 \text{ g/mol}}{22.4 \text{ L/mol}} = 1.25 \text{ g/L}$$

Problem 9.6

Calculate the density of helium under standard conditions.

A second corollary of Avogadro's Hypothesis is that, at constant temperature and pressure, the volume of a gas sample depends on the number of molecules (or moles) the sample contains. Stated a little differently, if the pressure and temperature are constant, the ratio between the volume of a gas sample and the number of molecules the sample contains is a constant. Stating this ratio as an equation,

$$\frac{\text{Volume of sample 1}}{\text{Volume of sample 2}} = \frac{\text{Number of molecules in sample 1}}{\text{Number of molecules in sample 2}}$$

Example 9.7

A gas sample containing 5.02×10^{23} molecules has a volume of 19.6 L. At the same temperature and pressure, how many molecules will be contained in 7.9 L of the gas?

Solution

If the temperature and pressure are kept constant, the volume of a gas is directly proportional to the number of molecules it contains. Substituting values into the equation:

$$\frac{19.6 \text{ L}}{7.9 \text{ L}} = \frac{5.02 \times 10^{23} \text{ molecules}}{\text{Molecules in sample 2}}$$

Rearranging and solving:

$$\text{Molecules in sample 2} = \frac{(5.02 \times 10^{23} \text{ molecules})(7.9 \text{ L})}{(19.6 \text{ L})}$$

$$= 2.02 \times 10^{23} \text{ molecules}$$

Problem 9.7   What volume of helium at STP will contain 7.96×10^{10} molecules? (Molar volume = 22.4 L at STP.)

E. The Ideal Gas Equation

The various statements relating the pressure, volume, temperature, and number of moles of a gas sample can be combined into one statement: The volume (V) occupied by a gas is directly proportional to its Kelvin temperature (T) and the number of moles (n) of gas in the sample, and it is inversely proportional to its pressure (P). In mathematical form, this statement becomes:

$$V = \frac{nRT}{P}$$

where V = volume, n = moles of sample, P = pressure, T = temperature in K, and R = a proportionality constant known as the gas constant. This equation, called the **ideal gas equation,** is often seen in the form

$$PV = nRT$$

The term *ideal gas* means a gas that obeys exactly the gas laws. **Real gases,** those gases whose molecules do not follow exactly the postulates of the kinetic molecular theory, exhibit minor variations in behavior from those predicted by the gas laws.

The value of the **gas constant** R can be determined by substituting into the equation the known values for one mole of gas at standard conditions.

$$R = \frac{PV}{nT} = \frac{1 \text{ atm} \times 22.4 \text{ L}}{1 \text{ mol} \times 273 \text{ K}} = 0.0821 \frac{\text{L-atm}}{\text{mol-K}}$$

Table 9.3 shows the value of the gas constant R when the units are different from those shown here.

TABLE 9.3 Several values of the gas constant R	
Value	*Units*
0.0821	L-atm/mol-K
8.31×10^3	L-Pa/mol-K
62.4	L-torr/mol-K
8.31	m^3-Pa/mol-K

Example 9.8   What volume is occupied by 5.50 g of carbon dioxide at 25°C and 742 torr?

Solution   1. Identify the variables in the equation, and convert the units to match those of the gas constant. We will use the gas constant 0.0821 L-atm/mol-K. This value establishes the units of volume (L), of pressure (atm), of moles, and of temperature (K) to be used in solving the problem.

$$P = 742 \text{ torr} \times \frac{1 \text{ atm}}{760 \text{ torr}} = 0.976 \text{ atm}$$

$$V = ? \text{ L}$$

$$n = 5.50 \text{ g} \times \frac{1 \text{ mol}}{44.0 \text{ g}} = 0.125 \text{ mol}$$

$$R = 0.0821 \text{ L-atm/mol-K}$$

$$T = 25 + 273 = 298 \text{ K}$$

2. Substituting these values into the ideal gas equation:

$$V = \frac{nRT}{P}$$

$$= 0.125 \text{ mol} \times \frac{(0.0821 \text{ L-atm/mol-K})(298 \text{ K})}{0.976 \text{ atm}}$$

$$= 3.13 \text{ L}$$

The units cancel; the answer is reasonable. The amount of carbon dioxide is about one-eighth mole. The conditions are not far from STP. The answer (3.13 L) is about one eighth of the molar volume (22.4 L).

Problem 9.8   What mass of oxygen will occupy 1.23 L at 37°C and 0.752 atm?

Example 9.9   Laughing gas is dinitrogen oxide, N_2O. What is the density of laughing gas at 30°C and 745 torr?

Solution   **Wanted**
Density (that is, mass/volume) of N_2O at 30°C and 745 torr

Strategy
The mass of one mole at STP is known. Using the ideal gas equation, we can calculate the volume of one mole at the given conditions. The density at the given conditions can be calculated.

Data

$$1 \text{ mol } N_2O = 44.0 \text{ g}$$

$$R = 0.0821 \text{ L-atm/mol-K}$$

$$T = 30 + 273 = 303 \text{ K}$$

$$P = 745 \text{ torr} \times \frac{1 \text{ atm}}{760 \text{ torr}} = 0.980 \text{ atm}$$

Substituting into the ideal gas equation,

$$V = \frac{nRT}{P} = \frac{(1 \text{ mol})(0.0821 \text{ L-atm/mol-K})(303 \text{ K})}{0.980 \text{ atm}} = 25.4 \text{ L}$$

Calculating the density:

$$\text{Density at } 30°C = \frac{44.0 \text{ g}}{25.4 \text{ L}} = 1.73 \text{ g/L}$$

Problem 9.9   Dry ice is solid carbon dioxide. Dry ice becomes a gas at $-78°C$. What is the density of carbon dioxide at this very low temperature and a pressure of 1.00×10^5 Pa?

Molar volume is often used to determine the molecular mass of a low-boiling liquid. The compound becomes gaseous at a measured temperature and pressure, and the mass of a measured volume of the vapor is determined. Example 9.10 illustrates this process.

Example 9.10   What is the molecular mass of a compound if 0.556 g of this compound occupies 255 mL at 9.56×10^4 Pa and 98°C?

Solution   **1.** Determine the moles n of sample using the ideal gas equation.

Data
The gas constant 0.0821 L-atm/mol-K will be used; the data given must be changed to these units.

$$P = 9.56 \times 10^4 \text{ Pa} \times \frac{1 \text{ atm}}{1.013 \times 10^5 \text{ Pa}} = 0.944 \text{ atm}$$

$$V = 255 \text{ mL} = 0.255 \text{ L}$$

$$R = 0.0821 \text{ L-atm/mol-K}$$

$$T = 98 + 273 = 371 \text{ K}$$

Substitute into the ideal gas equation:

$$n = \frac{PV}{RT} = \frac{(0.944 \text{ atm})(0.255 \text{ L})}{(0.0821 \text{ L-atm/mol-K})(371 \text{ K})} = 0.00790 \text{ mol}$$

2. Next determine the molecular mass of the compound. The mass of the sample was given as 0.556 g. Calculations have shown that this mass is

0.00790 mol. A simple ratio will determine the molecular weight of the substance.

$$\frac{grams}{mole} = \frac{0.556\ g}{0.00790\ mol} = 70.4\ g/mol$$

Problem 9.10 Calculate the molecular mass of a compound if 0.926 g of the compound occupies 0.251 L at 725 torr and 85°C.

9.6 Mixtures of Gases; Partial Pressures

We have already noted that the composition of a gas does not affect the validity of the gas laws. It follows then that mixtures of gases must follow those laws in the same way that a single gas does. They do and, when Boyle's Law is applied to a mixture of gases, there is a relationship between the composition of a gas sample and its total pressure. This relationship is known as Dalton's Law of Partial Pressures (the same Dalton who proposed the atomic theory described in Section 3.1). **Dalton's Law of Partial Pressures** states: (1) Each gas in a mixture of gases exerts a pressure, known as its **partial pressure,** that is equal to the pressure the gas would exert if it were the only gas present; (2) the total pressure of the mixture is the sum of the partial pressures of all the gases present. This law is based on the postulate of the kinetic molecular theory (Section 9.3), which states that a gas sample is mostly empty space. The gas molecules are so far apart from one another that each acts independently. A mathematical expression of the Law of Partial Pressures is:

$$P_{Total} = P_1 + P_2 + P_3 + \cdots$$

where P_{Total} equals the total pressure of the mixture, and $P_1, P_2, P_3, \ldots$ are the partial pressures of the gases present in the mixture.

Suppose we have 1 L oxygen at 1 atm pressure in one container, 1 L nitrogen at 0.5 atm pressure in a second container, and 1 L hydrogen at 3 atm pressure in a third container (Figure 9.11). If we combine the samples in a single

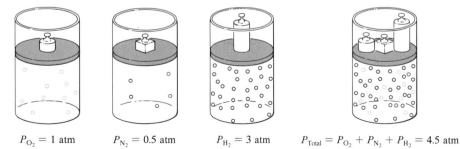

$P_{O_2} = 1\ atm$ $P_{N_2} = 0.5\ atm$ $P_{H_2} = 3\ atm$ $P_{Total} = P_{O_2} + P_{N_2} + P_{H_2} = 4.5\ atm$

FIGURE 9.11 The total pressure of a mixture of gases equals the sum of the individual gas pressures.

1-L container, the total pressure is 4.5 atm (1 atm + 0.5 atm + 3 atm). The partial pressure of oxygen, P_{O_2}, is 1 atm (the pressure it alone exerted in its container). Similarly, the partial pressure of nitrogen, P_{N_2}, is 0.5 atm, and that of hydrogen, P_{H_2}, is 3.0 atm.

A corollary of this law is that, in a mixture of gases, the percent of each gas in the total volume is the same as the percent of each partial pressure in the total pressure. From the total pressure of a mixture of gases and its percent composition, we can calculate the partial pressure of the individual gases.

$$\frac{V_{gas}}{V_{Total}} = \frac{P_{gas}}{P_{Total}}$$

Example 9.11

Dry air contains 78.08% nitrogen, 20.95% oxygen, and 0.93% argon. Calculate the partial pressure of each gas in a sample of dry air at 760 torr. Calculate also the total pressure exerted by the three gases combined.

Solution

1. The equation is:

$$P_{Total} = P_{N_2} + P_{O_2} + P_{Ar}$$

2. Calculate the partial pressure of each gas by using the corollary of Dalton's Law, which states that each partial pressure is the same percent of the total pressure as the percent each gas is of the total volume.

$$P_{N_2} = \frac{78.08}{100} \times 760 \text{ torr} = 593.4 \text{ torr}$$

$$P_{O_2} = \frac{20.95}{100} \times 760 \text{ torr} = 159.2 \text{ torr}$$

$$P_{Ar} = \frac{0.93}{100} \times 760 \text{ torr} = 7.1 \text{ torr}$$

3. The total pressure is:

$$P_{Total} = 593.4 + 159.2 + 7.1 = 759.7 \text{ torr}$$

The difference between the total pressure of the three gases and the total pressure of the air sample is due to the partial pressure of other gases, such as carbon dioxide, present in dry air.

Problem 9.11

Air in the trachea (windpipe) contains 19.5% oxygen, 0.4% carbon dioxide, 6.2% water vapor, and 74.1% nitrogen. Assuming that the pressure in the trachea is 1 atm, what are the partial pressures of these gases in this part of the body?

Example 9.12   Gases insoluble in water can be purified by bubbling them through water. This process removes impurities that are soluble in water; but, at the same time, water vapor is picked up by the sample. A sample of nitrogen that was purified by this method had a volume of 6.523 L at 26°C and a total pressure of 746 torr. In any gas sample saturated with water vapor at 26°C, the partial pressure of the water is 25.2 torr. How many moles of nitrogen did the sample contain?

Solution   The solution to this problem requires the use of the ideal gas equation in the form

$$n = \frac{PV}{RT}$$

To use the gas constant $R = 0.0821$ L-atm/mol-K, we must first convert the values for each parameter in the equation to those of the gas constant:

$$T = 26 + 273 = 299 \text{ K}$$

The total pressure of the gas sample is the sum of the partial pressure of the nitrogen and the partial pressure of the water vapor ($P_{Total} = P_{N_2} + P_{H_2O}$). Rearranging this equation gives

$$P_{N_2} = P_{Total} - P_{H_2O} = 746 \text{ torr} - 25.2 \text{ torr} = 721 \text{ torr}$$

Converting to atmospheres to match the units of R,

$$P_{N_2} = 721 \text{ torr} \times \frac{1 \text{ atm}}{760 \text{ torr}} = 0.949 \text{ atm}$$

Substituting these values into the equation gives

$$n_{N_2} = \frac{(0.949 \text{ atm})(6.523 \text{ L})}{(0.0821 \text{ L-atm/mol-K})(299 \text{ K})} = 0.252 \text{ mol N}_2$$

Problem 9.12   Calculate the moles of carbon dioxide in a gas sample that has been bubbled through water and now occupies a volume of 0.876 L at 35°C and has a total pressure of 0.982 atm. The partial pressure of water vapor at this temperature is 45.2 torr.

9.7 Stoichiometry Involving Gases

Many chemical reactions involve gases. Section 9.5 gave several equations that relate the mass of a gas (in terms of moles) to its volume, temperature, and pressure. These relationships can be applied to stoichiometric problems involving gases.

Example 9.13   What volume of carbon dioxide at 37°C (body temperature) and 740 torr is produced by the metabolism of 1.0 g ethyl alcohol (C_2H_5OH)? The balanced equation is:

$$C_2H_5OH + 3\ O_2 \longrightarrow 2\ CO_2 + 3\ H_2O$$

Solution   To solve this problem, follow the steps listed in Section 8.4 to find the number of moles of CO_2 formed. Then use the ideal gas equation to convert from the number of moles of carbon dioxide to a volume.

Wanted
Liters (V) of CO_2 at 37°C and 740 torr

Given
1.0 g C_2H_5OH

Conversion factors

1 mol C_2H_5OH = 46.1 g C_2H_5OH
1 mol C_2H_5OH yields 2 mol CO_2
37°C = 310 K

Arithmetic equation

$$? \text{ mol } CO_2 = 1.0 \text{ g } C_2H_5OH \times \frac{1 \text{ mol } C_2H_5OH}{46.1 \text{ g } C_2H_5OH} \times \frac{2 \text{ mol } CO_2}{1 \text{ mol } C_2H_5OH}$$

$$= 0.043 \text{ mol } CO_2$$

Substituting this value into the ideal gas equation gives:

$$V = \frac{nRT}{P} = 0.043 \text{ mol } CO_2 \times \frac{(0.0821 \text{ L-atm/mol-K})(310 \text{ K})}{740 \text{ torr}} \times \frac{760 \text{ torr}}{1 \text{ atm}}$$

Answer
1.1 L CO_2

Problem 9.13   What volume of oxygen measured at 40°C and 1 atm reacts completely with 5.0 g propane? The balanced equation for the reaction is:

$$C_3H_8(g) + 5\ O_2(g) \longrightarrow 3\ CO_2(g) + 4\ H_2O(g)$$

9.8   Real Gases

Thus far in this chapter, we have assumed that all gases are ideal and behave in accordance with the postulates of the kinetic molecular theory and the ideal gas equation. Under standard conditions of temperature and pressure, and also at higher temperatures and lower pressures, the behavior of most real gases such as oxygen, nitrogen, and carbon dioxide is that predicted by the gas laws and the kinetic molecular theory. It is for this reason that we study the properties of ideal gases. However, as the temperature of a gas is decreased, the kinetic energy of the molecules decreases, their movement becomes more sluggish, and the attractive forces that exist between real molecules play a larger role in determining the behavior of the sample. Likewise, if the pressure is increased and the volume decreased until the volume of the space between the molecules approximates the volume of the molecules themselves, the molecules can no longer act as the wholly independent particles postulated by the kinetic molecular theory.

Under these conditions of low temperature and high pressure, any attractive forces that exist between the molecules of the gas come into play. These attractive forces are **dipole–dipole interactions** and dispersion forces. **Dispersion forces,** also called **London** or **Van der Waal's forces,** are weak forces of attraction that exist between all molecules without regard to the polarity of the molecules. To understand the nature of these forces, we need to remember that, even though a molecule may have no permanent dipole, it does have a cloud of rapidly moving electrons. If this cloud is distorted, no matter how briefly, the molecule will then have a temporary negative charge at one end and a temporary positive charge at the other end (Figure 9.12). In other words, the molecule

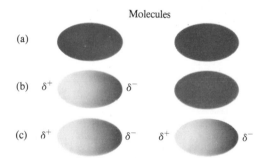

Molecules

(a)

(b) δ^+ δ^-

(c) δ^+ δ^- δ^+ δ^-

FIGURE 9.12 The development of temporary dipoles in molecules: (a) electron clouds with charge evenly dispersed; (b) temporary distortion of left cloud, causing a temporary dipole; (c) induced distortion of right cloud caused by presence of dipole in left cloud, also resulting in a temporary dipole.

has a temporary dipole. This temporary dipole can distort the electron clouds of nearby molecules so that they, too, have temporarily induced dipoles. The forces of attraction between the temporary partial positive charges on some molecules and the temporary partial negative charges on neighboring molecules are the dispersion forces.

Under standard conditions of temperature and pressure, molecules move freely without intermolecular attraction, as illustrated in Figure 9.13a. In Figure 9.13b, the molecules are moving more slowly, they are closer together, and

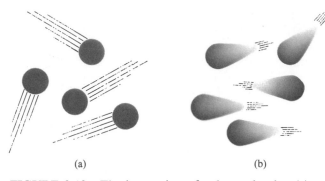

(a)   (b)

FIGURE 9.13   The interaction of polar molecules: (a) gas molecules move freely at STP, without interaction; (b) interaction occurs when the gas is at low temperature or high pressure, causing temporary dipoles.

they interact. The dipole in one molecule, regardless of whether it is real or temporary, interacts with the dipole of its neighbors. The lower the temperature and the closer the molecules are together (a result of higher pressure), the more effective are these dipole–dipole interactions in preventing the free movement of molecules required by the kinetic molecular theory. Gases that show these tendencies are said to be real gases, as opposed to ideal gases (those whose behavior is close to that predicted by the gas laws). Gases of low molecular weight and no polarity are the most ideal—for example, hydrogen and helium.

We expect the behavior of real gases to deviate more and more from the ideal as the polarity (either real or induced) and the molecular weight of the molecules increase. Molecular weight is a factor because the size and mass of a molecule increase as its molecular weight increases.

9.9 Summary

Samples of matter are either solid, liquid, or gaseous. Each state has characteristics of shape, compressibility, and density that are dependent on the bonds between the component particles of the sample. Each particle also has a kinetic energy ($KE = \frac{1}{2}mv^2$, where m = mass and v = velocity). A collection of particles will have a characteristic distribution of kinetic energy, the average of which is dependent on the temperature of the sample.

The kinetic molecular theory for gases describes the behavior of an ideal gas as a collection of very tiny molecules moving rapidly and freely in a very large volume. There are no attractive forces between these molecules. Collisions occur with no loss of kinetic energy whether the collision is with other molecules or with the walls of the container. The pressure a gas exerts is caused by collisions with the walls of the container. The temperature of a gas is proportional to the average kinetic energy of its molecules.

An ideal gas obeys certain laws that are predictable from the kinetic molecular theory. They are:

Boyle's Law: $P_1V_1 = P_2V_2$

Charles' Law: $V_1/T_1 = V_2/T_2$

Combined Gas Law: $P_1V_1/T_1 = P_2V_2/T_2$

Avogadro's Hypothesis and molar volume: 22.4 L at STP

Dalton's Law of Partial Pressures: $P_{Total} = P_1 + P_2 + P_3 + \cdots$

Ideal Gas Equation: $PV = nRT$

Application of these laws makes possible a wide variety of calculations.

Real gases differ from the ideal gas described by the kinetic molecular theory in that their molecules have real volume and there are attractive forces between the molecules due to permanent or temporary dipoles.

Key Terms

atmosphere (9.4)
Avogadro's Hypothesis (9.5D)
barometer (9.4)
Boyle's Law (9.5A)
Charles' Law (9.5B)
Combined Gas Law (9.5C)
compressibility (9.1D)
Dalton's Law of Partial Pressures (9.6)
dipole–dipole interactions (9.8)
dispersion forces (London, Van der Waal's) (9.8)
elastic collisions (9.3)
gas constant (9.5E)
gas density (9.5D)
ideal gas (9.3)
ideal gas equation (9.5E)

kinetic energy (9.2)
kinetic molecular theory (9.3)
manometer (9.4)
molar volume (9.5D)
most probable kinetic energy (9.2A)
normal boiling point (9.1)
normal melting point (9.1)
partial pressure (9.6)
Pascal (9.4)
pressure (9.4)
real gases (9.5E)
standard pressure (9.4)
standard temperature (9.4)
standard temperature and pressure (STP) (9.4)
torr (9.4)

Multiple-Choice Questions

For Questions 1–7 the following answers are possible.

    a. one mole of aluminum at STP

    b. one mole of neon at STP

    c. one mole of ammonia at STP

    d. one mole of carbon dioxide at STP

    e. All are the same.

MC1. Which sample occupies the smallest volume?

MC2. Which sample contains the most atoms?

MC3. In which sample do the particles have the highest average kinetic energy?

MC4. Which sample has the greatest mass?

MC5. Which sample has a density of 0.759 g/L?

MC6. In which sample do the molecules (or atoms in a monatomic sample) have the least attraction for one another?

MC7. Which sample has the highest melting point?

MC8. What volume of hydrogen measured at 0.95 atm and 27°C can be prepared by the reaction of 0.500 mol zinc with an excess of hydrochloric acid?
  a. 1.06 L   **b.** 11.2 L   **c.** 11.7 L   **d.** 13.0 L   **e.** 22.4 L

MC9. A 2.63-L sample of neon is collected at 730 torr and 26°C. What volume will the gas occupy at 1.10 atm and 10°C?
  a. 3.18 L   **b.** 16.5 L   **c.** 0.850 L   **d.** 646 L   **e.** 2.17 L

MC10. The volume of a gas sample at constant pressure is related to temperature T (in Kelvin) by the graph:

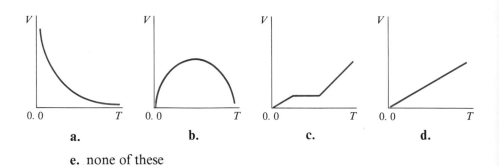

a.   **b.**   **c.**   **d.**

e. none of these

Problems

9.1 Characteristics of the Solid, Liquid, and Gaseous States

9.14. Why is it difficult to compress a liquid or a solid?

9.15. In describing the properties of gases, why must we speak of the "average" kinetic energy instead of just kinetic energy?

9.3 The Kinetic Molecular Theory

9.16. Use the kinetic molecular theory to explain:
  a. why a foul smelling substance is added to natural gas to aid in detecting leaks.
  b. why a baked potato sometimes explodes in the oven.

c. how an aerosol can works and why it sometimes ceases to work even though shaking indicates it still contains a liquid.
d. how hot-air ballooning is related to Charles' Law.

9.17. State the postulates of the kinetic molecular theory of gases. Explain how they predict Charles' and Boyle's Laws.

9.4 Measuring Gas Samples

9.18. The pressure exerted by a column of water 10 cm high with a cross section of 0.5 cm² is twice that of a column 5 cm high with a 1.0-cm² cross section. Explain why the two pressures differ.

9.19. In Section 9.4 we described how a barometer is made. What would happen to the height of the mercury column in Figure 9.6 if the closed end of the glass tube were cut off? Why?

2.0 atm, then to 4.0 atm, to 8.0 atm, and finally to 16.0 atm. Calculate the volume of the sample at each pressure. If the pressure continues to increase, will the volume ever reach zero?

9.23. Calculate the final pressure if 2.63 L of gas at 25°C and 1.00 atm pressure is allowed to expand to 8.45 L at the same temperature.

9.24. Why do aerosol cans carry the warning "do not incinerate"?

9.25. Calculate the density of ethane (C_2H_6) at STP.

*__9.26.__ What is the density of hydrogen at standard conditions? at 25°C and 0.925 atm?

9.27. Does the density of a gas increase or decrease as the pressure increases at constant temperature? as the temperature increases at constant pressure?

9.28. Complete the following table:

V_1	T_1	P_1	V_2	T_2	P_2
546 L	43°C	6.5 atm	_____	65°C	1.9 atm
43 mL	−56°C	865 torr	47.5 mL	_____	1.5 atm
4.2 L	234 K	0.87 atm	3.2 L	29°C	_____
1.3 L	25°C	1.89×10^4 Pa	_____	0°C	1.0 atm

9.29. Complete the following table:

Initial conditions			Final conditions		
V_1	T_1	P_1	V_2	T_2	P_2
6.35 L	10°C	0.75 atm	_____	0°C	1 atm
75.6 L	0°C	1 atm	_____	35°C	735 torr
1.06 L	75°C	0.55 atm	0.76 L	0°C	_____

9.20. Convert 0.895 atm to torr and to Pascals.

9.21. What is the pressure in atmospheres on a scuba diver at a depth of 50 ft? (Each 33 ft of water adds a pressure of 1 atm.)

9.5 The Gas Laws

9.22. A sample of gas has a volume of 1.0 L at 1.0 atm. The temperature is kept constant and the pressure is raised first to

9.30. A balloon filled with a 1.2-L sample of air at 25°C and 0.98 atm pressure is submerged in liquid nitrogen at −196°C. Calculate the final volume of the gas in the balloon. (The pressure remains constant.)

*__9.31.__ a. A 1.65-L sample of a gas is collected at 25°C and 450 mm Hg pressure. What

will be its volume at 40°C and 550 mm Hg pressure?

b. How many moles of gas are contained in the gas sample?

9.32. The density of liquid octane, C_8H_{18}, is 0.7025 g/mL. If 1.00 mL of liquid octane is changed to a gas at 100°C and 725 torr, what volume will the vapor occupy?

*9.33. Calculate the molecular weight of a gas if 3.03 g of the gas occupy 660 mL at 735 mm Hg and 27°C.

9.34. Calculate the volume that 1.1 g oxygen occupy at 2.0 atm and 5.0°C.

9.35. If a 156-g block of dry ice, $CO_2(s)$, is changed to a gas at 25°C and 740 torr, what volume will the gas occupy?

9.7 Stoichiometry Involving Gases

*9.36. Oxygen can be prepared in the laboratory by heating potassium chlorate. The balanced equation is:

$$2\ KClO_3 \longrightarrow 2\ KCl + 3\ O_2$$

What volume of oxygen would be prepared at 140°C and 737 torr by the decomposition of 635 g potassium chlorate?

9.37. What volume of hydrogen at 55°C and 0.95 atm can be prepared by the reaction of 1.5 g magnesium with excess hydrochloric acid? The balanced equation is:

$$Mg + 2\ HCl \longrightarrow H_2 + MgCl_2$$

Review Problems

9.38. A sample of gas collected at 38°C and 740 torr weighs 0.0630 g and occupies 26.2 mL. What is its volume at STP? What is its molecular weight?

9.39. An average pair of lungs have a volume of 6.5 L. If the air they contain is 21% oxygen, how many molecules of oxygen do they contain at 1 atm and 37°C?

9.40. What volume of air (21% oxygen) measured at 25°C and 0.975 atm is required to completely burn 3.42 g aluminum?

*9.41. A 0.673-g sample of aluminum is treated with an excess of hydrochloric acid. If

0.765 L hydrogen is produced at 20°C and 0.987 atm, what is the percent purity of the aluminum?

9.42. Gaseous hydrogen sulfide at 36°C and 1.113×10^5 Pa is bubbled through a solution of tin(II) chloride. Only 45% of the hydrogen sulfide reacts as the gas is bubbled through the solution. What volume of hydrogen sulfide must be bubbled through a solution of tin(II) chloride to form 3.46 g SnS [tin(II) sulfide]?

9.43. How many grams of sodium chloride must be decomposed to yield 65 L chlorine measured at 32°C and 723 torr?

· 10 ·

The Condensed States of Matter

When steam cools, it condenses to a liquid. When liquid water cools, it becomes solid ice. These two states, liquid and solid, are known as the condensed states. The characteristic properties of matter in these states were discussed briefly in Section 9.1. In this chapter, we will see how the behavior of matter in condensed states is related to, but is not the same as, the behavior of gases as described in Chapter 9. We will consider:

1. How the kinetic molecular theory of gases is modified to explain the behavior of molecules in the solid and liquid states.
2. The attractive forces that operate between molecules in these states.
3. The vapor pressure of liquids and solids; how it relates to the distribution of kinetic energy in a collection of molecules in these two states and to the boiling point and freezing point of a substance.
4. How changes in kinetic and potential energy are related to temperature changes and changes of state of a sample.
5. The unusual properties of water.

10.1 The Kinetic Molecular Theory Applied to the Condensed States

As we saw in Section 9.3, the kinetic molecular theory of gases has five postulates: (1) the volume occupied by a gas is largely empty space; by comparison, the real volume of the gas molecules is negligible; (2) there are no attractive forces between the molecules of a gas; (3) the gaseous molecules are in constant, random, rapid, straight-line motion; (4) the molecules collide frequently with one another and with the walls of the container, but no energy is lost by these collisions; and (5) the average kinetic energy of the molecules is proportional to the temperature of the sample.

These postulates do not hold completely for liquids and solids. We have already seen that, when the average kinetic energy of a gas sample decreases, the sample eventually reaches a temperature at which this kinetic energy is no longer great enough to overcome the attractive forces between the molecules, and they start to interact with one another. These forces become more effective as the temperature lowers, and the sample becomes first a liquid and then, at a lower temperature, a solid. In these states, the sample has the properties described in Section 9.1. For condensed states, the kinetic molecular theory must be modified. The properties of molecules in these states can be described as follows:

1. Instead of negligible volume, the molecules of a liquid or a solid occupy most of the volume of the sample. This fact is apparent from the lack of compressibility of matter in these two states.

2. There are appreciable attractive forces between the molecules; these forces keep the liquid or the solid together.

3. Although not moving as rapidly as in a gas, the molecules are still in constant, rapid motion. In a liquid this movement can be seen by gently placing a drop of ink (or dye) on the surface of a large amount of water. Without any stirring, the colored ink will be rapidly dispersed throughout the entire container as the colored molecules are buffeted about by the constantly moving water molecules (Figure 10.1). In Section 10.4, we will discuss movement of particles in a solid.

4. In a liquid the molecules collide without loss of kinetic energy; in a solid the particles are held so rigidly in place that they do not collide.

5. Just as in gases, each molecule has some kinetic energy and the average of these energies is proportional to the temperature of the sample. There is a distribution of kinetic energies in the sample similar to that in a gas sample (see Section 9.2).

10.2 Attractive Forces between Particles

The magnitude of the attraction of one particle for another is important in determining whether the substance containing those particles is a solid, a liquid, or a gas under normal conditions (20°C, 1 atm). These attractive forces also represent a part of the potential or stored energy of a sample. We know that when the **kinetic energy** of a sample changes, its temperature changes. When the **potential energy** of a sample changes, the temperature does not change. Instead the energy that is added or subtracted breaks or forms bonds. For example, when 6.02 kJ of energy are added to 1 mol (18 g) of ice at 0°C, there is no change in temperature, only a change in state from solid to liquid. The added energy counteracts the forces that held the water molecules in the rigid ice structure. It does not break the bonds within molecules, only those between molecules.

$$H_2O(s) + energy \longrightarrow H_2O(l)$$

The bonds between the atoms, ions, or molecules in a substance range from very strong to comparatively weak. The melting and boiling points of substances of similar molecular weight are a measure of the relative strength of their intermolecular, interatomic, or interionic bonds. The higher-melting substance contains stronger bonds between its particles; the lower-melting substance has comparatively weaker interparticle bonds.

A. Intermolecular Forces in Liquids

Low-melting solids and compounds that are liquids at room temperature are usually covalent. Ionic compounds almost always melt well above room temperature. (The melting point of sodium chloride, a typical ionic solid, is 801°C.) In samples of covalent compounds that are either low-melting or liquid, three types of **intermolecular forces** are possible. Two of these have already been

FIGURE 10.1 When a drop of a dye is placed on the surface of some water, the molecules of dye are buffeted by the moving molecules of water and dispersed throughout.

discussed: **dispersion forces** (Section 9.8) and **dipole–dipole interactions** (Sections 7.4 and 9.8). The third type of intermolecular force is known as hydrogen bonding.

Hydrogen bonding is a special kind of dipole–dipole interaction. It represents the attraction between a small partially negative atom (usually nitrogen, oxygen, or fluorine) and a partially positive hydrogen (usually bonded to another very electronegative atom like nitrogen, oxygen, or fluorine) in another molecule. We can illustrate the existence of hydrogen bonding by comparing the boiling points of the hydrides of several elements. (A **hydride** of an element is the compound that element forms with hydrogen; for example, ammonia is the hydride of nitrogen.) We know that the boiling point of a substance is a measure of its intermolecular forces. Figure 10.2 plots the boiling points of the hydrides of the elements in Groups IV, V, VI, and VII of the periodic table.

Elements of Group IV show the regular increase in boiling point that would be expected for a regular increase in molecular weight. For Groups V, VI,

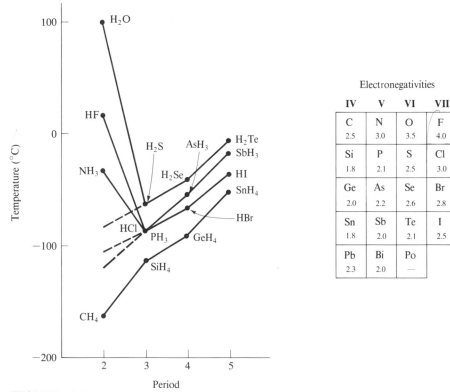

Electronegativities			
IV	**V**	**VI**	**VII**
C	N	O	F
2.5	3.0	3.5	4.0
Si	P	S	Cl
1.8	2.1	2.5	3.0
Ge	As	Se	Br
2.0	2.2	2.6	2.8
Sn	Sb	Te	I
1.8	2.0	2.1	2.5
Pb	Bi	Po	
2.3	2.0	—	

FIGURE 10.2   Molecular weight versus boiling point of the hydrides of Groups IV, V, VI, and VII. The solid line shows the actual boiling points; the dashed line, the predicted boiling point of the hydride of the lightest member of each group if there were no hydrogen bonding.

and VII, the hydrides of the three heaviest members of each group show a regular increase in boiling point, but the boiling point of the hydride of the lightest member of each group is much higher than would be predicted. Their expected boiling points are shown by the dashed continuation of the line plotting boiling points against molecular weight. For Group VI, the graph shows that the boiling point of water is approximately 200°C higher than would be predicted on the basis of molecular weight. Ammonia and hydrogen fluoride also boil at a higher temperature than expected. We postulate that the abnormally high boiling points of ammonia, water, and hydrogen fluoride are due to strong interactions between the molecules, so strong that these compounds behave more as aggregates of molecules than as single molecules. These aggregates are held together by hydrogen bonds. Within this group, the hydrogen bond strength is greatest in H—F, less in H—O, and even less in H—N bonds. Notice that this order corresponds to the order of decreasing electronegativity of F, O, and N.

Let us consider the nature of hydrogen bonding in water at the molecular level. We know that the O—H bond is polar covalent. The oxygen atom bears a partial positive charge.

When two water molecules come close to each other, a partially positive hydrogen atom of one water molecule interacts with the partially negative oxygen atom of the other to form a hydrogen bond. The hydrogen atom is apparently bonded to oxygen atoms in both molecules. We say that the two molecules of water are held together by a hydrogen bond. Figure 10.3 shows several water molecules. The hydrogen bonds between neighboring molecules are shown as dashed lines. The overall effect is a network of bonds with no separate molecules.

In closing this discussion of hydrogen bonding, we must emphasize that, even though hydrogen bonds are stronger than dipole–dipole interactions or dispersion forces, they are still much weaker than intermolecular or interionic bonds.

Hydrogen bonds play an important role in chemistry, particularly in the chemistry of biochemical molecules such as proteins and nucleic acids.

B. Interparticle Forces in Solids

High-melting solids are metals, ionic compounds, or network covalent compounds. This latter group includes diamonds, quartz, and silica, the hardest materials known. The interparticle forces in high-melting solids also fall into three categories:

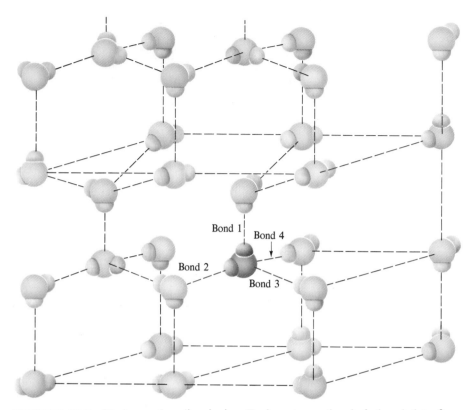

FIGURE 10.3 Hydrogen bonding in ice. Each water molecule is bonded to four others. Its two hydrogen atoms are attracted to the oxygen atoms in two other water molecules, and its oxygen atom attracts hydrogens in two more water molecules.

1. Network covalent solids

In Chapter 7 we discussed covalent bonds in small molecules. Most of these are low-melting, but the interatomic bonds in these small molecules are very strong. **Network covalent solids** (Figure 10.4a) can be thought of as huge covalent molecules of enormous molecular weight. In these molecules every atom is covalently bonded to several other atoms so as to form a network of covalent bonds (hence the name). Such compounds have very high melting points. Diamonds (mp 3550°C) and quartz (SiO_2; mp 1610°C) are typical network covalent solids. When these compounds melt, a regular covalent bond of shared electrons is broken, the same kind of bond that is broken when water is decomposed to hydrogen and oxygen.

2. Ionic bonds

In Chapter 7 we discussed ionic compounds, pointing out that the particles in an ionic compound are not atoms or molecules but ions. In the solid state, the ions are held in a regular, rigid structure (Figure 10.4b) by the electrostatic

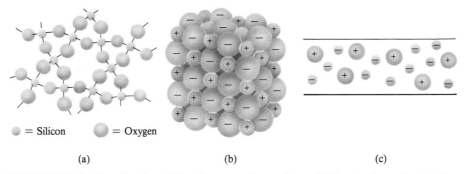

= Silicon   = Oxygen

(a)   (b)   (c)

FIGURE 10.4   Bonding in solids: (a) a network covalent solid (notice how closely this structure resembles the structure of hydrogen bonding in ice); (b) an ionic solid; (c) a metal.

forces of attraction between oppositely charged particles, as noted in Section 7.1A. Ionic solids usually have very high melting points, indicating the great strength of the electrostatic forces between ions.

3. Metallic bonding

The bonding in metals differs from that in ionic or covalent solids. We know that metals typically have one, two, or three valence electrons. One picture of a metallic solid shows these valence electrons as a fluid within which float the nuclei surrounded by their inner electrons (Figure 10.4c). The conductivity of metals is due to the movement of this "sea" of electrons. Hammering a metal into a thin sheet spreads out the fluid; similarly, drawing a metal into a thin wire is a rearrangement of the sea of electrons containing the nuclei into a thin stream. Metal fatigue, a problem in the aircraft industry and in nuclear power plants, is associated with metals losing their "fluid" nature and assuming a rigid structure. The melting points of metals vary over a wide range. Mercury is the lowest-melting ($-39°C$); tungsten is one of the highest-melting ($3410°C$).

Example 10.1   Two sets of compounds are given. Arrange each in order of increasing strength of interparticle force.

> **a.** CH_3CH_2OH $CH_3CH_2CH_2CH_3$ Ni   **b.** PH_3 MgO CH_3CH_3

Solution   **a.** The correct order is $CH_3CH_2CH_2CH_3$, CH_3CH_2OH, Ni. The first compound, $CH_3CH_2CH_2CH_3$, is a hydrocarbon. It contains no polar bonds. Between its molecules are only weak dispersion forces. The second compound, CH_3CH_2OH, contains hydrogen bonded to the very electronegative oxygen; its molecules would be held together by comparatively strong hydrogen bonds. The last substance, nickel, is a metal. Interatomic bonds in metals are strong.

b. The order is CH_3CH_3, PH_3, MgO. CH_3CH_3 is a hydrocarbon containing no polar bonds. Its intermolecular forces are very weak. In PH_3, the phosphorus is not sufficiently electronegative to allow the formation of hydrogen bonds but the phosphorus–hydrogen bond is still polar; there would be weak dipole–dipole interaction between the molecules. MgO is essentially ionic; the difference in electronegativity between magnesium and oxygen is appreciable. There will be strong interionic bonds between Mg^{2+} and O^{2-}.

Problem 10.1   Arrange the following substances in order of increasing interparticle forces. Name the forces.

        a. NaCl CH_3OH Cl_2 **b.** He   HCl   LiCl

10.3   Physical Properties of Liquids

A liquid is a collection of molecules held together by attractive forces strong enough to keep the molecules in a single unit but not strong enough to hold them in a rigid structure. The physical properties of liquids result from this structure.

A. Vapor Pressure

In Section 9.2 we discussed the kinetic energy in a collection of molecules and showed (in Figure 9.3) how the kinetic energies of the molecules were distributed. Figure 10.5 is similar in shape to Figure 9.3 but refers to the distribution of kinetic energies in a liquid.

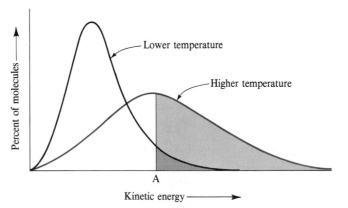

FIGURE 10.5 Distribution of kinetic energies among the molecules of a liquid at two different temperatures.

Let us assume that a molecule with kinetic energy greater than A in Figure 10.5 has enough energy to overcome the attraction of its neighboring molecules. If the molecule is on the surface of the liquid, it can escape from the body of the liquid and become a gaseous molecule. The number of molecules that have at least this much energy is represented by the area under the curve to the right of A. When these molecules escape the liquid, we say that they have evaporated or **vaporized;** the process is called **evaporation.** We can show this process by the equation

Liquid + energy $\longrightarrow$ gas

If these gaseous molecules are confined in the space above the liquid — that is, if the liquid sample is in a closed container — some of them, in their random gaseous motion, strike the surface of the liquid and are recaptured by it. This recapturing is called **condensation.**

When the rate of condensation equals the rate of evaporation, the molecules are in a state of dynamic equilibrium. When a system is in a state of **dynamic equilibrium,** two opposing processes are going on at the same rate. In this equilibrium, the opposing processes are evaporation and condensation, and

Rate of evaporation = Rate of condensation

For this equilibrium to be established, the system (the closed flask containing both liquid and vapor) must be at constant pressure. The molecules that have escaped are in the atmosphere over the liquid and, like all gases, exert a pressure. The partial pressure (see Section 9.6) they exert is known as the **vapor pressure** of the liquid.

Note that this equilibrium can take place only in a closed container or **closed system.** The vapor must accumulate enough to allow a normal distribution of energies before the rate of condensation will equal the rate of vaporization. In an open container, the vapor can escape and equilibrium will not be reached (see Figure 10.6). We recognize this fact when we store liquids in closed containers. Even ethyl ether, with a boiling point (bp 34.5°C) slightly above room temperature, can be stored at room temperature in a tightly closed container. Figure 10.6 shows the two processes of evaporation from an open container and equilibrium in a closed container.

Vapor pressure increases with temperature. Figure 10.5 shows why. At higher temperatures, more molecules have sufficient energy to escape the body of the liquid. Figure 10.7 illustrates how the vapor pressure of a liquid changes with temperature. In that figure, the vapor pressure of water (bp 100°C) and of carbon tetrachloride (bp 78°C) is plotted against temperature (°C). Notice how rapidly the vapor pressure increases. Notice too that, for each, the vapor pressure equals 760 torr at the boiling point. This fact then defines the **normal boiling point** of a liquid: the temperature at which the vapor pressure of a liquid equals 1 atm.

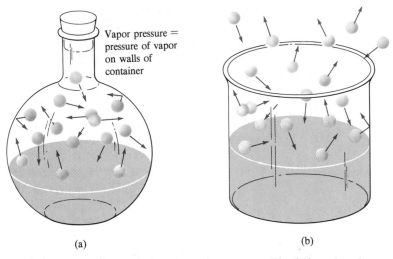

Vapor pressure =
pressure of vapor
on walls of
container

(a)   (b)

FIGURE 10.6   (a) Equilibrium between vapor and liquid in a closed container. (b) Nonequilibrium (evaporation) in an open container; equilibrium cannot be established because the vapor does not collect.

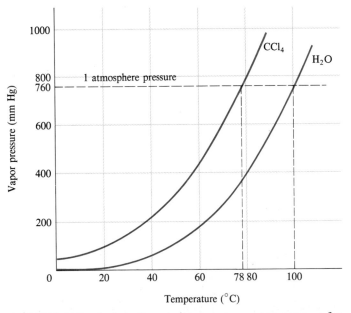

FIGURE 10.7   A plot of vapor pressure versus temperature for carbon tetrachloride (CCl_4) and for water (H_2O). Notice that the vapor pressure of these two liquids equals 760 torr (1 atm) at their normal boiling points (78°C for carbon tetrachloride and 100°C for water).

Although the normal boiling point of a substance is defined as the temperature at which its vapor pressure equals 1 atm, a liquid will boil whenever its vapor pressure equals the gaseous pressure in the space over the liquid. At high altitudes, such as on a mountain, liquids boil at lower temperatures than normal because the atmospheric pressure is less than 1 atm. Foods take longer to cook because the water boils at a lower temperature. Conversely, in a pressure cooker, the vapor is confined and builds up a pressure greater than that of the atmosphere. The boiling point of water in a pressure cooker is above 100°C because of this greater pressure, which means that the water can be hotter than 100°C and still remain a liquid. Foods cook in a shorter time because they are cooking in water that is at a temperature above 100°C.

B. The Specific Heat of Liquids

At the molecular level, the temperature of a liquid is proportional to the average kinetic energy of the molecules within the liquid; any change in temperature corresponds to a change in the average kinetic energy of the molecules.

The **specific heat** of a liquid is the amount of energy necessary to change the temperature of a one-gram sample by one degree Celsius (review Section 2.5B). The amount of energy necessary to cause this change differs between liquids. This difference means that each liquid has a unique specific heat. Given data for the mass of the sample, its temperature change (ΔT), and the energy added, the specific heat of the sample can be calculated:

$$\text{Specific heat} = \frac{\text{energy change}}{(\text{mass of sample})(\Delta T)}$$

Example 10.2 Calculate the specific heat of ethyl alcohol if 402 J are required to change the temperature of a 9.63-g sample from 21°C to 38°C.

Solution **Wanted**
Specific heat (in J/g°C) of ethyl alcohol
Given
9.63 g alcohol, 402 J, $\Delta T = 17$°C (temperature change from 21°C to 38°C)
Equation

$$\text{Specific heat} = \frac{402 \text{ J}}{(9.63 \text{ g})(17°\text{C})}$$

Answer
2.46 J/g°C

Problem 10.2 Calculate the specific heat of octane (C_8H_{18}) if 2.11×10^3 J are required to change the temperature of a 16.32-g sample of octane from 25°C to 83°C.

Example 10.3 . Suppose 655 J of energy are added to 16.5 g water at 25.0°C. Calculate the final temperature of the sample.

Solution   **Wanted**
Final temperature. [*Note:* The final temperature will be (25.0°C + ΔT), therefore we must calculate ΔT.]

Given
16.5 g water, 655 J

Conversion factors

Specific heat of water = 4.184 J/g°C

$$\text{Specific heat} = \frac{\text{energy}}{(\text{mass})(\Delta T)}$$

Rearranging this equation gives:

$$\Delta T = \frac{\text{energy}}{(\text{mass})(\text{specific heat})}$$

Equation

$$\Delta T = \frac{655 \text{ J}}{(16.5 \text{ g})(4.184 \text{ J/g°C})} = 9.49°C$$

Final T = 25.0°C + 9.49°C = 34.5°C

Problem 10.3   Calculate the final temperature when 1.573×10^3 J are added to 21.3 g water at 36.0°C.

C. The Molar Heat of Vaporization

When energy is added to a liquid, its temperature rises to the boiling point at a rate that is dependent on its specific heat. Energy added to a liquid at its boiling point does not change the temperature. Instead, this added energy counteracts the intermolecular forces in the liquid, and the molecules break apart one by one to vaporize and become gaseous. The amount of energy required to vaporize one mole of a liquid at its boiling point is its **molar heat of vaporization** (ΔH_{vap}). Table 10.1 shows the molecular weight, boiling point, specific heat, and molar heat of vaporization for several liquids. The liquids are listed in order of increasing polarity. Notice that this order is independent of molecular weight.

The molar heat of vaporization of a liquid can be calculated from experimental data, as the following example shows.

TABLE 10.1 Physical properties of liquids

Liquid	Molecular weight	bp (°C)	ΔH_{vap} (kJ/mol)	Specific heat (J/g°C)
carbon disulfide (CS$_2$)	76.1	46°	28.4	1.000
chloroform (CHCl$_3$)	119.4	61.7°	31.4	0.966
ethyl alcohol (C$_2$H$_5$OH)	46.1	78.5°	40.4	2.45
water (H$_2$O)	18	100°	40.7	4.18

Example 10.4   To vaporize 13.6 g benzene at its boiling point, 535 J are required. Calculate the molar heat of vaporization of benzene.

Solution   **Wanted**
The molar heat of vaporization in joules per mole ($\Delta H_{vap} = ?$ J/mol)

Given
7.47 kJ are required to vaporize 13.6 g benzene.

Conversion factor

1 mol benzene = 78.1 g benzene

Equation

$$? \text{ J/mol} = \frac{7.47 \text{ kJ}}{13.6 \text{ g}} \times \frac{78.1 \text{ g benzene}}{1 \text{ mol benzene}}$$

Answer
42.9 kJ/mol

Problem 10.4   Calculate the molar heat of vaporization of butane (C$_4$H$_{10}$) if 10.17 kJ are required to vaporize 26.5 g butane at its boiling point.

Example 10.5   The molar heat of vaporization of octane (C$_8$H$_{18}$) is 38.8 kJ/mol. Calculate the energy in joules necessary to vaporize 83.7 g octane at its boiling point.

Solution   **Wanted**
? J

Given
83.7 g octane

Conversion factors

1 mol octane = 114.2 g octane

ΔH_{vap} of octane = 38.8 kJ/mol = 38.8 × 10^3 J/mol

Equation

$$? \, J = 83.7 \text{ g octane} \times \frac{1 \text{ mol octane}}{114.2 \text{ g octane}} \times \frac{38.8 \times 10^3 \text{ J}}{1 \text{ mol octane}}$$

Answer
$28.4 \times 10^3 \text{ J}$

Problem 10.5   Calculate the heat required to vaporize 95 g of water at 100°C.

D. Surface Tension and Viscosity

The surface tension of liquid is the property that causes a liquid to have the smallest possible surface area. Within a liquid, attractive forces operate all around a molecule; on the surface, attraction is only into the body of the liquid (see Figure 10.8). This attraction into the liquid causes the molecules on the surface to minimize surface area, becoming more tightly bound together than those within the liquid.

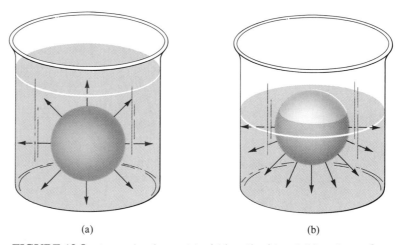

(a)   (b)

FIGURE 10.8   Attractive forces (a) within a liquid and (b) at the surface.

Surface tension measures this surface binding. Between different liquids of similar molecular weight, the difference in surface tension can be observed by comparing the different-sized drops formed by the liquids being compared. Large drops mean large intermolecular forces and large surface tension. Surface tension decreases as the temperature and kinetic energy of the molecules increase.

The resistance to flow, or the **viscosity** of a liquid, is another measure of the strength of its intermolecular forces. All liquids flow; some (like gasoline) flow easily, others (like diesel oil) flow very slowly. Resistance to flow means high viscosity and strong intermolecular forces just as did high surface tension. Viscosity decreases as temperature increases; recall the difference in rate of flow between cold and warm diesel oil.

10.4 Physical Properties of Solids

In Section 9.1 and again in Section 10.2B, we discussed the characteristics of solids. These properties suggest that the particles (ions, molecules, or atoms) in a solid occupy fixed positions from which they cannot easily move. This orderly arrangement is called the crystal structure or **crystal lattice** of the solid. Strong attractive forces between the particles keep them in this arrangement. Even so, as suggested in Figure 10.9, each particle in a solid has some kinetic energy and

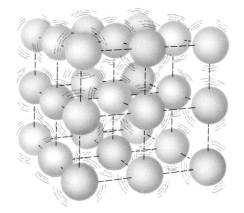

FIGURE 10.9 The crystal structure of a solid, showing the movement of the component ions, atoms, or molecules within their assigned space.

is in constant motion within its space in the solid (unless the solid is at a temperature of absolute zero, at which all motion ceases). The ions, atoms, or molecules of which the solid is composed vibrate, rotate, and even move around within their assigned space in the crystal structure.

As in gas and liquid samples, particles in a solid have a distribution of kinetic energies that depends on the temperature of the sample. At every temperature, some particles have enough energy to overcome the forces that hold them in place and escape from the solid as a vapor. Hence each solid, like each liquid, has a vapor pressure that increases as the temperature increases. The process by which atoms or molecules go directly from the solid state to the gaseous state is called **sublimation.** Solids that sublime have unusually high vapor pressures.

Dry ice (solid carbon dioxide) is a familiar example of a solid that sublimes. Solid water (ice) sublimes. If you live in a cold, dry climate, you may have noticed that ice disappears (actually it sublimes to a colorless gas) at temperatures below 0°C. Moth balls sublime; they disappear after protecting woolens during a hot summer.

As a solid is heated, its change in temperature depends on its specific heat. The solution of specific heat problems is the same whether for a liquid (Section 10.3B) or a solid (Section 2.5B).

A solid melts (changes to a liquid) when the average kinetic energy of its particles is high enough to overcome the attractive forces between them. By definition, the **melting point** of a solid is the temperature at which the liquid and solid states are in equilibrium at a pressure of 1 atm. We show this equilibrium in equation form by writing "solid" as a reactant and "liquid" as a product and connecting them with a double (equilibrium) arrow:

$$\text{solid} \underset{\text{freezing}}{\overset{\text{melting}}{\rightleftharpoons}} \text{liquid}$$

During melting, both solid and liquid are present until all the solid is converted to liquid. Conversely, during freezing, solid and liquid are present until all the liquid is converted to solid. These processes are indicated by writing "melting" over the forward equilibrium arrow and "freezing" under the reverse equilibrium arrow. Because freezing is the reverse of melting, the freezing point of a substance is the same as its melting point.

The **molar heat of fusion**, ΔH_{fus}, of a substance is the amount of heat that must be supplied to convert one mole of that substance from a solid to a liquid at its melting point. Calculations involving heats of fusion are similar to those involving heats of vaporization.

Example 10.6 The molar heat of fusion of carbon tetrachloride (CCl_4) is 3.26 kJ/mol. How many joules must be supplied to 34 g solid carbon tetrachloride at its melting point to change the sample to a liquid?

Solution

Wanted
? J

Given
34 g CCl_4

Conversion factors

$$\Delta H_{fus} = \frac{3.26 \text{ kJ}}{\text{mol}}$$

1 mol CCl_4 = 153.8 g CCl_4

Equation

$$? \text{ J} = 34 \text{ g CCl}_4 \times \frac{1 \text{ mol CCl}_4}{153.8 \text{ g CCl}_4} \times \frac{3.26 \text{ kJ}}{1 \text{ mol CCl}_4}$$

Answer
0.72 kJ

Problem 10.6   How many joules are required to melt 15 g ice at 0°C? The molar heat of fusion of water is 6.02 kJ/mol.

10.5 ## Transitions from the Solid through the Liquid to the Gaseous State

We have now discussed the properties of matter in its three states: solid, liquid, and gas. In each case, we have talked about the average kinetic energy of the molecules in that state and about changes in average kinetic energy as the temperature is changed. We have also talked about changes from one state to another.

The energy necessary to transform a substance from a solid below its melting point to a gas above its boiling point can be determined by calculating the energy required for each change involved and summing these energies. Figure 10.10 plots the energy required for such a series of changes. In this change, pressure was held constant at 1 atm, so the melting and boiling points of the substance are the normal ones.

Note carefully the shape of the graph. Although the length and slopes of the segments differ between compounds, the general shape will stay the same. The temperatures for the horizontal segments are the melting point and the boiling point of the substance, and the lengths of these segments depend on its molar heat of fusion and molar heat of vaporization. The slopes of the other segments of the graph depend on the specific heats of the substance: the first segment on the specific heat of the solid, the second on the specific heat of the liquid, and the third on the specific heat of the gas.

Note that the sloping segments represent a change in temperature and therefore a change in kinetic energy, whereas the horizontal segments represent no change in temperature and therefore no change in kinetic energy. These segments do represent a change in the potential energy of the sample as bonds are broken.

The amount of energy required to change a sample from one state and temperature to another state and temperature can be calculated if the various physical constants are known for that substance.

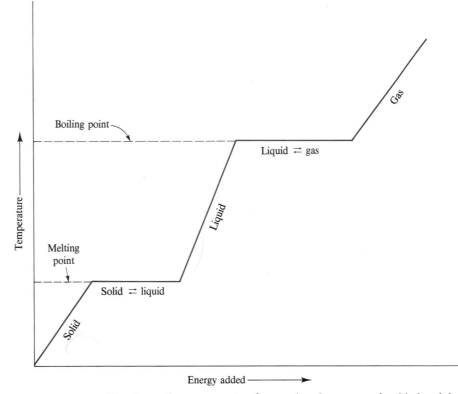

FIGURE 10.10   The change in temperature of a sample as heat energy is added and the sample changes from a solid to a liquid to a gas.

Example 10.7

How much energy (in joules) is needed to change 55.0 g ice at −5°C to steam at 110°C?

Specific heat of ice = 2.05 J/g°C

Specific heat of liquid water = 4.184 J/g°C

Specific heat of steam = 2.01 J/g°C

Molar heat of fusion of water = 6.02 kJ/mol

Molar heat of vaporization of water = 40.7 kJ/mol

Solution

The answer to this problem is obtained by calculating the energy change for each segment of the graph in Figure 10.10 and then finding the total energy change. A scaled-down version of the graph is shown here (Figure 10.11). In this solution, the letter of each step corresponds with a lettered segment of the figure.

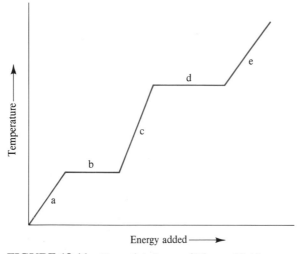

FIGURE 10.11 Essential shape of Figure 10.10.

a. To raise the temperature of ice from $-5\,^{\circ}$C to $0\,^{\circ}$C, a 5-degree change in the temperature of the solid, the equation is

$$? J = 55.0\ g \times \frac{2.05\ J}{g\,^{\circ}C} \times 5\,^{\circ}C = 0.564\ kJ$$

b. To melt the ice at $0\,^{\circ}$C:

$$? J = 55.0\ g \times \frac{1\ mol}{18.0\ g} \times \frac{6.02\ kJ}{1\ mol} = 18.4\ kJ$$

c. To raise the temperature of liquid water from $0\,^{\circ}$C to $100\,^{\circ}$C:

$$? J = 55.0\ g \times \frac{4.184\ J}{g\,^{\circ}C} \times 100\,^{\circ}\ C = 23.0\ kJ$$

d. To vaporize liquid water at $100\,^{\circ}$C:

$$? J = 55.0\ g \times \frac{1\ mol}{18.0\ g} \times \frac{40.7\ kJ}{1\ mol} = 124.4\ kJ$$

e. To heat water vapor (steam) from $100\,^{\circ}$C to $110\,^{\circ}$C:

$$? J = 55.0\ g \times \frac{2.01\ J}{g\,^{\circ}C} \times 10\,^{\circ}C = 1106\ J = 1.1\ kJ$$

The sum of these energy changes is the total amount of energy required by the change:

$$? J = 0.564\ kJ + 18.4\ kJ + 23.0\ kJ + 124.4\ kJ + 1.1\ kJ$$
$$= 167.5\ kJ$$

Correcting to three significant figures:

$$? J = 168 \text{ kJ}$$

Problem 10.7   Calculate the number of joules released when 60 g ice melt at 0°C and are raised to room temperature (25°C).

Example 10.8   Using the physical constants given in Example 10.7, calculate the final temperature if 8.11 kJ are added to 15.9 g ice at 0°C.

Solution   **Step 1.** How much of the available energy will be required to melt the ice? The ΔH_{fus} of water is 6.02 kJ/mol. To melt the ice, the equation is:

$$? \text{ kJ} = 15.9 \text{ g} \times \frac{1 \text{ mol}}{18.0 \text{ g}} \times \frac{6.02 \text{ kJ}}{1 \text{ mol}} = 5.32 \text{ kJ}$$

Step 2. How much energy remains to warm the liquid?

$$8.11 \text{ kJ} - 5.32 \text{ kJ} = 2.79 \text{ kJ}$$

Step 3. By how many degrees will 2.79 kJ raise the temperature of 15.9 g water?

$$°C = \frac{2.79 \text{ kJ}}{(15.9 \text{ g})(4.184 \text{ J/g°C})} \times \frac{1000 \text{ J}}{1 \text{ kJ}} = 41.9°C$$

Step 4. Calculate the final temperature:

$$°C = 0°C + 41.9°C = 41.9°C$$

Problem 10.8   Calculate the final temperature if 7.93×10^3 J are added to 19.3 g ice at −10°C.

Both of these examples show clearly that much more energy is involved in a change of state than in a change in temperature.

10.6   The Uniqueness of Water

Before closing this chapter, let us consider briefly the properties of water. Table 10.2 summarizes these properties. The same table also lists the properties of methane, a compound of similar molecular weight (therefore expected to be similar to water in many physical properties), and of hydrogen sulfide (the hydride of sulfur), an element in the same column of the periodic table as oxygen. Our study of the periodic table showed that the compounds water and hydrogen sulfide can be expected to be quite similar. Notice that both methane and hydrogen sulfide are gaseous under normal conditions. Water, on the other hand, is a liquid under normal conditions and a solid or gas at temperatures not far from normal.

TABLE 10.2   A comparison of water with methane (similar molecular weight) and hydrogen sulfide (a hydride of an element similarly located in the periodic table)

	Methane, CH_4	Water, H_2O	Hydrogen sulfide, H_2S
Formula weight	16.0	18.0	34.1
Melting point	$-182°C$	$0°C$	$-85°C$
Boiling point	$-16°C$	$100°C$	$-61°C$
Density: solid (g/cm^3)	—	0.999	—
liquid (g/mL)	—	1.000	—
gas (g/L)	0.71	0.80	1.5
Specific heat: solid ($J/g°C$)	—	2.05	—
liquid ($J/g°C$)	—	4.18	—
gas ($J/g°C$)	2.1	2.01	—
Heat of fusion (kJ/mol)	0.934	6.02	2.37
Heat of vaporization (kJ/mol)	8.90	40.7	18.7
Toxicity	high	none	very high
Odor	bad	none	bad
Usefulness as solvent	none*	excellent	none*

* Both would be useful solvents below their boiling points but, as both melt well below room temperature and are highly toxic, they have limited use as solvents.

The physical state of water at normal temperature can by and large be attributed to the polarity of its molecules and the strength of its hydrogen bonds. Methane is neither polar nor hydrogen bonded. Hydrogen sulfide is polar, but much less so than water, and is not hydrogen bonded.

What are some of the special properties of water? It is an excellent solvent. It dissolves ionic compounds, polar compounds, and low-molecular-weight nonpolar compounds. The implications of this property, particularly in biological processes, are enormous.

The high specific heat and heat of vaporization of water keep most of our planet temperate. Consider how much less the daily temperature fluctuates near large bodies of water than in deserts. Our bodies evaporate water (sweat) in order to maintain their constant temperature.

The small difference in density between ice and water is of far-reaching importance. Ice is slightly less dense and floats on water. Lakes freeze from the top down, not only allowing a haven for fish in freezing weather but allowing the ice to melt in summer.

10.7   Summary

Substances are liquids or solids if the attractive forces between their particles are strong enough to overcome their motion (their kinetic energy). These forces are, in order of increasing strength, dispersion forces (London forces), dipole–dipole interaction, and hydrogen bonds. In addition to these forces, the particles in solids may be bonded to one another by ionic bonds, network covalent bonds, or metallic bonds.

A liquid has a vapor pressure that increases with temperature. When the vapor pressure of a liquid equals atmospheric pressure, the liquid boils. When the pressure of the atmosphere is 1 atm, the temperature at which a liquid boils is its normal boiling point.

Vapor pressure in the solid state is less than that in the liquid state. When the vapor pressures are equal, the solid melts. Within a solid, the particles are joined by rigid bonds that define the crystal lattice of the solid.

Each substance has a specific heat for each physical state in which it occurs. Specific heat measures the rate of change of kinetic energy within the sample. Each substance also has a heat of fusion and a heat of vaporization that measure the strength of the attractive forces between the particles. The energy necessary to cause a change in temperature and/or a change in state can be calculated.

Key Terms

closed system (10.3A)
condensation (10.3A)
crystal lattice (10.4)
dipole–dipole interaction (10.2A)
dispersion forces (10.2A)
dynamic equilibrium (10.3A)
evaporation (10.3A)
hydride (10.2A)
hydrogen bonding (10.2A)
intermolecular forces (10.2A)
ionic bonds (10.2B2)
kinetic energy (10.2)
melting point (10.4)

metallic bonding (10.2B3)
molar heat of fusion (10.4)
molar heat of vaporization (10.3C)
network covalent solids (10.2B1)
normal boiling point (10.3A)
potential energy (10.2)
specific heat (10.3B, 10.4)
sublimation (10.4)
surface tension (10.3D)
vaporized (10.3A)
vapor pressure (10.3A)
viscosity (10.3D)

Multiple-Choice Questions

MC1. Which of the following statements correctly describe the normal boiling point of a liquid?

1. It is the same temperature everywhere on the Earth.

2. It depends on the attractive forces between the particles (molecules, ions, or atoms) of which the substance is composed.

3. It is the temperature at which the vapor pressure of the liquid equals 1 atm.

4. No molecule (or ion or atom) of the liquid becomes gaseous until the liquid boils.

a. none   **b.** all   **c.** 1, 2, and 3   **d.** 3 and 4   **e.** 2 and 3

The possible answers to Questions 2, 3, and 4 are:

        a. solid   **b.** liquid   **c.** real gas   **d.** ideal gas
        e. The statement is true for all states of matter.

MC2. At 20°C and 1 atm, in which state do the particles making up a sample occupy the smallest percent of the total volume?

MC3. In which is the average kinetic energy of the particles proportional to the Kelvin temperature?

MC4. At 20°C and 1.0 atm, in which state would you expect to find calcium chloride?

For Questions 5 and 6, five substances are listed in one column and their melting points are listed in a second column, but not in the same order.

        a. sodium bromide, $NaBr$ 3652°C
        b. heptane, C_7H_{16} -46°C
        c. graphite, C -91°C
        d. hexyl alcohol, $C_6H_{13}OH$ 321°C
        e. cadmium, Cd 747°C

MC5. Which substance melts at -91°C?

MC6. How many are solids at room temperature (20°C)?
        a. 0   **b.** 1   **c.** 2   **d.** 3   **e.** 4

MC7. The specific heat of silver is 0.238 J/g°C. How much energy is needed to raise the temperature of 1.5 g silver from 22°C to 91°C?
        a. 430 J   **b.** 25 J   **c.** 320 J   **d.** 240 J   **e.** 43 J

MC8. The postulates of the kinetic molecular theory for gases are:
    1. Molecules have no effective volume.
    2. There are no attractive forces between molecules.
    3. All collisions between molecules are elastic.
    4. Molecules are in constant, straight-line motion.
    5. The average kinetic energy of the molecules is proportional to the Kelvin temperature of the sample.
    Which of these postulates are true of a molecular liquid?
        a. all   **b.** 2, 3, 4, and 5   **c.** 4 and 5   **d.** 1, 4, and 5
        e. 2, 3, and 5

MC9. The normal boiling point of benzene is 80.1°C. At 0.85 atm, benzene will boil at a temperature:
        a. above 80.1°C.   **b.** below 80.1°C.   **c.** at 80.1°C.
        d. Not enough information is given to know.

MC10. Energy is added to 100 g water at -5°C until the temperature of the sample is 200°C. The physical constants of water involved in this change are:

Specific heat (solid) = 2.05 J/g°C
Specific heat (liquid) = 4.18 J/g°C
Specific heat (gas) = 2.01 J/g°C

Molar heat of fusion = 6.02 kJ/mol
Molar heat of vaporization = 40.7 kJ/mol

Which of the following steps will require the largest amount of energy?
a. warming the solid
b. melting the solid at its melting point
c. warming the liquid
d. vaporizing the liquid at its boiling point
e. warming the vapor

Problems

10.2 Attractive Forces between Particles

*10.9. The melting point of several substances at 1 atm is given below. Suggest what type of intermolecular force may be operating in each substance.

Substance	mp (°C)
propane, C_3H_8	−190
gallium, Ga	29.8
potassium nitrate, KNO_3	334
sulfur dioxide, SO_2	−73
nitrogen oxide, NO	−164
chlorine, Cl_2	−101
lanthanum oxide, La_2O_3	2307

10.10. Molecular weights and boiling points are given for three pairs of gases. For each pair, explain why the boiling points differ and why the direction of that difference is as reported.

	Formula	Mol. wt.	bp (°C)
fluorine	F_2	38.0	−188
bromine	Br_2	159.8	58.8
hydrogen chloride	HCl	36.5	−84.9
chlorine	Cl_2	70.9	−34.6
ammonia	NH_3	17.0	−33.3
methane	CH_4	16.0	−164

10.11. Why do low-formula-weight ionic compounds usually have higher melting points than low-formula-weight covalent compounds?

10.3 Physical Properties of Liquids

10.12. The vapor pressure of tribromomethane is 10 mm Hg at 34°C, 40 mm Hg at 63°C, 100 mm Hg at 86°C, and 600 mm Hg at 143°C. Plot these values and predict the boiling point of tribromomethane.

*10.13. Calculate the specific heat of dichlorodifluoromethane if it requires 247 J to change the temperature of 36.6 g dichlorodifluoromethane from 0°C to 25°C.

10.14. Calculate the final temperature of 48.5 g water at 15°C to which 6.57 kJ are added.

10.15. The heat of vaporization of dichlorodifluoromethane is 19.7 kJ/mol. Calculate the energy necessary to vaporize 39.2 g of this compound. (The formula weight of this compound is 120.9.)

10.16. Calculate the energy necessary to vaporize 13.9 g chloromethane (formula weight = 50.5) if the molar heat of vaporization is 5.20 kJ/mol.

10.17. The specific heat of mercury is 0.0332 cal/g°C. Calculate the number of joules necessary to raise the temperature of one mole of mercury by 36°C.

10.4 Physical Properties of Solids
10.5 Transitions from the Solid through the Liquid to the Gaseous State

*10.18. How much energy (kJ) is needed to melt 25 g ice at 0°C and to raise its temperature to 85°C?

10.19. Compare the number of joules absorbed when 100 g ice at 0°C are changed to water at 37°C with the number absorbed when 100 g water are warmed from 0°C to 37°C.

10.20. Temperature is a measure of kinetic energy. At the melting point of a substance,
   a. is there a change in its kinetic energy? Explain.
   b. is there a change in temperature? Explain.

10.21. The specific heat of aluminum is 16.7 J/g°C. Calculate the energy necessary to change the temperature of 56 g aluminum from 36°C to 79°C.

10.22. Calculate the final temperature if 5783 J of energy are added to 54 g ice at 0°C.

10.23. Calculate the final temperature if 46.2 kJ are added to 74 g ice at −10°C. (See Table 10.2 for physical constants needed.)

10.24. Calculate the amount of energy required to change 12 g liquid water at 65°C to steam at 100°C. (See Table 10.2 for the necessary physical constants.)

10.25. In another text, the heat of fusion of methanol (CH_3OH) is given as 16 cal/g and the heat of vaporization as 262 cal/g. Calculate the molar heat of fusion and the molar heat of vaporization in joules per mole for methanol.

10.26. Using the data in Table 10.2, calculate the energy needed to vaporize 6.3 mol methane at its boiling point and to raise the temperature of the sample to 0°C.

10.27. a. An ice-cube tray is filled with water at 25°C and placed in the freezing compartment. The volume of the tray is 0.653 L. Calculate the energy change when this water freezes.
   b. Is this energy released or absorbed by the water?
   c. The tray is aluminum and weighs 245 g. The specific heat of aluminum is 16.7 J/g°C. How do these data affect your results in part a of this problem?

10.28. Energy of 167 J is required to change the temperature of 5.0 g solid acetone a total of 7 degrees. Calculate the specific heat of acetone.

*10.29. A piece of zinc weighing 75.1 g at a temperature of 100°C is dropped into 100 mL water at 20°C. What is the final temperature of the water? (The specific heat of zinc is 0.388 J/g°C.)

10.30. Butane, C_4H_{10}, propanol, C_3H_7OH, and sodium chloride are very close in molecular weight but very different in melting and boiling points. Predict which will be lowest and which will be highest. Justify your answer.

10.31. The density of liquid water at 0°C is 0.9999 g/cm³. The density of solid water (ice) at 0°C is 0.9168 g/cm³. How does the volume of a sample of water change as it freezes? Why do water pipes burst when the water in them freezes?

10.32. How many joules are released when 10 g steam at 100°C are condensed and cooled to body temperature (37°C)? How many joules are released when 10 g water at 100°C are cooled to 37°C? Why are steam burns more painful than hot-water burns?

Review Problems

10.33. A compound contains 92.2% carbon and the rest hydrogen. A 5.00-g sample of this compound occupies a volume of 1.90 L at 120°C and 0.811 atm. What is the molecular formula of this compound?

10.34. A metal X forms a chloride with the formula XCl_3. In forming this compound, 5.00 g of the metal react completely with 10.6 g hydrogen chloride. What is the metal?

10.35. What volume of oxygen measured at 86°C and 545 torr would be formed by the complete decomposition of 15.6 g potassium chlorate, $KClO_3$, to potassium chloride and oxygen?

10.36. Given the following physical constants for acetone, C_3H_6O, calculate the amount of energy necessary to raise the temperature of 8.5 g acetone from −110°C to +100°C.

Melting point = −95°C
Boiling point = +56°C
Specific heat (solid) = 2.21 J/g°C
Specific heat (liquid) = 2.13 J/g°C
Specific heat (gas) = 1.28 J/g°C
Molar heat of fusion = 97.9 J/mol
Molar heat of vaporization
= 32.0 kJ/mol

10.37. What is the final temperature of 345 g water to which 1.56 kJ energy are added? The water is originally at 25°C.

▪ **11** ▪

Solutions

Our discussions thus far have concentrated on the properties of pure substances: elements, compounds formed by the combination of elements, and the reactions of these pure substances. In actual experience, we do not often encounter pure substances. More often, the matter we see and use is a mixture. In this chapter we will discuss one particular kind of mixture: solutions.

In this chapter we will discuss the following:

1. Definitions of the components of a solution—the solute and the solvent.
2. The meaning of solubility, the equilibrium present in a saturated solution, and the effect of intermolecular forces, changing temperature, and changing pressure on solubility.
3. The quantitative ways of describing concentration—in particular, molarity.
4. The use of concentrations in stoichiometric calculations.
5. The reactions of ionic compounds in solution; net ionic equations and spectator ions.

11.1   The Characteristics of Solutions

A **solution** is a homogeneous mixture. One substance (the **solute**) has dissolved in another (the **solvent**). Although the term *solution* usually calls to mind a liquid solvent, both solute and solvent can be in any of the three states of matter. For example, a solid can be dissolved in a liquid (salt water), a liquid dissolved in a liquid (rubbing alcohol), a gas dissolved in a liquid (carbonated beverages), or a solid dissolved in another solid (alloys). As for terminology, if the solution is liquid and the solute is solid or gas, the liquid component is called the solvent. If both solute and solvent are liquids, the component present in the larger amount is usually considered to be the solvent.

Remember that a solution is **homogeneous.** Only one **phase** or physical state is visible. A cloudy liquid is not a solution, for the cloudiness means that a substance insoluble in the solvent has been suspended in the solvent. Milk is a suspension, not a solution. Tea is a solution.

11.2   Solubility

The **solubility** of a substance is the amount of that substance that will dissolve in a given amount of solvent. Solubility is a quantitative term. Solubilities vary enormously (recall Table 8.2). The terms *soluble* and *insoluble* are relative. A substance is said to be **soluble** if more than 0.1 g of that substance dissolves in 100 mL solvent. If less than 0.1 g dissolves in 100 mL solvent, the substance is said to be **insoluble** or, more exactly, **sparingly soluble.** The terms *miscible* and *immiscible* may be encountered when considering the solubility of one liquid in another. **Miscible** means soluble without limits; for example, alcohol is miscible with water. **Immiscible** and insoluble mean the same; oil is immiscible with water, as in oil and vinegar salad dressing (see Figure 11.1).

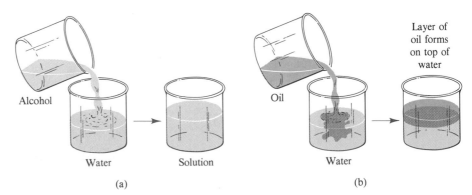

FIGURE 11.1   Soluble and insoluble. Alcohol is soluble in water; when added to water, it forms a clear solution. Oil is insoluble in water; when added to water, the two liquids form separate layers.

A. Determining Solubility

How is the solubility of a substance determined? A known amount of the solvent—for example, 100 mL—is put in a container. Then the substance whose solubility is to be determined is added until, even after vigorous and prolonged stirring, some of that substance does not dissolve. Such a solution is said to be saturated because it contains as much solute as possible at that temperature. In this saturated solution, the amount of solute is the solubility of that substance at that temperature in that solvent. Doing this experiment with water as the solvent and sodium chloride as the solute, we find that, at 20°C, 35.7 g of the salt dissolve in 100 mL water. The solubility of sodium chloride is, then, 35.7 g/100 mL water at 20°C. Sodium chloride is a moderately soluble salt. The solubility of sodium nitrate is 92.1 g/100 mL water at 20°C; sodium nitrate is a very soluble salt. At the opposite end of the scale is barium sulfate, which has a solubility of 2.3×10^{-4} g/100 mL water at 20°C. Barium sulfate is an insoluble salt. See Table 8.2 for the solubility of other compounds and Table 8.3 for the rules predicting the solubility of ionic compounds.

B. Saturated Solutions

A **saturated solution** is one in which the dissolved solute is in equilibrium with the undissolved solute. The container in Figure 11.2a holds a saturated solution of sucrose (cane sugar); at the bottom of the container is some undissolved sucrose. If we could see the individual molecules of sucrose in this solution, we would see that some molecules of sucrose are moving away from the solid crystals at the bottom of the container as they dissolve. The same number of molecules are coming out of solution to become part of the undissolved sucrose.

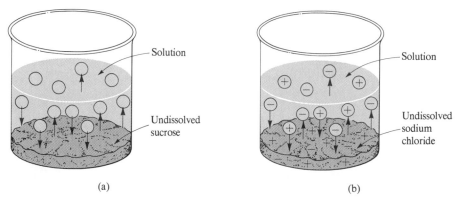

FIGURE 11.2 Equilibrium in solutions: (a) the equilibrium in a sucrose solution between dissolved and undissolved molecules; (b) the equilibrium in an ionic solution between dissolved ions and undissolved sodium chloride.

Dissolution (dissolving) and **precipitation** are occurring at the same rate, thereby satisfying the requirement for a dynamic equilibrium. This requirement was set forth in Section 10.3, where we discussed the equilibrium between liquid and vapor. We can express the equilibrium in the sucrose solution with the equation

$$\text{sucrose}(s) \rightleftharpoons \text{sucrose}(aq)$$

Two statements can be made about this solution: (1) the two processes dissolution and precipitation are going on at the same time, and (2) the number of molecules in the solution remains constant.

A saturated solution of an ionic compound is slightly different from that of a covalent compound like sucrose. The ionic compound dissolves and exists in solutions as ions; the covalent compound dissolves and exists in solutions as molecules. The equilibrium of sodium chloride with its ions in a saturated solution would be shown by the equation

$$\text{NaCl}(s) \rightleftharpoons \text{Na}^+(aq) + \text{Cl}^-(aq)$$

These equilibria are diagramed in Figure 11.2.

An **unsaturated solution** contains less solute than does a saturated solution. No equilibrium is present. When additional solute is added to an unsaturated solution, it dissolves. When additional solute is added to a saturated solution, the amount of dissolved solute does not increase, because the limit of solubility has already been reached. Adding more solute to a saturated solution simply increases the amount of undissolved solute.

Remember in discussing these solutions that solubility changes with temperature. (This factor will be discussed further in Section 11.2C2.) A solution that is saturated at one temperature may be unsaturated at a different temperature.

This spot is perhaps appropriate for a few comments on preparing solutions. Dissolution requires interaction between the molecules (or ions) of the solute and the molecules of the solvent. A finely divided solute will dissolve more rapidly than one that is in large chunks, because more contact is made between solute and solvent. Constant stirring increases the rate of dissolution, because stirring changes the particular solvent molecules that are in contact with undissolved solute. Because the solubility of solids and liquids generally increases with temperature, solids and liquids are often dissolved in warm solutes.

C. Factors Affecting Solubility

Many factors affect the solubility of one substance in another. We will discuss a few, in particular.

1. Forces between particles

One factor that affects solubility is the nature of the **intermolecular forces or interionic forces** in both the solute and the solvent. In Chapter 10, we listed the kinds of forces operating in solids and liquids. When one substance dissolves in another, the attractive forces in both must be overcome. The dissolving solute must be able to break up the aggregation of molecules in the solvent (that is, the hydrogen bonds between molecules or the dispersion forces between molecules in a **nonpolar solvent**), and the molecules of the solvent must have sufficient attraction for the solute particles to remove them one by one from their neighbors in the undissolved solute. If the solute is ionic, only a very **polar solvent** like water provides enough interaction to effect dissolution (see Figure 7.17). In those ionic compounds like barium sulfate that are called insoluble, the interaction between the ions is greater than can be overcome by interaction with the polar water molecules. If the solute particles are polar molecules, polar solvents such as alcohols can usually effect dissolution. If the solute is nonpolar, it may dissolve only in nonpolar solvents, not because polar solvent molecules are unable to overcome the weak dispersion forces between the solute molecules, but because these dispersion forces are too weak to overcome the dipole–dipole interaction between the solvent molecules.

A general rule is that *like dissolves like*. Ionic and polar compounds are most apt to be soluble in polar solvents like water or liquid ammonia. Nonpolar compounds are most apt to be soluble in nonpolar solvents, such as carbon tetrachloride, and hydrocarbon solvents like gasoline.

The solubility of gases in water depends a great deal on the polarity of the gas molecules. Those gases whose molecules are polar are much more soluble in water than are nonpolar gases. Ammonia, a strongly polar molecule, is very soluble in water (89.9 g/100 g H_2O); so is hydrogen chloride (82.3 g/100 g H_2O). Helium and nitrogen are nonpolar molecules. Helium is only slightly soluble (1.8×10^{-4} g/100 g H_2O), as is nitrogen (2.9×10^{-3} g/100 g H_2O).

Table 11.1 shows specific examples of different kinds of compounds and their relative solubilities in water, a polar solvent; in alcohol, a less polar solvent; and in benzene, a nonpolar solvent.

TABLE 11.1   Solubility and interparticle bonds

		Solubility in		
Kind of bonds	*Example*	*Water*	*Alcohol*	*Benzene*
ionic	sodium chloride	very soluble	slightly soluble	insoluble
polar covalent	sucrose (sugar)	very soluble	soluble	insoluble
nonpolar covalent	naphthalene	insoluble	soluble	very soluble

2. Temperature

Table 11.2 shows the solubility of several substances in water at 20°C and 100°C. Notice that the solubilities of solids and liquids usually increase as the temperature increases, but the solubilities of gases decrease with increasing temperature. This property of gases causes our concern for the fish population of lakes, oceans, and rivers threatened with thermal (heat) pollution. Fish require dissolved oxygen to survive. If the temperature of the water increases, the concentration of dissolved oxygen decreases, and the survival of the fish becomes questionable.

TABLE 11.2 Solubility and temperature

Compound	Type of bonding	Solubility (g/100 mL water)	
		At 20°C	At 100°C
sodium chloride	ionic	35.7	39.1
barium sulfate	ionic	2.3×10^{-4}	4.1×10^{-4}
sucrose	polar covalent	179	487
ammonia	polar covalent	89.9	7.4
hydrogen chloride	polar covalent	82.3	56.1
oxygen	nonpolar covalent	4.5×10^{-3}	3.3×10^{-3}

3. Pressure

The pressure on the surface of a solution has very little effect on the solubilities of solids and liquids. It does have an enormous effect on the solubility of gases. As the partial pressure (Section 9.6) of a gas in the atmosphere above the surface of a solution increases, the solubility of that gas increases. A carbonated beverage is bottled and capped under a high partial pressure of carbon dioxide so that a great deal of carbon dioxide dissolves. When the bottle is uncapped, the partial pressure of carbon dioxide above the liquid drops to that in the atmosphere, the solubility of carbon dioxide decreases, and the gaseous CO_2 comes out of solution as fine bubbles.

This same phenomenon is of concern in scuba diving. As divers descend, they are under ever-increasing pressure due to the increasing mass of water above them. Each 33 ft of water exert a pressure of 1 atm. As the diver descends, the pressure on the air in the lungs increases rapidly, and the blood dissolves more than normal amounts of both nitrogen and oxygen. As the diver comes up, the pressure decreases and the gases come out of solution. If the change in pressure is too swift, the bubbles of gas are released into the blood stream and tissues, instead of into the lungs. The bubbles cause sharp pain wherever they occur, most notably in the joints and ligaments. This dangerous and sometimes fatal condition is called *bends,* the name deriving from the temporary deformities caused by the affected diver's being unable to straighten his or her joints. Because helium, at all pressures, is less soluble than nitrogen, divers often breathe a mixture of helium and oxygen instead of air, a nitrogen–oxygen mixture.

The term **hyperbaric** means greater-than-normal atmospheric pressure. In hyperbaric medical treatments, patients are subjected to a pressure greater than atmospheric. This pressure increases the amount of oxygen dissolved in the blood. Treatment in hyperbaric, or high-pressure, chambers is of particular value for patients who are suffering from severe anemia, hemoglobin abnormalities, or carbon monoxide exposure or undergoing skin grafts, for such treatment increases the amount of oxygen transported by the blood to the tissues.

11.3 Expressing Concentrations of Solutions

A complete description of a solution states what the solute is and how much solute is dissolved in a given amount of solvent or solution. The quantitative relationship between solute and solvent is the **concentration** of the solution. This concentration may be expressed using several different methods, as discussed next.

A. Concentration by Mass

The concentration of a solution may be given as the mass of solute in a given amount of solution, as in the following statements: The northern part of the Pacific Ocean contains 35.9 g salt in each 1000 g seawater. The North Atlantic Ocean has a higher salt concentration, 37.9 g salt/1000 g seawater.

B. Concentration by Percent

The concentration of a solution is often expressed as **percent concentration** by mass or percent by volume of solute in solution. Percent by mass is calculated from the mass of solute in a given mass of solution. A 5%-by-mass aqueous solution of sodium chloride contains 5 g sodium chloride and 95 g water in each 100 g solution.

$$\text{Percent by mass} = \frac{\text{mass of solute}}{\text{mass of solution}} \times 100\%$$

Example 11.1   How many grams of glucose and of water are in 500 g of a 5.3%-by-mass glucose solution?

Solution   We know that 5.3% of the solution is glucose:

$$\frac{5.3 \text{ g glucose}}{100 \text{ g solution}} \times 500 \text{ g solution} = 26.5 \text{ g glucose}$$

The remainder of the 500 g is water:

$$500 \text{ g} - 26.5 \text{ g} = 437.5 \text{ g water}$$

Problem 11.1   What mass of sodium chloride is needed to prepare 315 g of a 0.9%-by-mass solution of sodium chloride in water?

If both solute and solvent are liquids, the concentration may be expressed as percent by volume. Both ethyl alcohol and water are liquids; the concentration of alcohol–water solutions is often given as percent by volume. For example, a 95% solution of ethyl alcohol contains 95 mL ethyl alcohol in each 100 mL solution.

$$\text{Percent by volume} = \frac{\text{volume of solute}}{\text{volume of solution}} \times 100\%$$

Example 11.2   Rubbing alcohol is an aqueous solution containing 70% isopropyl alcohol by volume. How would you prepare 250 mL rubbing alcohol from pure isopropyl alcohol?

Solution   We know that 70% of the volume is isopropyl alcohol:

$$\frac{70 \text{ mL isopropyl alcohol}}{100 \text{ mL solution}} \times 250 \text{ mL solution} = 175 \text{ mL isopropyl alcohol}$$

To prepare the solution, enough water is added to 175 mL isopropyl alcohol to form 250 mL solution.

Problem 11.2   Using pure ethyl alcohol and water, how would you prepare 1.0 L of a 40%-by-volume ethyl alcohol solution?

Because the density of liquids changes slightly as the temperature changes, a concentration given in percent by mass is accurate over a wider range of temperatures than is a concentration given in percent by volume. Sometimes a combination of mass and volume is used to express the concentration — the mass of solute dissolved in each 100 mL solution. Using this method, a 5% (wt/vol) solution of sodium chloride contains 5 g sodium chloride in each 100 mL solution.

C. Concentration in Parts per Million (ppm) and Parts per Billion (ppb)

The terms **parts per million (ppm)** and **parts per billion (ppb)** are encountered more and more frequently as we become aware of the effects of substances present in trace amounts in water and air, and as we develop instruments sensitive enough to detect substances present in such low concentrations. In discussing mass, parts per million means concentration in grams per 10^6 grams, or micrograms per gram. In discussing volume, parts per million may mean milliliters per cubic meter, or the mixed designation of milligrams per cubic meter. For parts per billion, the general trend is toward the use of micrograms per liter when discussing water contaminants, micrograms per cubic meter for air, and micrograms per kilogram for soil concentrations.

D. Concentration in Terms of Moles

The concentration of a solution may be stated as **molarity** (M), which is the number of moles of solute per liter of solution or the number of **millimoles** (mmol) (1 millimole $= 10^{-3}$ mole) per milliliter of solution.

$$\text{Molarity } (M) = \frac{\text{moles solute}}{\text{volume (liter) solution}} = \frac{\text{millimoles solute}}{\text{milliliter solution}}$$

A 6 M (say "six molar") solution of hydrochloric acid contains 6 mol hydrochloric acid in 1 L solution. A 0.1 M solution of sodium iodide contains 0.1 mol sodium iodide in 1 L solution.

The molarity of a solution gives a ratio between moles of solute and volume of solution. It can be used as a conversion factor between these two units in calculations involving solutions. As a conversion factor, it can be used two ways:

1. Moles/volume (L) states the number of moles in one liter of solution. This conversion factor is used in calculating the number of moles of solute in a given volume of solution.

2. Volume (L)/moles states that one liter contains some number of moles of solute. This conversion factor is used to calculate the volume of a solution that contains a given quantity of solute.

Example 11.3 How many moles of hydrochloric acid are in 200 mL of 0.15 M HCl?

Solution   **Wanted**
? mol HCl

Given
200 mL of 0.15 M HCl

Conversion factors

1 L of 0.15 M HCl contains 0.15 mol HCl—that is,

$$\frac{0.15 \text{ mol HCl}}{1.0 \text{ L solution}}$$

Equation

$$? \text{ mol HCl} = 200 \text{ mL of } 0.15 \ M \text{ HCl} \times \frac{1 \text{ L}}{1000 \text{ mL}} \times \frac{0.15 \text{ mol HCl}}{1 \text{ L of } 0.15 \ M \text{ HCl}}$$

Answer
0.30 mol HCl

Note that, each time the volume of a solution is stated, the concentration of the solution is given. This form may look confusing, but without this marking it is easy to forget which solution you are referring to.

Problem 11.3 How many moles of glucose are in 450 mL of 0.125 M glucose?

Example 11.4 What mass of sodium hydroxide (NaOH) is needed to prepare 100 mL of 0.125 M sodium hydroxide?

Solution   **Wanted**
? g NaOH

Given
100 mL of 0.125 M NaOH

Conversion factors

1 L of 0.125 M NaOH contains 0.125 mol NaOH—that is,

$$\frac{0.125 \text{ mol NaOH}}{1 \text{ L solution}}$$

1 mol NaOH = 40.0 g NaOH

Equation

$$? \text{ g NaOH} = 100 \text{ mL of } 0.125 \ M \text{ NaOH} \times \frac{1 \text{ L}}{1000 \text{ mL}}$$

$$\times \frac{0.125 \text{ mol NaOH}}{1 \text{ L solution}} \times \frac{40.0 \text{ g NaOH}}{1 \text{ mol NaOH}}$$

Answer
0.500 g NaOH

Problem 11.4 What mass of sodium chloride is needed to prepare 1.50 L of 0.125 M NaCl?

Example 11.5 What volume of 3.25 M sulfuric acid is needed to prepare 0.500 L of 0.130 M H_2SO_4?

Solution We are to prepare 0.500 L of 0.130 M H_2SO_4 by adding an amount of water to an amount of 3.25 M H_2SO_4. The moles of sulfuric acid in the final (more dilute) solution will be the same as the moles of sulfuric acid in the portion of the more concentrated solution. We can calculate the moles of sulfuric acid in the final dilute solution:

? mol H_2SO_4 in 0.500 L of 0.130 M H_2SO_4

$$= 0.500 \text{ L of } 0.130 \ M \ H_2SO_4 \times \frac{0.130 \text{ mol } H_2SO_4}{1 \text{ L solution}}$$

$$= 0.065 \text{ mol } H_2SO_4$$

This answer gives the moles of acid needed. We can calculate the volume of 3.25 M H_2SO_4 that would contain 0.065 mol H_2SO_4.

$$? \text{ L of } 3.25 \ M \ H_2SO_4 = 0.065 \text{ mol } H_2SO_4 \times \frac{1.0 \text{ L of } 3.25 \ M \ H_2SO_4}{3.25 \text{ mol } H_2SO_4}$$

$$= 0.020 \text{ L of } 3.25 \ M \ H_2SO_4$$

This answer gives the volume of concentrated acid that contains the moles of acid needed for the dilute solution. This volume of 3.25 M H_2SO_4 would be dissolved in 480 mL (500 mL − 20 mL) water to prepare 0.500 L of 0.130 M H_2SO_4. This problem is diagramed in Figure 11.3.

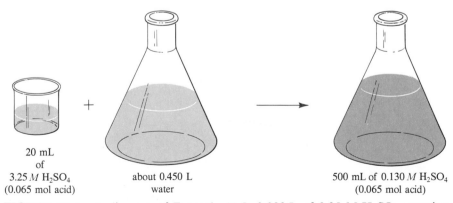

20 mL
of
3.25 M H_2SO_4 about 0.450 L   500 mL of 0.130 M H_2SO_4
(0.065 mol acid)   water   (0.065 mol acid)

FIGURE 11.3 A diagram of Example 11.5: 0.020 L of 3.25 M H_2SO_4 contains 0.065 mol acid. This solution is added to about 450 mL water, and enough water is added to increase the final volume to exactly 0.500 L. The resulting solution is 0.130 M H_2SO_4.

To prepare the solution, we would add 0.020 L of 3.25 M H_2SO_4 to an amount of water somewhat less than 480 mL and then add enough more water to raise the volume to exactly 0.500 L. Why do we first add acid to some water and then add more water until the correct volume is reached? First, when acid dissolves in water, a great deal of heat is generated. If the acid is added to a volume of water, the acid (being more dense than water) falls toward the bottom of the container; the heat is dissipated through the solution and does no harm. If the water is added to the acid, the water (being lighter than the acid) stays on top of the acid; the heat, generated in the small interface between the acid and the water, may cause an explosion. Having mixed the acid and water, we then add more water. Second, when two liquids are combined, the final volume is frequently not the arithmetic sum of the added volumes. Therefore, we start with a smaller volume than required and then dilute the solution up to the needed volume.

Problem 11.5   What volume of 6.0 M HCl is needed to prepare 275 mL of 0.255 M HCl?

Example 11.6   What volume of 6.39 M sodium chloride contains 51.2 mmol sodium chloride?

Solution

Wanted
? mL of 6.39 M NaCl

Given
51.2 mmol NaCl

Conversion factors

1 L of 6.39 M NaCl contains 6.39 mol NaCl

1 mL of 6.39 M NaCl contains 6.39 mmol NaCl

Equation

$$? \text{ mL of } 6.39\ M \text{ NaCl} = 51.2 \text{ mmol NaCl} \times \frac{1 \text{ mL of } 6.39\ M \text{ NaCl}}{6.39 \text{ mmol NaCl}}$$

Answer
8.01 mL of 6.39 M NaCl

Problem 11.6   What volume of 0.195 M HCl contains 50.0 mmol HCl?

Example 11.7   How do we prepare 75.0 mL of 0.96 M sulfuric acid from 18 M acid?

Solution

This problem is similar to Example 11.5. We are to prepare 75.0 mL of 0.96 M sulfuric acid by diluting 18 M sulfuric acid with water. We can calculate the millimoles of sulfuric acid in the final solution:

$$? \text{ mmol } H_2SO_4 = 75 \text{ mL} \times \frac{0.96 \text{ mmol } H_2SO_4}{1 \text{ mL of } 0.96\ M\ H_2SO_4}$$

$$= 72 \text{ mmol } H_2SO_4$$

We can calculate the volume of 18 M H_2SO_4 that will contain 72 mmol H_2SO_4:

$$? \text{ mL of } 18 \; M \; H_2SO_4 = 72 \text{ mmol } H_2SO_4 \times \frac{1 \text{ mL of } 18 \; M \; H_2SO_4}{18 \text{ mmol } H_2SO_4}$$

$$= 4.0 \text{ mL of } 18 \; M \; H_2SO_4$$

The solution would be prepared by adding 4.0 mL of 18 M H_2SO_4 to about 50 mL of water and then diluting that solution to exactly 75 mL.

Problem 11.7   What volume of 15 M acetic acid must be used to prepare 48 mL of 0.245 M acetic acid?

Table 11.3 lists several of the commonly used ways of expressing concentrations.

TABLE 11.3   Common units of concentration

	Solute		*Solvent*		*Solution*	*Comments*
Percent by weight	? g	+	? g	$\longrightarrow$	100 g	accurate, independent of temperature
Percent by volume	? mL	+	? mL	$\longrightarrow$	100 mL	used when solute is liquid; concentration varies slightly with temperature
Percent, weight/ volume	? g	+	100 mL	$\longrightarrow$	—	used in technical labs
Molarity (M)	moles		—		1 liter ⎫	used in chemical
Millimole/liter	10^{-3} mole		—		1 liter ⎬	calculations
Millimole/milliliter	10^{-3} mole		—		10^{-3} liter ⎭	
Parts per million (ppm)	mg		—		kg ⎫	used in
Parts per billion (ppb)	μg		—		kg ⎬	environmental studies

11.4   **Calculations Involving Concentrations**

The stoichiometry calculations done in Chapter 8 involved pure substances and samples whose mass could be determined by weighing. In Chapter 9, the reactant was often a gas. In these cases the mass of the sample could be determined if we knew its pressure, volume, and temperature by applying the ideal gas equation and the concept of molar volume. Now, by using the concentration of a solution as a conversion factor, we can extend stoichiometric calculations to include solutions.

Example 11.8   What weight of barium sulfate is precipitated by the addition of an excess of sulfuric acid to 55.6 mL of 0.54 M barium chloride?

Solution   **Equation**

$$BaCl_2 + H_2SO_4 \longrightarrow BaSO_4 + 2\ HCl$$

Given
55.6 mL of 0.54 M $BaCl_2$

Conversion factors

0.54 mol $BaCl_2$ in 1 L of 0.54 M $BaCl_2$
0.54 mmol $BaCl_2$ in 1 mL of 0.54 M $BaCl_2$
1 mol $BaCl_2$ forms 1 mol $BaSO_4$
1 mol $BaSO_4$ weighs 233.4 g

Arithmetic equation

$$? \text{ g } BaSO_4 = 55.6 \text{ mL of } 0.54\ M\ BaCl_2 \times \frac{0.54 \text{ mmol } BaCl_2}{1 \text{ mL of } 0.54\ M\ BaCl_2}$$

$$\times \frac{1 \text{ mmol } BaSO_4}{1 \text{ mmol } BaCl_2} \times \frac{10^{-3} \text{ mol } BaSO_4}{1 \text{ mmol } BaSO_4} \times \frac{233.4 \text{ g } BaSO_4}{1 \text{ mol } BaSO_4}$$

Answer
7.0 g $BaSO_4$

Problem 11.8   What mass of magnesium will react completely with 125 mL of 1.25 M HCl?

In the next example, two solutions are used. The concentration and volume of the first are known; only the concentration of the second is known. Its volume can be calculated.

Example 11.9   What volume of 0.154 M sodium hydroxide will completely react with 25.0 mL of 0.0952 M hydrochloric acid?

Solution   **Equation**

$$NaOH + HCl \longrightarrow NaCl + H_2O$$

Wanted
? mL of 0.154 M NaOH

Given
25.0 mL of 0.0952 M HCl

Conversion factors

> 1 mL of 0.0952 M HCl contains 0.0952 mmol HCl
>
> 1 mol NaOH reacts with 1 mol HCl
>
> 1 mL of 0.154 M NaOH contains 0.154 mmol NaOH

Arithmetic equation

> ? mL of 0.154 M NaOH = 25.0 mL of 0.0952 M HCl
>
> $$\times \frac{0.0952 \text{ mmol HCl}}{1 \text{ mL of } 0.0952 \ M \text{ HCl}} \times \frac{1 \text{ mmol NaOH}}{1 \text{ mmol HCl}}$$
>
> $$\times \frac{1 \text{ mL of } 0.0154 \ M \text{ NaOH}}{0.154 \text{ mmol NaOH}}$$

Answer

15.5 mL of 0.154 M NaOH

Problem 11.9 Calculate the molarity of a solution of hydrochloric acid if 15.0 mL of this solution reacts exactly with 26.2 mL of 0.126 M potassium hydroxide.

▬■▬

The next example illustrates the use of molarity in stoichiometric problems involving gases.

▬■▬

Example 11.10 What is the concentration of an aqueous solution of hydrochloric acid if 25.0 mL of the acid reacts with an excess of solid calcium carbonate to yield 0.307 L carbon dioxide, measured at 0.95 atm and 25°C?

Solution **Analysis of the problem**

1. The product of the reaction is a gas whose volume was not measured at standard conditions of temperature and pressure. We can calculate the number of moles of gas produced by using the ideal gas equation: $PV = nRT$.

2. Using the number of moles of carbon dioxide produced, we can calculate the number of moles of hydrochloric acid used. The moles of HCl are in 25.0 mL solution. Molarity is a ratio between moles of solute and volume of solution. By dividing the number of moles of HCl by the volume (L) of solution in which it was dissolved, we will obtain the molarity of the acid solution.

Calculations

1. Using the ideal gas equation, calculate the moles of CO_2 produced.

$$PV = nRT$$

$$0.95 \text{ atm} \times 0.307 \text{ L} = n \times 0.0821 \frac{\text{L-atm}}{\text{mol-K}} \times (25 + 273) \text{ K}$$

$$n = \frac{0.95 \text{ atm} \times 0.307 \text{ L}}{0.0821 \frac{\text{L-atm}}{\text{mol-K}} \times 298 \text{ K}} = 0.012 \text{ mol CO}_2$$

2. Carry out the stoichiometric calculation.

Equation

$$2 \text{ HCl} + \text{CaCO}_3 \longrightarrow \text{CO}_2 + \text{CaCl}_2 + \text{H}_2\text{O}$$

Wanted

The molarity of the acid solution or mol HCl/L solution

Given

0.012 mol CO_2

Conversion factor

2 mol HCl yield 1 mol CO_2

Arithmetic equation

$$? \, M \text{ HCl} = \frac{? \text{ mol HCl}}{1 \text{ L solution}}$$

$$= 0.012 \text{ mol CO}_2 \times \frac{2 \text{ mol HCl}}{1 \text{ mol CO}_2} \times \frac{1}{25 \text{ mL}} \times \frac{1000 \text{ mL}}{1 \text{ L}}$$

Answer

0.96 M HCl

Problem 11.10 What volume of carbon dioxide measured at 27°C and 0.93 atm is formed by the reaction of 22.5 mL of 0.105 M HCl with an excess of solid magnesium carbonate? The balanced equation for the reaction is:

$$\text{MgCO}_3 + 2 \text{ HCl} \longrightarrow \text{MgCl}_2 + \text{CO}_2 + \text{H}_2\text{O}$$

11.5   Titration

Laboratories, whether medical or industrial, are frequently asked to determine the exact concentration of a particular substance in a solution. For example, what is the concentration of acetic acid in a sample of vinegar? What are the concentrations of iron, calcium, and magnesium ions in a hard-water sample? Such determinations may be made using a technique known as titration.

In a **titration,** a known volume of a solution of unknown concentration is reacted with, or titrated by, a known volume of a solution of known concentration. By knowing the titration volumes and the mole ratio in which the solutes react, the concentration of the second solution can be calculated. The method used is similar to that used in Examples 11.8 and 11.9. The solution of unknown concentration may contain an acid (such as stomach acid), a base (such as ammonia), an ion (such as iodide ion), or any other substance whose concentration must be determined.

There are several requirements for analytical titrations:

1. The equation for the reaction must be known, so that the stoichiometric ratio can be used in calculations.

2. The reaction must be fast and complete.

3. When the reactants have combined exactly, there must be a clear-cut change in some measurable property of the reaction mixture. The occurrence of this change is called the **endpoint** of the reaction.

4. There must be a way of measuring accurately the amount of each reactant, whether that reactant is initially in solution or is a solid to be dissolved.

Let us discuss these requirements as they apply to a particular titration, that of a solution of sulfuric acid of known concentration with a sodium hydroxide solution of unknown concentration. The balanced equation for this acid–base reaction is:

$$2\ NaOH + H_2SO_4 \longrightarrow Na_2SO_4 + 2\ H_2O$$

Sodium hydroxide ionizes in water to form sodium ions and hydroxide ions; sulfuric acid ionizes to form hydrogen ions and sulfate ions. The reaction between hydroxide ions and hydrogen ions is rapid and complete; thus, the second requirement for an analytical titration is met.

What clear-cut change in property will occur when the reaction is complete? Suppose the sodium hydroxide solution is slowly added to the acid solution. As each hydroxide ion is added, it reacts with a hydrogen ion to form a water molecule. As long as unreacted hydrogen ions remain in solution, the solution is acidic. When the number of added hydroxide ions exactly equals the original number of hydrogen ions, the solution becomes neutral. If any extra hydroxide ions are added, the solution becomes basic. How will the experimenter know when the solution becomes basic? In Section 5.7D you learned about organic compounds called **indicators** that have one color in acidic solutions and another in basic solutions. If such an indicator is present in acid–base titration, it changes color when the solution changes from acidic to basic. Phenolphthalein is an acid–base indicator that is colorless in acidic solutions and pink in basic solutions. If phenolphthalein is added to the original sample of sulfuric acid, the solution is colorless and will remain so while hydrogen ions are in excess. After enough sodium hydroxide solution has been added to react with all of the hydrogen ions, the next drop of base will provide a slight excess of hydroxide ions and the solution will turn pink. Thus, there will be a visible and clear-cut indication of the occurrence of the endpoint. Table 11.4 lists three indicators that could be used in acid–base titrations.

TABLE 11.4   Common acid–base indicators

Indicator	Color in acid	Color in base
phenolphthalein	colorless	pink
methyl orange	red	yellow
bromothymol blue	yellow	blue

The requirement to measure accurately the volumes of solutions used is met by the use of **volumetric glassware**—in particular, burets—to measure the volumes of the solutions. Recall from Section 2.2 that the precision of a buret is one part per thousand. Figure 11.4 shows a typical titration setup.

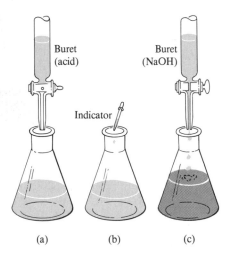

Buret (acid)   Buret (NaOH)

Indicator

(a)   (b)   (c)

FIGURE 11.4   A typical acid–base titration: (a) An exact volume of a solution of known concentration of acid is measured into a flask. (b) A few drops of phenolphthalein indicator are added. (c) The solution containing sodium hydroxide in an unknown concentration is added until a faint pink color is visible for a few seconds.

Data and calculations for a typical acid–base titration are shown in Table 11.5. Notice that three trials were run—a standard procedure to check the precision of titration.

Titration reactions are not always acid–base reactions. They may be oxidation–reduction reactions, precipitation reactions, or combination reactions. The endpoint may be determined by a change in color of an added indicator or by a change in color of the solution when the reaction is complete. In other

---

TABLE 11.5   Data from the titration of 0.108 M sulfuric acid with a solution of sodium hydroxide of unknown concentration

---

Equation

$$2\ NaOH + H_2SO_4 \longrightarrow Na_2SO_4 + 2\ H_2O$$

Data

Molarity of H_2SO_4:   0.108 M = 0.108 mmol/mL

	Trial I	*Trial II*	*Trial III*
Volume of 0.108 M H_2SO_4:	25.0 mL	25.0 mL	25.0 mL
Volume of NaOH solution:			
Buret readings: finish	34.12 mL	39.61 mL	35.84 mL
start	0.64 mL	6.15 mL	2.34 mL
Volume of NaOH at endpoint:	33.48 mL	33.46 mL	33.50 mL

Average volume used: 33.48 mL.

Arithmetic equation*

$$? \ M \ NaOH = 25.0 \ mL \ H_2SO_4 \times \frac{0.108 \ mmol \ H_2SO_4}{1 \ mL \ H_2SO_4} \times \frac{2 \ mmol \ NaOH}{1 \ mmol \ H_2SO_4} \times \frac{1}{33.48 \ mL \ NaOH}$$

Answer

0.161 M NaOH

---

* The first three factors in the equation give the millimoles of H_2SO_4 used. Dividing by the volume of acid used gives the molarity (mol/L). The 1 in the last factor is a dimensionless number.

titrations, the endpoint may be marked by a change in electrical conductivity of the reaction mixture, by the formation of a precipitate, or by a variety of other means.

Example 11.11   Three 25.00-mL samples of a sulfuric acid solution of unknown concentration were titrated to a phenolphthalein endpoint with 0.129 M potassium hydroxide solution. The buret readings for each of the trials were as follows:

	Trial I	*Trial II*	*Trial III*
Final reading:	34.32 mL	33.35 mL	35.15 mL
Initial reading:	1.13 mL	0.20 mL	2.07 mL

Calculate the molarity of the sulfuric acid solution.

Solution   The solution to this problem should be set up in the form of Table 11.5.

Equation

$$2\ KOH + H_2SO_4 \longrightarrow K_2SO_4 + 2\ H_2O$$

Data

Molarity of KOH: 0.129 M (0.129 mmol KOH/mL solution)

	Trial I	*Trial II*	*Trial III*
Volume of H_2SO_4 used:	25.00 mL	25.00 mL	25.00 mL
Volume of KOH used:	33.19 mL	33.15 mL	33.08 mL

(These data are obtained by subtracting the buret readings given in the body of the problem.)

Average volume of 0.129 *M* KOH used: 33.14 mL

Arithmetic equation

$$? \, M \, H_2SO_4 = \frac{\text{mmol } H_2SO_4}{\text{mL}}$$

$$= 33.14 \text{ mL of } 0.129 \, M \text{ KOH} \times \frac{0.129 \text{ mmol KOH}}{1 \text{ mL of } 0.129 \, M \text{ KOH}}$$

$$\times \frac{1 \text{ mmol } H_2SO_4}{2 \text{ mmol KOH}} \times \frac{1}{25.0 \text{ mL } H_2SO_4}$$

Answer

0.0855 *M* H_2SO_4

Problem 11.11   Calculate the concentration of an acetic acid solution using the following data. Three 25.0-mL samples of acid were titrated to a phenolphthalein endpoint with 0.121 *M* KOH. The buret readings were recorded:

	Trial I	*Trial II*	*Trial III*
Final reading:	21.31 mL	40.94 mL	30.72 mL
Initial reading:	1.35 mL	21.21 mL	10.93 mL

11.6   **Ionic Reactions in Solution**

Ionic compounds are different from those containing only covalent bonds. The latter exist as molecules whether in the pure state (crystals, liquids) or in solution. The former are a combination of two different kinds of particles. One kind (cations) carry a positive charge, and the second kind (anions) carry a negative charge. Sodium chloride is a typical ionic compound. It exists not as sodium chloride units but as sodium ions and chloride ions. In the solid state, these ions are arranged in a three-dimensional lattice, one kind of ion alternating with the other. In solution, the sodium ions and the chloride ions behave independently of each other.

Consider a particular example of this independent behavior. Suppose you have two solutions. One contains silver ions and an anion whose silver salt is soluble. The second contains chloride ions and a cation whose chloride is soluble. When the two solutions are combined, a white precipitate of silver chloride forms. We can write an equation for the reaction:

$$Ag^+ + Cl^- \longrightarrow AgCl(\downarrow)$$

The other ions are still present in the solution surrounding the precipitate (the supernatant liquid), but they do not participate in the reaction (see Figure 11.5a).

Earlier we would have written the equation differently. Assuming that nitrate was the anion with silver and sodium the cation with chloride, the equation would have been

$$AgNO_3 + NaCl \longrightarrow AgCl(\downarrow) + NaNO_3$$

Another way to write the equation shows the ionic compounds in solution as separate ions:

$$Ag^+ + NO_3^- + Na^+ + Cl^- \longrightarrow AgCl(\downarrow) + Na^+ + NO_3^-$$

This last equation shows clearly that only the silver and the chloride ions take part in the reaction; the sodium and nitrate ions do nothing. These nonparticipating ions are called **spectator ions,** meaning that they do not participate in the reaction — just as people in the stands at a football game are spectators, meaning that they do not participate in the game. The best equation for a reaction is one that shows only the participating ions; such an equation is called the **net ionic equation.** The net ionic equation for the reaction of silver ion with chloride is

$$Ag^+ + Cl^- \longrightarrow AgCl(\downarrow)$$

Consider another situation. When an acid reacts with a hydroxide, water and a salt are formed. All acids release hydrogen ions in solution. Hydroxides furnish hydroxide ions. Hydrogen ions and hydroxide ions react to form water, a covalent un-ionized molecule (see Figure 11.5b). The equation for this reaction is

$$H^+ + OH^- \longrightarrow H_2O$$

Previously we would have written this neutralization reaction using complete formulas. For example, the equation for the reaction of hydrochloric acid with sodium hydroxide would be

$$HCl + NaOH \longrightarrow NaCl + H_2O$$

Rewriting this equation to show the acid, the hydroxide, and the salt as ions — which is the way these compounds exist in solution — would give the equation

$$H^+ + Cl^- + Na^+ + OH^- \longrightarrow Na^+ + Cl^- + H_2O$$

This equation shows clearly that the hydrogen and the hydroxide ions react and that the sodium and chloride ions are spectator ions. They would not appear in the net ionic equation.

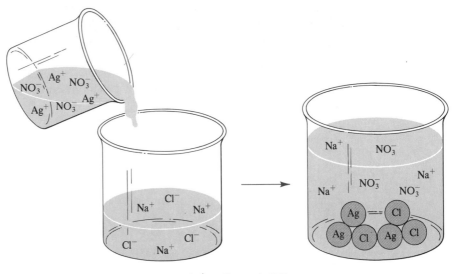

(a) $Ag^+ + Cl^- \longrightarrow AgCl(\downarrow)$

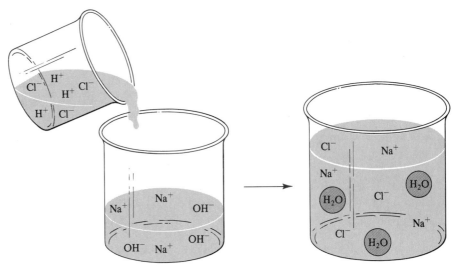

(b) $H^+ + OH^- \longrightarrow H_2O$

FIGURE 11.5   (a) When a solution of silver nitrate is added to a solution of sodium chloride, the silver ions combine with the chloride ions to form a precipitate of silver chloride. The sodium and the nitrate ions are nonparticipating spectator ions. (b) When hydrochloric acid is added to a solution of potassium nitrate, the hydrogen ions of the acid combine with the hydroxide ions of the potassium hydroxide to form molecules of water. The chloride and potassium ions are nonparticipating spectator ions.

Net ionic equations are also useful in showing displacement reactions. When magnesium is added to an acid, hydrogen gas and magnesium ions are formed. This statement gives us enough information to write the equation

$$Mg + 2\,H^+ \longrightarrow Mg^{2+} + H_2$$

Again, the identity of the anion of the acid is not important. It is present as a spectator ion and need not be shown in the equation.

Like all equations, net ionic equations must be balanced. Previously we have been concerned only with balancing equations by numbers and kinds of atoms. Now we must also be certain that equations are balanced by charge; that is, the total charge on the reactants must equal the total charge on the products.

Example 11.12   **a.** Write the net ionic equation for the formation of the insoluble compound copper(II) sulfide.

b. Write the net ionic equation for the formation of a precipitate of calcium phosphate.

For each reaction name the appropriate spectator ions.

Solution   **a.** The formula of copper(II) sulfide is CuS. The cation is Cu^{2+}; the anion is S^{2-}. The net ionic equation for the formation of CuS would show only these two ions.

$$Cu^{2+} + S^{2-} \longrightarrow CuS$$

Whatever anion is with Cu^{2+} and whatever cation is with S^{2-} in the reactants do not take part in the reaction and are spectator ions. Appropriate spectator ions might be nitrate as cation, sodium as anion. We chose these ions because the reactants must be soluble and both nitrates and sodium salts are always soluble.

b. The formula of calcium phosphate is $Ca_3(PO_4)_2$. The cation is Ca^{2+}; the anion is PO_4^{3-}. The net ionic equation is

$$3\,Ca^{2+} + 2\,PO_4^{3-} \longrightarrow Ca_3(PO_4)_2$$

Note that the net charge on the reactants is zero as is that on the product. Appropriate spectator ions would be nitrate as anion, sodium as cation.

Problem 11.12   Write the net ionic equations for the formation of the following insoluble compounds. Name the appropriate spectator ions. Be sure the equations are balanced by mass and charge.

   a. cobalt(II) sulfide   **b.** iron(II) hydroxide   **c.** strontium sulfate

Net ionic equations can be derived from complete equations. The complete equation is written first. It is examined to determine whether there has been a change in oxidation numbers. If there has been such a change, those ions or molecules that contain the atoms changing oxidation number are isolated for the net ionic equation. If no oxidation numbers change, the products are examined to find whether a covalent compound was formed. If so, the ions that combined to form that molecule are isolated for the net ionic equation. If all the products are ionic, and no oxidation number change occurred, one of the products must be a precipitate. The solubility rules in Table 8.3 are used to determine which product is the precipitate, and the ions it contains are isolated for the net ionic equation.

■

Example 11.13   Write the net ionic equation for the following:

    a. the reaction between silver nitrate and potassium sulfate

    b. the reaction of sulfuric acid with sodium hydroxide

    c. the displacement of bromine from sodium bromide by chlorine

Solution   **a.** The complete balanced equation for the reaction is

$$2\,AgNO_3 + K_2SO_4 \longrightarrow Ag_2SO_4 + 2\,KNO_3$$

Oxidation #: $+1, +5, -2$ $+1, +6, -2$ $+1, +6, -2$ $+1, +5, -2$

There have been no changes in oxidation numbers. Neither product is covalent. According to the solubility rules, silver sulfate is insoluble. Isolating the ions it contains gives the net ionic equation

$$2\,Ag^+ + SO_4^{2-} \longrightarrow Ag_2SO_4$$

This equation can be checked against the complete ionic equation

$$2\,Ag^+ + 2\,\cancel{NO_3^-} + 2\,\cancel{K^+} + SO_4^{2-} \longrightarrow Ag_2SO_4 + 2\,\cancel{K^+} + 2\,\cancel{NO_3^-}$$

in which we have crossed out the nonparticipating or spectator ions.

b. The complete balanced equation for the reaction is

$$H_2SO_4 + 2\,NaOH \longrightarrow Na_2SO_4 + 2\,H_2O$$

Oxidation #: $+1, +6, -2$ $+1, -2, +1$ $+1, +6, -2$ $+1, -2$

The oxidation numbers are shown and none has changed. Water is covalent and the other compounds are ionic. The net ionic equation for the reaction should be

$$H^+ + OH^- \longrightarrow H_2O$$

Writing the complete equation in ionic form gives:

$$2\,H^+ + \cancel{SO_4^{2-}} + 2\,\cancel{Na^+} + 2\,OH^- \longrightarrow 2\,H_2O + 2\,\cancel{Na^+} + \cancel{SO_4^{2-}}$$

and crossing out the spectator ions shows that our choice for the net ionic equation was correct.

c. The complete equation for the reaction is

$$2\ NaBr + Cl_2 \longrightarrow Br_2 + 2\ NaCl$$

Oxidation #: $+1, -1 \qquad 0 \qquad\qquad 0 \qquad +1, -1$

From the oxidation numbers, we see that both bromine and chloride have changed oxidation number. Isolating these ions gives the net ionic equation

$$2\ Br^- + Cl_2 \longrightarrow Br_2 + 2\ Cl^-$$

The complete ionic equation is

$$2\ \cancel{Na^+} + 2\ Br^- + Cl_2 \longrightarrow Br_2 + 2\ \cancel{Na^+} + 2\ Cl^-$$

Crossing out the spectator sodium ions gives the same net ionic equation.

Problem 11.13   Write the net ionic equations for the following reactions:

a. the reaction of aluminum chloride with sodium hydroxide to form a precipitate

b. the reaction of nitric acid with potassium hydroxide

c. the displacement reaction of metallic zinc with nickel(II) nitrate

---

11.7 Physical Properties of Solutions

The physical properties of a solution are different from those of the pure solvent. Many differences in physical properties are predictable if the solute in the pure state is **nonvolatile** — that is, if it has a very low vapor pressure. Sugar, sodium chloride, and potassium nitrate are examples of nonvolatile solutes. **Colligative properties** are those physical properties of solutions of nonvolatile solutes that depend only on the number of particles present in a given amount of solution, not on the nature of those particles. We will consider four colligative properties: vapor pressure lowering, boiling point elevation, freezing point depression, and osmotic pressure.

A. Vapor Pressure Lowering

At any given temperature, the vapor pressure of a solution containing a nonvolatile solute is less than that of the pure solvent (see Section 10.3A for a discussion of vapor pressure). This effect is called **vapor pressure lowering.** The solid line in Figure 11.6 is a plot of the vapor pressure of pure water versus temperature. The break in the curve at 0°C is the intersection of the curve of the vapor pressure of the solid with the curve of the vapor pressure of the liquid. The dashed line in Figure 11.6 is a plot of the vapor pressure of an aqueous solution

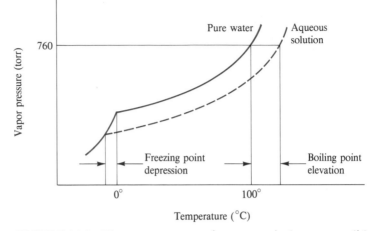

FIGURE 11.6   The vapor pressure of pure water is shown as a solid line; the vapor pressure of an aqueous solution is shown as a dashed line. Note the differences between the solution and the pure substance in melting point and boiling point.

of sugar versus temperature. Notice that the vapor pressure of the solution is always less than that of the pure solvent. What causes this difference?

The surface of a pure solvent (Figure 11.7a) is populated only by solvent molecules. Some of these molecules are escaping from the surface, and others are returning to the liquid state (see Section 10.3A). The surface of a solution is

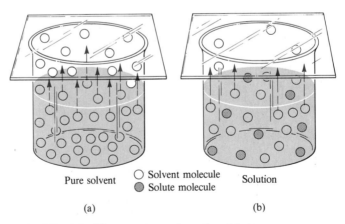

FIGURE 11.7   Vapor pressure lowering: (a) the vapor pressure of a pure liquid; (b) the vapor pressure of a solution. In (b), the number of solvent molecules on the surface of the liquid has been decreased by the presence of the solute molecules. Fewer solvent molecules can vaporize, and the vapor pressure is lower.

populated by two kinds of molecules; some are solvent molecules, others are solute molecules. Only the solvent molecules are volatile. They alone can escape to build up the vapor pressure of the solution. There are fewer solvent molecules on the surface of the solution than on the surface of the pure liquid. Fewer will vaporize and, as a consequence, the vapor pressure of the solution will be less than that of the pure liquid at the same temperature (see Figure 11.7b).

B. Boiling Point Elevation

The boiling point of a substance is the temperature at which the vapor pressure of the substance equals atmospheric pressure. A solution containing a nonvolatile solute, having a lower vapor pressure than the pure solvent, must be at a higher temperature before its vapor pressure equals atmospheric pressure and it boils. Thus, the boiling point of a solution containing a nonvolatile solute is higher than that of the pure solvent (see Figure 11.6). This effect is called **boiling point elevation.**

C. Freezing Point Depression

Recall that freezing point and melting point are two terms that describe the same temperature, the temperature at which the vapor pressure of the solid equals the vapor pressure of the liquid and at which the solid and the liquid are in equilibrium. Remember, too, that vapor pressure decreases as the temperature decreases. The vapor pressure of a solution is lower than that of the solvent, so the vapor pressure of a solution will equal that of the solid at a lower temperature than in the case of the pure solvent. Thus, the freezing point will be lower for a solution than for the pure solvent (see Figure 11.6). This effect is called **freezing point depression.** Remember that, just as it is the solvent that vaporizes when a solution boils, it is the solvent, not the solution, that becomes solid when a solution freezes. When a salt solution freezes, the ice is pure water (solid); the remaining solution contains all the salt.

Application of this principle leads us to add antifreeze (a nonvolatile solute) to the water in the radiators in our cars. We thus lower the freezing point of the solvent (water), and the solution remains a liquid even at subfreezing temperatures.

D. Osmosis and Osmotic Pressure

Osmosis and **osmotic pressure** depend on the ability of small molecules to pass through **semipermeable membranes** like a thin piece of rubber, a cell membrane, or a thin piece of plastic wrap. Think of the membrane as a sieve with very tiny holes. Solvent particles are small and can very easily pass through

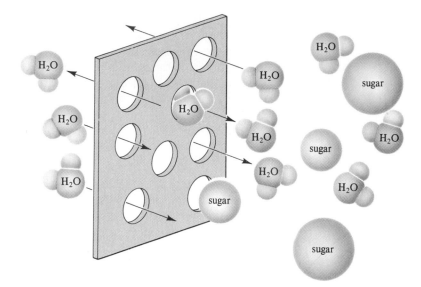

FIGURE 11.8   A semipermeable membrane allows small solvent molecules to pass through but prevents the passage of larger particles like those of a nonvolatile solute.

these holes; solute particles are larger and cannot pass through (Figure 11.8). When a semipermeable membrane separates a solution from pure solvent, solvent molecules move back and forth through the membrane, but not in equal numbers. More move from the pure solvent into the solution than from the solution into the solvent.

The movement of solvent molecules will continue to be uneven until the number of solvent molecules is the same on both sides of the membrane. The process is called osmosis. Figure 11.9 illustrates these points. In Figure 11.9a, different amounts of pure solvent (water) are separated by a semipermeable membrane. Water molecules from both sides move through the membrane until the pressure of solvent on both sides of the membrane is equal. This equality is indicated by the equal heights of the columns (Figure 11.9b).

In Figure 11.9c, solute molecules have been added on one side of the membrane, creating a solution. Now there are fewer solvent molecules next to this side of the membrane than there are on the side of the pure solvent. To overcome this difference, solvent molecules move more rapidly from the solvent side than from the solution side in an effort to equate these numbers. (Notice the bigger arrow meaning migration from the solvent side.) In Figure 11.9d, the height of the solution is greater than that of the solvent, but now the rate at which the solvent molecules pass through the membrane is the same from both sides. The difference in heights of the columns is proportional to the osmotic pressure of the original solution.

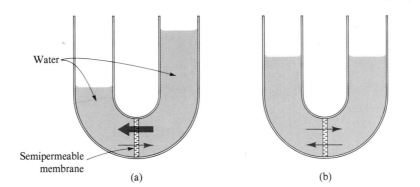

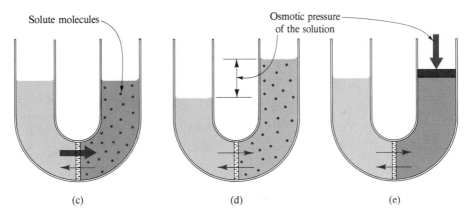

FIGURE 11.9   Osmosis and osmotic pressure. Differing amounts of pure solvent on either side of a semipermeable membrane (a) will, through osmosis, become equally divided on either side of the membrane (b). However, if solute molecules are added to one side (c), some of the solvent will migrate into the solution side, causing a difference in osmotic pressure (d). The difference in pressure can be counteracted by increased surface pressure on the solution side (e).

Osmotic pressure is also being measured in Figure 11.9e. Here, pressure is being applied to the surface of the solution. The osmotic pressure of the solution is the pressure that must be applied to the solution to prevent migration of solvent molecules from the more dilute (or pure solvent) side into the solution.

E. Differences between Colligative Properties of Solutions of Ionic and Molecular Compounds

For any solution, the amount that the vapor pressure is lowered, the freezing point depressed, or the boiling point elevated with respect to the properties of the pure solvent depends on the number of solute particles in solution, not on

TABLE 11.6   Colligative properties of some ionic and nonionic solutions

Solution	g solute/ 1000 g water	mp, °C	bp, °C	Osmotic pressure at 25°C
water	0	0	100	0
glucose solutions	18 (0.1 mol)	−0.19	100.05	2.4 atm
	36 (0.2 mol)	−0.36	100.10	4.8 atm
	180 (1.0 mol)	−1.8	100.52	24 atm
	360 (2.0 mol)	−3.6	101.04	73 atm
NaCl solution	6 (0.1 mol)	−0.33	100.08	4.1 atm
Na_2SO_4 solution	14 (0.1 mol)	−0.43	100.17	5.4 atm

the nature of those particles. Similarly, the osmotic pressure of a solution is dependent only on the number of solute particles, not on their nature. Table 11.6 shows the melting point (freezing point), boiling point, and osmotic pressure of several glucose solutions. The number of moles of glucose (therefore, the number of glucose molecules in a given amount of water) differs among these solutions. The greater the number of molecules of solute, the greater the difference between the properties of the pure solvent and those of the solution.

Table 11.6 shows the colligative properties of a solution of 0.1 mol sodium chloride in 1000 g water and of 0.1 mol sodium sulfate in 1000 g water. Notice that they differ from those of a solution with a molecular solute. One mole of glucose, a molecular compound, dissolves to yield one mole of particles; however, one mole of sodium chloride, an ionic compound, dissolves to yield two moles of particles — one mole of sodium ions and one mole of chloride ions. One mole of sodium sulfate (Na_2SO_4) dissolves to yield three moles of particles — two of sodium ions and one of sulfate ions. Because the number of particles determines the colligative properties of a solution, one mole of sodium chloride dissolved in a given amount of water causes approximately twice the change in colligative properties than does one mole of glucose dissolved in the same volume of water. One mole of sodium sulfate dissolved in the same amount of water causes approximately three times the change.

11.8   Summary

A solution is a homogeneous mixture. Although both solute and solvent can be in any physical state, the solvent is usually a liquid. The solubility of one substance in another depends on the attractive forces between its particles as well as those between the molecules of the solvent. Solubility also depends on temperature and, in the case of gases, on the partial pressure of the gas above the solution.

Concentration measures the amount of solute in a given amount of solvent or solution. Concentration can be expressed in terms of relative weights or volumes of the solute and solvent, in percent, or as molarity. Molarity means moles per liter of solution or millimoles per milliliter of solution. In environmental studies, concentration is usually given as parts per million (or billion). Molarity can be used in stoichiometric calculations, as for example in titration problems.

When reactions take place between ions in solution, the nonparticipating ions are called spectator ions. The reaction can be described by a net ionic equation.

The physical properties of a solution are different from those of the pure solvent because of the change in the vapor pressure caused by the presence of the solute. The number of dissolved particles rather than their mass or composition determines the extent to which the vapor pressure and freezing point are lowered and the boiling point and osmotic pressure of the solution are increased.

Key Terms

boiling point elevation (11.7B)
colligative properties (11.7)
concentration (11.3)
dissolution (11.2B)
endpoint (11.5)
freezing point depression (11.7C)
homogeneous (11.1)
hyperbaric (11.2C3)
immiscible (11.2)
indicator (11.5)
insoluble (11.2)
interionic forces (11.2C1)
intermolecular forces (11.2C1)
millimole (11.3D)
miscible (11.2)
molarity (11.3D)
net ionic equation (11.6)
nonpolar solvents (11.2C1)
nonvolatile solute (11.7)
osmosis (11.7D)

osmotic pressure (11.7D)
parts per billion (11.3C)
parts per million (11.3C)
percent concentration (11.3B)
phase (11.1)
polar solvent (11.2C1)
precipitation (11.2B)
saturated solution (11.2B)
semipermeable membrane (11.7D)
solubility (11.2)
soluble (11.2)
solute (11.1)
solution (11.1)
solvent (11.1)
sparingly soluble (11.2)
spectator ions (11.6)
titration (11.5)
unsaturated solution (11.2B)
vapor pressure lowering (11.7)
volumetric glassware (11.5)

Multiple-Choice Questions

Questions 1 and 2 have the following possible answers:

    a. solid carbon (diamonds)   **b.** sodium nitrate, $NaNO_3$
    c. glycerol, $C_3H_5(OH)_3$ **d.** butane, C_4H_{10} **e.** iron

MC1. Which substance is most likely to be soluble in a nonpolar solvent like gasoline?

MC2. Which substance will dissolve in water to form a solution that will carry an electric current?

MC3. When a solution of sulfuric acid is added to a potassium hydroxide solution, a reaction takes place. Which of the following is an acceptable equation for the reaction?
    a. $H_2SO_4 + 2\ KOH \longrightarrow KSO_4 + 2\ H_2O$
    b. $2\ HSO_4 + K(OH)_2 \longrightarrow K(SO_4)_2 + 2\ H_2O$
    c. $H_2SO_4 + 2\ KOH \longrightarrow K_2SO_4 + H_2O$
    d. $H_2SO_4 + 2\ KOH \longrightarrow K_2SO_4 + 2\ H_2O$
    e. $H_2SO_4 + 2\ POH \longrightarrow P_2SO_4 + 2\ H_2O$

MC4. Which of the following are true of the mixture formed when 4.6 g Na_2SO_4 are added to 100 mL water?
    1. The name of the solvent is sodium sulfate.
    2. The solute is an electrolyte.
    3. A precipitate will form.
    4. The solution formed will be 0.32 M.
    a. none   **b.** all   **c.** 1, 2, and 3   **d.** 2, 3, and 4
    e. 2 and 4

MC5. What volume of 18.0 M sulfuric acid is needed to prepare 5.00 L 0.105 M sulfuric acid?
    a. 9.45 mL   **b.** 29.2 mL   **c.** 0.290 L   **d.** 9.45×10^{-2} L
    e. 29.2 L

MC6. When compared with pure water, a 0.5 M solution of potassium chloride will have:
    a. a lower boiling point   **b.** a higher freezing point
    c. a higher vapor pressure
    d. a greater ability to conduct an electrical current
    e. none of these

MC7. What volume of 0.20 M sodium hydroxide will react exactly with 25 mL 0.20 M sulfuric acid?
    a. 12.5 mL   **b.** 20 mL   **c.** 25 mL   **d.** 50 mL   **e.** 100 mL

MC8. Which of the following is not required for an acid–base titration?
    a. The salt formed by the reaction must be soluble.
    b. The volumes of solutions used must be accurately measured.
    c. There must be a way of determining when the endpoint of the titration is reached.

d. The concentration of one of the solutions must be known.

e. All these conditions are required.

MC9. Which of the following affect the ability of a solid to dissolve in a particular liquid?

1. the temperature of the liquid and the solid

2. the atmospheric pressure

3. the intermolecular forces in the solvent

4. the interparticle forces in the solid

a. none   **b.** all   **c.** 1 and 2   **d.** 1, 3, and 4   **e.** 3 and 4

MC10. What is the net ionic equation for the reaction of magnesium with hydrochloric acid?

a. $Mg + 2\,H^+ \longrightarrow Mg^{2+} + H_2$

b. $Mg + 2\,HCl \longrightarrow MgCl_2 + H_2$

c. $2\,Mg + 2\,H^+ \longrightarrow 2\,Mg^+ + H_2$

d. $Mn + 2\,H^+ \longrightarrow Mn^{2+} + H_2$

e. None of these is the correct net ionic equation for this reaction.

Problems

11.1 The Characteristics of Solutions

11.14. Give examples in which

a. a gas is dissolved in a liquid

b. a solid is dissolved in a liquid

c. a liquid is dissolved in a liquid

11.15. Identify the solute and solvent in the following solutions:

a. aqueous potassium nitrate

b. iodine in carbon tetrachloride

c. chlorine water

d. potassium hydroxide in ethyl alcohol

11.16. Which of the following compounds would you expect to be quite soluble in water?

a. potassium chloride (KCl)

b. benzene (C_6H_6)

c. chloroform $(CHCl_3)$

d. hydrogen iodide (HI)

e. sodium acetate $(NaC_2H_3O_2)$

f. formaldehyde (HCHO)

*11.17. For each of the following, predict its relative solubility in water and in gasoline, a nonpolar solvent. Remember the solubility rules in Chapter 8.

a. barium sulfate

b. lithium nitrate

c. octane (C_8H_{18})

d. bromine

e. carbon tetrafluoride (CF_4)

11.18. Describe the equilibrium present in:

a. a saturated solution of potassium nitrate and some undissolved potassium nitrate.

b. a solution from which barium sulfate has been precipitated.

c. a saturated solution of starch, a molecular compound.

11.3 Expressing Concentrations of Solutions

11.19. Carry out the following conversions:

a. 5.0 g $NaHCO_3$ in 1 L solution to molarity

b. 12 M HCl to g HCl/1 L solution

c. 0.15 M sodium hydroxide to g NaOH/100 mL solution

d. 1.33 g silver nitrate in 100 mL solution to molarity

*11.20. Describe how to prepare:
   a. 5.0 L of 0.15 M sulfuric acid from 18 M H_2SO_4.
   b. 400 mL of 0.10 M KOH from solid potassium hydroxide.
   c. 100 mL of 0.25 M HCl from 6 M HCl.
   d. 500 mL of 50% alcohol from 95% alcohol (% by volume).
   e. 450 mL of 3% (wt/vol) glucose in water.

11.21. What is the mass of the solute in the following?
   a. 1.5 L of 0.10 M $AgNO_3$
   b. 0.500 L of 3.0 M HNO_3
   c. 25.00 mL of 0.155 M NaOH

11.22. What volume of each of the following solutions contains 0.10 mol solute?
   a. 0.15 M barium chloride
   b. 0.25 M copper(II) sulfate
   c. 5.0 M ammonium nitrate
   d. 15 M nitric acid
   e. 0.30 M iron(II) chloride
   f. 0.55 M zinc(II) nitrate

11.23. What are the differences between a saturated solution and a 1 M solution?

11.24. The solubility of sodium bicarbonate is 6.9 g/100 g water at 25°C. What weight of sodium bicarbonate will dissolve in 250 g water at that temperature? What is the molarity of the saturated solution?

11.25. A solution contains 2.6 g glucose in 150 mL solution. Calculate the percent concentration (wt/vol) and the molarity of this solution.

11.26. Copper sulfate is obtained as a pentahydrate, $CuSO_4 \cdot 5H_2O$. In the solid form, each unit of $CuSO_4$ is associated with five molecules of H_2O. Starting with copper sulfate pentahydrate, describe how you would prepare 500 mL of a solution that is 0.25 M in copper ion.

11.27. An aqueous solution contains one part per million by weight of fluoride ion (μg/g). How would you prepare 10 L of this solution using sodium fluoride? What is the molarity of this solution?

*11.28. A solution of vitamin C, ascorbic acid (mol. wt 176.1), contains 1.0 g ascorbic acid per 200 mL solution. What is the molarity of this solution?

11.29. Beer is 12% ethyl alcohol by volume.
   a. Calculate the volume of alcohol in one liter of beer.
   b. Calculate the mass of alcohol in one liter of beer (density of alcohol = 0.789 g/mL).
   c. Calculate the molarity of ethyl alcohol in beer. (The molecular formula of ethyl alcohol is C_2H_5OH.)

11.30. Phenol, C_6H_5OH, is a mild antiseptic used in several nonprescription mouthwashes. In two of these mouthwashes, the concentration of phenol is 1.4% (wt/vol). Calculate the molarity of phenol in these solutions.

11.31. A commercial liquid noncaloric sweetener contains 1.62% (wt/vol) of the calcium salt of saccharin. The molecular formula of this calcium salt is $Ca(C_7H_4NSO_3)_2$. Calculate the molarity of this solution.

11.32. Calculate the number of millimoles of sodium ion in 11.65 mL of 0.150 M sodium carbonate.

11.33. Calculate the volume of 0.256 M hydrochloric acid that would contain 36.5 mmol acid.

*11.34. What volume of 15M HNO_3 is needed to prepare 1.65 L of 6.0 M HNO_3?

11.35. What volume of 0.351 M sodium hydroxide can be prepared from 35.0 mL of 4.15 M NaOH?

11.36. What is the molarity of the final solution if 58.3 mL of 12 *M* HCl is diluted to 11.5 L?

11.4 Calculations Involving Concentrations

*11.37. Calculate the concentration of a hydrochloric acid solution if 25.0 mL of this solution react completely with 33.5 mL of 0.103 *M* silver nitrate.

11.38. Calculate the concentration of a solution of sulfuric acid if 15.0 mL of this solution react completely with 26.2 mL of 0.125 *M* potassium hydroxide solution.

11.39. Calculate the molarity of a sodium hydroxide solution if 23.90 mL of this solution react completely with 25.0 mL of 0.215 *M* hydrochloric acid.

11.40. What weight of silver chloride is precipitated by the reaction of 35.0 mL of 0.15 *M* silver nitrate with excess 0.110 *M* sodium chloride?

11.41. Calculate the concentration of iodide ion in a solution of sodium iodide if 24.2 mL of the solution react completely with 16.7 mL of 0.176 *M* silver nitrate solution. The balanced equation is:

$$NaI + AgNO_3 \longrightarrow AgI + NaNO_3$$

11.42. What mass of bromine will be formed if an excess of chlorine is bubbled through 1.0 L of 5.0 *M* sodium bromide?

11.5 Titration

11.43. What are the requirements of a titration?

11.44. Using the following data, calculate the molarity of the hydrochloric acid solution.

Concentration of NaOH: 0.132 *M*
Volume of HCl samples: 25.0 mL
Buret readings for NaOH:

Trial I	Trial II	Trial III
26.14 mL	34.56 mL	44.25 mL
1.98	10.36	20.13

*11.45. Given the following data, calculate the concentration of the sulfuric acid solution.

Concentration of KOH: 0.987 *M*
Volume of H_2SO_4 samples: 10.0 mL
Buret readings for KOH:

Trial I	Trial II	Trial III
35.62 mL	27.89 mL	23.76 mL
13.87	6.01	1.87

11.6 Ionic Reactions in Solution

*11.46. Write net ionic equations for the preparation of the following compounds (all insoluble).

  a. chromium(III) hydroxide
  b. lead(II) iodide
  c. zinc(II) sulfide
  d. calcium carbonate

*11.47. Write net ionic equations to describe the following reactions.
  a. hydrochloric acid and sodium hydroxide
  b. sodium chloride and lead(II) nitrate
  c. sulfuric acid and silver(I) nitrate
  d. sulfuric acid and sodium hydroxide
  e. zinc and hydrochloric acid
  f. chlorine and potassium iodide

11.7E Colligative Properties

11.48. What is a colligative property? Explain why the vapor pressure of a pure liquid is greater than that of a solution in which that liquid is the solvent.

*11.49. Which of the following would you expect to have the highest boiling point: 0.1 *M* glucose, 1.0 *M* glucose, or 10.0 *M* glucose? Why?

11.50. The following solutions are all 0.1 M. Which would have the highest boiling point?
   a. sodium chloride
   b. potassium sulfate
   c. ethyl alcohol

*11.51. Which of the following 0.20 M solutions would have the lowest freezing point? Explain your choice.
   a. sodium sulfate
   b. sucrose (covalent)
   c. ammonium phosphate

Review Problems

11.52. A solution is prepared by dissolving 16.0 g sodium hydroxide in 750 mL water. A 25-mL sample of this solution reacts with 40.9 mL of a solution of hydrochloric acid. What is the molarity of the hydrochloric acid solution?

11.53. What volume of carbon dioxide measured at 25°C and 0.752 atm is obtained by the reaction of 45.0 g calcium carbonate with 165 mL of 0.215 M hydrochloric acid?

11.54. What mass of silver is obtained when 5.0 g copper is added to 125 mL of

0.555 M silver nitrate? The equation for the reaction is:

$$Cu + 2\ Ag^+ \longrightarrow Cu^{2+} + Ag$$

11.55. A 75-g sample of pure sodium chloride is added to 125 mL water at 25°C. Is the resulting solution saturated or unsaturated? The mixture is filtered and added to 1.5 L of 0.125 M silver nitrate. What weight of silver nitrate is precipitated?

11.56. What mass of sodium hydroxide is needed to prepare a solution in which the hydroxide ion concentration is 0.205 M?

▪12▪

Acids and Bases

Acids and bases have been discussed many times already in this text. In Section 5.7D acids and bases (hydroxides) were introduced in the discussion of differences between metals (hydroxides typically contain metals) and nonmetals (acids typically contain nonmetals). The nomenclature of acids is in Section 6.2. In Section 8.2D, acid–base reactions were discussed and identified as neutralization reactions; the reactions of acids with some metals were discussed in the section on displacement reactions (Section 8.2C). The net ionic equations for these two types of reactions were described in Section 11.6. In addition, our discussion of titrations (Section 11.5) centered on acid–base titrations. There is very little more about acids and bases that needs to be introduced, but, because they are such important compounds, it is advisable to summarize their properties.

In this chapter our summary will cover, in particular:

1. The various definitions of acids and bases, including the Arrhenius definitions and the Brønsted–Lowry definitions.
2. A review of the nomenclature of acids.
3. A review of acid–base reactions.
4. The differences in the extent to which acids ionize and the implications of these differences.
5. The equilibrium that exists in solutions of slightly ionized acids.
6. The equilibrium constant and related calculations.
7. The use of pH and pK_a in describing acid solutions.

12.1   Definitions of Acids and Bases

A. The Arrhenius Definitions

In Chapter 5, we defined an acid as a substance that releases hydrogen ions in aqueous solutions and a base as a substance that releases hydroxide ions in aqueous solutions. Because this behavior depends on dissociation into ions, and because the theory of ionization was first proposed by the Swedish chemist Svante Arrhenius (1859–1927), these definitions are frequently referred to as the **Arrhenius definitions.**

Table 12.1, a reproduction of Table 5.11, lists several familiar acids and bases.

TABLE 12.1   Common hydroxides and acids

Common hydroxides		*Common acids*	
sodium hydroxide	NaOH	hydrochloric acid	HCl
potassium hydroxide	KOH	acetic acid	$HC_2H_3O_2$
calcium hydroxide	$Ca(OH)_2$	nitric acid	HNO_3
aluminum hydroxide	$Al(OH)_3$	sulfuric acid	H_2SO_4
ammonium hydroxide	NH_4OH	carbonic acid	H_2CO_3
		phosphoric acid	H_3PO_4

B. The Brønsted–Lowry Definitions

The Arrhenius definitions of acids and bases describe the characteristics of aqueous solutions of acids and bases. In 1923, T. M. Lowry in England and J. M. Brønsted in Denmark proposed a system that defines acids and bases in terms of the mechanism by which they react. According to the **Brønsted–Lowry definitions:**

An acid is a proton (H^+) donor.

A base is a proton (H^+) acceptor.

Because a hydrogen ion consists of a nucleus containing a single proton, the terms *hydrogen ion* and *proton* are synonymous. These definitions somewhat broaden the category of substances that are acids or bases. The category of acids now includes those shown in Table 12.1 as well as ions such as ammonium ion, NH_4^+, and bicarbonate ion, HCO_3^-. Among Brønsted–Lowry bases are the hydroxide ion, OH^-; the anion of any acid; and ammonia, NH_3. Many substances such as water, bicarbonate ion, and ammonia can act as either an acid or a base.

In the Brønsted–Lowry system, an acid reacts by donating a proton to a base. In doing so, the acid becomes its **conjugate base.** The formula of the conjugate base is the formula of the acid less one hydrogen. The reacting base becomes its **conjugate acid.** The formula of the conjugate acid is the formula of the base plus one hydrogen ion. Let us illustrate this system using the neutralization of hydrochloric acid with sodium hydroxide. When hydrochloric acid reacts with a hydroxide ion, water and the chloride ion are formed. In the equation for the reaction each acid–base pair has the same subscript. Acid$_1$ is HCl, its conjugate base is base$_1$; hydroxide ion is base$_2$, and its conjugate acid (water) is acid$_2$.

$$\underset{\text{acid}_1}{\text{HCl}} + \text{Na}^+ + \underset{\text{base}_2}{\text{OH}^-} \longrightarrow \underset{\text{acid}_2}{\text{H}_2\text{O}} + \text{Na}^+ + \underset{\text{base}_1}{\text{Cl}^-}$$

Chloride ion is the conjugate base of hydrochloric acid. Water is the conjugate acid of the hydroxide ion. In this equation the sodium ion is a spectator ion.

The equation for the reaction of hydrochloric acid with ammonia is

$$\underset{\text{acid}_1}{\text{HCl}} + \underset{\text{base}_2}{\text{NH}_3} \longrightarrow \underset{\text{acid}_2}{\text{NH}_4^+} + \underset{\text{base}_1}{\text{Cl}^-}$$

When water reacts with ammonia, it is acting as an acid:

$$\underset{\text{acid}_1}{\text{H}_2\text{O}} + \underset{\text{base}_2}{\text{NH}_3} \longrightarrow \underset{\text{acid}_2}{\text{NH}_4^+} + \underset{\text{base}_1}{\text{OH}^-}$$

Hydroxide ion is the conjugate base of water. When water reacts with an acid, it is acting as a base:

$$\underset{\text{base}_1}{\text{H}_2\text{O}} + \underset{\text{acid}_2}{\text{HCl}} \longrightarrow \underset{\text{acid}_1}{\text{H}_3\text{O}^+} + \underset{\text{base}_2}{\text{Cl}^-}$$

The conjugate acid of water is the **hydronium ion,** H_3O^+, an ion formed by the association of a hydrogen ion with a water molecule.

Example 12.1   In the following list, group A contains Brønsted–Lowry acids and group B contains Brønsted–Lowry bases. Show by equation how each substance in group A acts as an acid using water as a base. Show by equation how each substance in group B acts as a base using acetic acid as an acid.

Group A: HSO_4^-, HNO_3, H_2S **Group B:** OH^-, HS^-, CO_3^{2-}

Solution   **Group A**

1. $\underset{\text{acid}_1}{\text{HSO}_4^-} + \underset{\text{base}_2}{\text{H}_2\text{O}} \longrightarrow \underset{\text{acid}_2}{\text{H}_3\text{O}^+} + \underset{\text{base}_1}{\text{SO}_4^{2-}}$

   The conjugate base of the bisulfate ion is the sulfate ion. The conjugate acid of water is the hydronium ion.

2. $\underset{\text{acid}_1}{\text{HNO}_3} + \underset{\text{base}_2}{\text{H}_2\text{O}} \longrightarrow \underset{\text{base}_1}{\text{NO}_3^-} + \underset{\text{acid}_2}{\text{H}_3\text{O}^+}$

3. $\underset{\text{acid}_1}{\text{H}_2\text{S}} + \underset{\text{base}_2}{\text{H}_2\text{O}} \longrightarrow \underset{\text{base}_1}{\text{HS}^-} + \underset{\text{acid}_2}{\text{H}_3\text{O}^+}$

Notice that the formula of the conjugate base of the acid is the formula of the acid less one hydrogen ion.

Group B

1. $\underset{\text{base}_1}{OH^-} + \underset{\text{acid}_2}{HC_2H_3O_2} \longrightarrow \underset{\text{acid}_1}{H_2O} + \underset{\text{base}_2}{C_2H_3O_2^-}$

2. $\underset{\text{base}_1}{HS^-} + \underset{\text{acid}_2}{HC_2H_3O_2} \longrightarrow \underset{\text{acid}_1}{H_2S} + \underset{\text{base}_2}{C_2H_3O_2^-}$

3. $\underset{\text{base}_1}{CO_3^{2-}} + \underset{\text{acid}_2}{HC_2H_3O_2} \longrightarrow \underset{\text{acid}_1}{HCO_3^-} + \underset{\text{base}_2}{C_2H_3O_2^-}$

Notice that the formula of the conjugate acid of a base contains one more hydrogen ion than the formula of the base.

Problem 12.1   Show by equation how the substances in group A act as Brønsted–Lowry acids and those in group B act as Brønsted–Lowry bases. Name all reactants and products. Use water as a base for group A and as an acid for group B.

Group A: $NH_4^+, H_2CO_3, HC_2H_3O_2$ **Group B:** NH_3, H_2O, HPO_4^{2-}

All of the reactions in Example 12.1 fit the general equation for a Brønsted–Lowry acid–base reaction:

$$\underset{\text{acid}_1}{HA} + \underset{\text{base}_2}{B^-} \longrightarrow \underset{\text{base}_1}{A^-} + \underset{\text{acid}_2}{HB}$$

The Brønsted–Lowry definitions greatly expand the classes of substances called acids and bases. Members of the acid group need not be compounds but can also be ions, and a base need not contain in its formula the hydroxide ion. The usefulness of these definitions is emphasized by considering the Lewis structures of typical members of the groups. Following are Lewis structures for some bases.

$$\left[:\!\overset{..}{\underset{..}{O}}\!-\!H \right]^- \qquad \underset{\overset{|}{H}}{H-\overset{..}{N}-H} \qquad \left[:\!O\!=\!C\!-\!\overset{..}{\underset{..}{O}}: \right]^{2-} \qquad H-\overset{..}{\underset{..}{O}}:$$

$$\qquad\qquad\qquad\qquad\qquad\qquad\qquad \underset{:\overset{..}{O}:}{|} \qquad\qquad\qquad\qquad \overset{|}{H}$$

$$\qquad OH^- \qquad\qquad\qquad NH_3 \qquad\qquad\qquad CO_3^{2-} \qquad\qquad\qquad H_2O$$

Each of these bases has at least one unshared pair of electrons on a very electronegative atom. When a proton or hydrogen ion is added to these structures, it forms a covalent bond with one of these unshared electron pairs, as shown below:

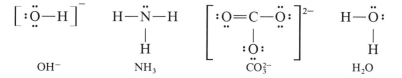

$$:\!\overset{..}{O}\!-\!H \qquad \left[H-\overset{\overset{\displaystyle H}{|}}{\underset{\overset{|}{H}}{N}}-H \right]^+ \qquad \left[:\!O\!=\!C\!-\!\overset{..}{O}\!-\!H \right]^- \qquad \left[H\!=\!\overset{..}{O}\!-\!H \right]^+$$

$$\overset{|}{H} \qquad\qquad\qquad\qquad\qquad\qquad\qquad \underset{:\overset{..}{O}:}{|} \qquad\qquad\qquad\qquad \overset{|}{H}$$

$$H_2O \qquad\qquad\qquad NH_4^+ \qquad\qquad\qquad HCO_3^- \qquad\qquad\qquad H_3O^+$$

Example 12.2   Use the Lewis structure for the formate ion, HCO_2^-, to show how it acts as a Brønsted–Lowry base.

Solution

$$\left[\begin{array}{c} :\overset{..}{\overset{\displaystyle O}{\|}} \\ H-C-\overset{..}{\underset{..}{O}}: \end{array} \right]^- + H_3O^+ \longrightarrow H-C-\overset{..}{\underset{..}{O}}-H + H_2O$$

New covalent bond

Unshared pair of electrons on a very electronegative atom

Problem 12.2   Draw the Lewis structure of the bisulfide ion, HS^-, and show how it acts as a Brønsted–Lowry base.

Acid–base definitions are summarized in Table 12.2.

TABLE 12.2   Definitions of acids and bases

System	Acid	Base
Arrhenius	in aqueous solution produces H^+	produces OH^-
Brønsted–Lowry	proton donor	proton acceptor

12.2 Nomenclature of Acids

The **nomenclature of acids** was described in Table 6.6 and in Sections 6.2B3 and 6.2C1. That material is summarized here in Table 12.3.

An acid containing only hydrogen and a nonmetal is named *hydro–ic* acid; for example, HCl is hydrochloric acid, and H_2S is hydrosulfuric acid.

Among the acids containing different numbers of oxygen atoms with the same nonmetal, the names of the most common include the root of the element and the ending *ic*. Thus, H_2SO_4 is sulfuric acid; HNO_3 is nitric acid. These names and formulas must be memorized.

The acid containing one less oxygen than the most common is named with the root of the element and the ending *ous*. Thus, H_2SO_3 is sulfurous acid; HNO_2 is nitrous acid.

Occasionally one encounters an acid containing two less oxygen atoms than the most common. These are named *hypo–ous* acids. Thus, HClO is hypochlorous acid.

Acids containing carbon are named by a wholly different system. The system need not be learned, but the formulas and names of a few of these acids should be memorized: $HC_2H_3O_2$ is acetic acid, H_2CO_3 is carbonic acid, and HCOOH is formic acid.

TABLE 12.3   Some common acids

Acid	Molecular formula	Anion formed in solution	Anion name
hydrochloric acid	HCl	Cl^-	chloride
sulfuric acid	H_2SO_4	HSO_4^-	hydrogen sulfate (bisulfate)
	HSO_4^-	SO_4^{2-}	sulfate
sulfurous acid	H_2SO_3	HSO_3^-	hydrogen sulfite (bisulfite)
	HSO_3^-	SO_3^{2-}	sulfite
nitric acid	HNO_3	NO_3^-	nitrate
nitrous acid	HNO_2	NO_2^-	nitrite
carbonic acid	H_2CO_3	HCO_3^-	hydrogen carbonate (bicarbonate)
	HCO_3^-	CO_3^{2-}	carbonate
phosphoric acid	H_3PO_4	$H_2PO_4^-$	dihydrogen phosphate
	$H_2PO_4^-$	HPO_4^{2-}	hydrogen phosphate
	HPO_4^{2-}	PO_4^{3-}	phosphate
Organic acids			
acetic	$HC_2H_3O_2$	$C_2H_3O_2^-$	acetate
formic	$HCOOH$	$HCOO^-$	formate

Example 12.3

Give the formula and name of the acids that have the following anions. Name the anions.

$$NO_2^-, F^-, HCOO^-$$

Solution

NO_2^- is the anion of HNO_2. This formula contains one less oxygen than nitric acid. HNO_2 is named nitrous acid, and the anion is nitrite. F^- is the anion of HF, an acid containing no oxygen atoms. The acid is named hydrofluoric acid, and the anion is fluoride. $HCOO^-$ is the anion of HCOOH, formic acid, and the anion is formate.

Problem 12.3

Give the formula and names of the acids that have the following anions. Name the anion.

$$CO_3^{2-}, I^-, ClO_3^-$$

12.3 Reactions of Acids

A. Reactions with a Base

Neutralization is the reaction of an acid with a base to form water and a salt (Section 8.2D1), as in the reaction of hydrochloric acid with sodium hydroxide:

$$\underset{\substack{\text{hydrochloric} \\ \text{acid}}}{HCl} + \underset{\substack{\text{sodium} \\ \text{hydroxide}}}{NaOH} \longrightarrow \underset{\substack{\text{sodium} \\ \text{chloride}}}{NaCl} + \underset{\text{water}}{H_2O}$$

The net ionic equation (Section 11.6) is

$$H^+ + OH^- \longrightarrow H_2O$$

We have seen acid–base reactions within the Brønsted–Lowry system as the reaction of a proton donor with a proton acceptor, as in the reaction of hydrochloric acid with the acetate ion:

$$\underset{\text{acid}_1}{HCl} + \underset{\text{base}_2}{C_2H_3O_2^-} \longrightarrow \underset{\text{acid}_2}{HC_2H_3O_2} + \underset{\text{base}_1}{Cl^-}$$

or with ammonia:

$$\underset{\text{acid}_1}{HCl} + \underset{\text{base}_2}{NH_3} \longrightarrow \underset{\text{acid}_2}{NH_4^+} + \underset{\text{base}_1}{Cl^-}$$

B. Reactions with a Metal

Acids react with some metals in displacement reactions (Section 8.2C) to form hydrogen gas and a salt, as in the reaction of zinc with sulfuric acid:

$$Zn(s) + H_2SO_4(aq) \longrightarrow ZnSO_4(aq) + H_2(g)$$

Writing this reaction as a net ionic equation emphasizes that it is an oxidation–reduction reaction:

$$2\,H^+ + Zn \longrightarrow Zn^{2+} + H_2$$

in which zinc is oxidized and hydrogen reduced.

Example 12.4

Write net ionic equations showing the reaction of hydrochloric acid with

a. sodium nitrite   **b.** ammonia   **c.** magnesium

Solution

The net ionic equation will show only the reaction of the proton of hydrochloric acid with the base provided by the reactants in parts **a** and **b** or with the metal magnesium in part **c**.

a. The base is the nitrite ion:

$$H^+ + NO_2^- \longrightarrow HNO_2 \quad \text{nitrous acid}$$

b. The base is ammonia:

$$H^+ + NH_3 \longrightarrow NH_4^+ \quad \text{ammonium ion}$$

c. The reactant is magnesium metal:

$$2\,H^+(aq) + Mg(s) \longrightarrow Mg^{2+}(aq) + H_2(g)$$

Problem 12.4   Write the net ionic equation for the reaction of hydrochloric acid with

a. potassium formate   **b.** bicarbonate ion   **c.** iron

12.4   Ionization of Acids

A. Weak and Strong Acids

Acids differ enormously in the extent to which they dissociate into ions in aqueous solution. Some acids, such as hydrochloric and nitric acids, are strong electrolytes, completely dissociated into ions; these acids are known as **strong acids.** Others, such as acetic and nitrous acids, are only partially dissociated in solution. These acids are weak electrolytes and are known as **weak acids.** The solution of a strong acid contains no acid molecules; the solution of a weak acid contains both molecules and ions (see Figure 12.1).

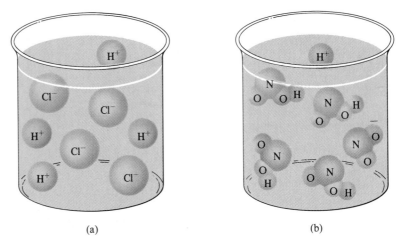

(a)   (b)

FIGURE 12.1   Strong and weak acids in solution: (a) the solution of a strong acid contains only ions; (b) the solution of a weak acid contains both molecules and ions.

The ions of a weak acid tend to recombine to re-form molecules of the acid. We can show this recombination in the ionization equation with double arrows, which mean that the reaction is reversible and goes both ways. Some molecules are dissociating into ions; some ions are recombining to form molecules. Acetic acid is a weak acid; we show its ionization with double arrows to indicate that a solution of acetic acid contains both molecules and ions. When using double arrows, it is customary to show a longer or heavier arrow in the direction of the predominant reaction. Thus, in the ionization of a weak acid, in which there are fewer ions than molecules, the longer arrow points toward the molecules. (The topic of weak acids is discussed again in Section 12.6B.)

$$HC_2H_3O_2 \rightleftharpoons H^+ + C_2H_3O_2^-$$

Because the ions of a strong acid do not recombine, we show its ionization with a single arrow, meaning that only the ions of the acid are present in its aqueous solution. For the strong acid nitric acid, the ionization would be shown as:

$$HNO_3 \longrightarrow H^+ + NO_3^-$$

In the Brønsted–Lowry system, a strong acid is one that easily donates a proton. The conjugate base of a strong Brønsted–Lowry acid is a weak base, one that has little affinity for a proton. Thus, for hydrochloric acid,

$$\underset{\text{base}_1}{H_2O} + \underset{\substack{\text{strong} \\ \text{acid}_2}}{HCl} \longrightarrow \underset{\substack{\text{weak conjugate} \\ \text{base}_2}}{Cl^-} + \underset{\text{acid}_1}{H_3O^+}$$

A weak acid does not readily donate a proton. Its conjugate base (or anion) is a relatively strong base, one with considerable attraction for a proton. Thus, for acetic acid,

$$\underset{\text{base}_2}{H_2O} + \underset{\text{weak acid}_1}{HC_2H_3O_2} \rightleftharpoons \underset{\substack{\text{strong conjugate} \\ \text{base}_1}}{C_2H_3O_2^-} + \underset{\text{acid}_2}{H_3O^+}$$

B. Polyprotic Acids

Molecules of some acids — such as sulfuric acid, H_2SO_4, and phosphoric acid, H_3PO_4 — have more than one ionizable hydrogen; these acids are called **polyprotic acids.** The ionization of these acids occurs in steps, with the molecule losing one proton at a time. The ionization of sulfuric acid, a diprotic acid, occurs as follows:

$$H_2SO_4 \longrightarrow H^+ + HSO_4^-$$
$$HSO_4^- \rightleftharpoons H^+ + SO_4^{2-}$$

Notice that, although the first ionization is that of a strong acid, the second is that of a weak acid.

The ionization of phosphoric acid, a triprotic acid, occurs as follows:

$$H_3PO_4 \rightleftharpoons H^+ + H_2PO_4^-$$
$$H_2PO_4^- \rightleftharpoons H^+ + HPO_4^{2-}$$
$$HPO_4^{2-} \rightleftharpoons H^+ + PO_4^{3-}$$

For sulfuric acid, and indeed for all polyprotic acids, the first ionization is much more complete than the subsequent ionizations.

Example 12.5   Show by equation the ionization in aqueous solution of

    a. chloric acid, $HClO_3$, a strong acid

    b. formic acid, $HCOOH$, a weak acid

    c. oxalic acid, $H_2C_2O_4$, a weak polyprotic acid

For oxalic acid, both ionizations are incomplete. Predict which anion is the stronger base.

Solution   **a.** A strong acid has only ions in solution, so we use a single arrow in the ionization equation.

$$HClO_3 \longrightarrow H^+ + ClO_3^-$$

b. A weak acid has both molecules and ions in solution; a double arrow is used in the ionization equation, with the longer arrow pointing toward the molecules.

$$HCOOH \rightleftharpoons H^+ + HCOO^-$$

c. A diprotic acid ionizes stepwise. The first anion is a weaker base than the second, for the first ionization is more complete than the second.

$$H_2C_2O_4 \rightleftharpoons H^+ + HC_2O_4^-$$
$$HC_2O_4^- \rightleftharpoons H^+ + C_2O_4^{2-}$$

Problem 12.5   Show by equation the ionization in aqueous solution of

    a. the strong acid, $HBrO_3$

    b. the weak acid, HNO_2

    c. the polyprotic weak acid, H_2SO_3. Both ionizations of sulfurous acid are incomplete.

Equilibrium in Solutions of Weak Acids

We have seen that equilibrium exists when two opposing reactions occur at the same rate. In Chapter 10, we discussed the equilibrium between a liquid and its vapor, with the opposing reactions being evaporation and condensation:

$$\text{Rate}_{\text{evaporation}} = \text{Rate}_{\text{condensation}}$$

In Chapter 11, we discussed the equilibrium between dissolved and undissolved solute in saturated solutions, in which the opposing reactions are precipitation and dissolution:

$$\text{Rate}_{\text{dissolution}} = \text{Rate}_{\text{precipitation}}$$

The ionization of a weak acid is also an equilibrium situation. Here the equilibrium is between molecules and ions. We have implied the existence of this equilibrium with double arrows in the ionization equations for weak acids. In a solution of acetic acid, a weak acid, the two ongoing reactions are dissociation of molecules into ions and recombination of ions into molecules. This equilibrium is shown by the equation

$$HC_2H_3O_2 \rightleftharpoons H^+ + C_2H_3O_2^-$$

The opposing reaction rates are

$$\text{Rate}_{\text{dissociation}} = \text{Rate}_{\text{recombination of ions}}$$

There is another criterion for equilibrium. For a system in equilibrium, a mathematical relationship exists between the concentrations of the components of the equilibrium. This relationship is known as the **equilibrium constant** K_{eq}. For an ionization equilibrium, this constant is called the **ionization constant.** If the ionization is that of a weak acid, the equilibrium constant is known as the **acid dissociation constant** and has the symbol K_a.

In general terms, for a reaction

$$a\text{A} + b\text{B} \rightleftharpoons c\text{C} + d\text{D}$$

the equilibrium constant expression is

$$K_{eq} = \frac{[\text{C}]^c[\text{D}]^d}{[\text{A}]^a[\text{B}]^b}$$

where the square brackets mean molar concentration. Note that the numerator of the equilibrium constant expression is the product of the concentrations of the products, each raised to a power equal to its coefficient in the balanced equation for the equilibrium reaction. The denominator is the product of the concentrations of the reactants, each raised to a power equal to its coefficient in the balanced equation for the equilibrium reaction. A weak acid with the formula HA would show in solution the equilibrium

$$HA \rightleftharpoons H^+ + A^-$$

The ionization constant expression for this equilibrium is

$$K_a = \frac{[H^+][A^-]}{[HA]}$$

By convention, these ionization equilibria are always written with the ions as products. Thus, in an ionization constant expression, the concentrations of the ions are always in the numerator, that of the un-ionized molecules in the denominator.

Therefore, for acetic acid, with the ionization equilibrium

$$HC_2H_3O_2 \rightleftharpoons H^+ + C_2H_3O_2^-$$

the equilibrium constant expression is

$$K_a = \frac{[H^+][C_2H_3O_2^-]}{[HC_2H_3O_2]}$$

We have already pointed out that a weak acid is only slightly ionized and that its solution contains mostly molecules and many fewer ions. Therefore, the concentration value of the un-ionized molecules in the denominator of the equilibrium constant is much larger than those of the ions in the numerator. Consequently, values of the ionization constants for weak acids are always much less than 1. Table 12.4 lists several weak acids and shows the equations for their ionization and the expression and value of their acid dissociation constants. Note several points in this table:

1. In regard to the acid dissociation constants of polyprotic acids, we have already noted (Section 12.4B) that polyprotic acids ionize stepwise and that each anion is less ionized than its conjugate acid. Each of the ionizations has an equilibrium equation and an equilibrium constant. Phosphoric acid has three ionizable hydrogens. The acid molecule ionizes to yield a hydrogen ion and a dihydrogen phosphate ion:

$$H_3PO_4 \rightleftharpoons H^+ + H_2PO_4^- \qquad K_a = 7.5 \times 10^{-3}$$

The dihydrogen phosphate ion ionizes to yield another hydrogen ion and the monohydrogen phosphate ion:

$$H_2PO_4^- \rightleftharpoons H^+ + HPO_4^{2-} \qquad K_a = 6.2 \times 10^{-8}$$

Finally, the monohydrogen phosphate ion ionizes to yield another hydrogen ion and the phosphate anion:

$$HPO_4^{2-} \rightleftharpoons H^+ + PO_4^{3-} \qquad K_a = 2.2 \times 10^{-13}$$

Notice that the acid dissociation constant becomes smaller with each ionization. For acids with more than one ionizable hydrogen, the first dissociation constant is always the largest; the value for each successive dissociation constant decreases. This progression goes along with the prediction in Section 12.4 that, in each successive ionization, the anion formed is a stronger base.

TABLE 12.4   Some common weak acids

Electrolyte	Equilibrium equation	Acid dissociation expression	K_a
acetic acid	$HC_2H_3O_2 \rightleftharpoons H^+ + C_2H_3O_2^-$	$\dfrac{[H^+][C_2H_3O_2^-]}{[HC_2H_3O_2]}$	1.8×10^{-5}
formic acid	$HCO_2H \rightleftharpoons H^+ + HCO_2^-$	$\dfrac{[H^+][HCO_2^-]}{[HCO_2H]}$	1.8×10^{-4}
nitrous acid	$HNO_2 \rightleftharpoons H^+ + NO_2^-$	$\dfrac{[H^+][NO_2^-]}{[HNO_2]}$	4.6×10^{-4}
hydrocyanic acid	$HCN \rightleftharpoons H^+ + CN^-$	$\dfrac{[H^+][CN^-]}{[HCN]}$	4.9×10^{-10}
carbonic acid	$CO_2 + H_2O \rightleftharpoons H^+ + HCO_3^-$	$\dfrac{[H^+][HCO_3^-]}{[CO_2]}$	4.3×10^{-7}
	$HCO_3^- \rightleftharpoons H^+ + CO_3^{2-}$	$\dfrac{[H^+][CO_3^{2-}]}{[HCO_3^-]}$	5.6×10^{-11}
phosphoric acid	$H_3PO_4 \rightleftharpoons H^+ + H_2PO_4^-$	$\dfrac{[H^+][H_2PO_4^-]}{[H_3PO_4]}$	7.5×10^{-3}
	$H_2PO_4^- \rightleftharpoons H^+ + HPO_4^{2-}$	$\dfrac{[H^+][HPO_4^{2-}]}{[H_2PO_4^-]}$	6.2×10^{-8}
	$HPO_4^{2-} \rightleftharpoons H^+ + PO_4^{3-}$	$\dfrac{[H^+][PO_4^{3-}]}{[HPO_4^{2-}]}$	2.2×10^{-13}
ammonium ion	$NH_4^+ \rightleftharpoons H^+ + NH_3$	$\dfrac{[H^+][NH_3]}{[NH_4^+]}$	5.5×10^{-10}

2. Carbonic acid (H_2CO_3) is a solution of carbon dioxide in water. Molecules of carbonic acid are not stable. The mixture of carbon dioxide and water ionizes stepwise like phosphoric acid:

$$CO_2 + H_2O \rightleftharpoons H^+ + HCO_3^-$$
$$HCO_3^- \rightleftharpoons CO_3^{2-} + H^+$$

3. Ammonium ion acts as a weak acid, ionizing to ammonia and a hydrogen ion:

$$NH_4^+ \rightleftharpoons H^+ + NH_3$$

Example 12.6   Ascorbic acid, vitamin C, is a weak electrolyte. Its molecular formula is $C_6H_8O_6$. In aqueous solution, ascorbic acid ionizes to form H^+ and ascorbate ion, $C_6H_7O_6^-$.

a. Write an equation for the equilibrium established in this ionization.

b. Write the acid dissociation constant expression for ascorbic acid.

Solution

a. The equation shows the loss of one hydrogen ion; the rest of the molecule is the ascorbate anion.

$$\underset{\text{ascorbic acid}}{C_6H_8O_6} \rightleftharpoons H^+ + \underset{\text{ascorbate ion}}{C_6H_7O_6^-}$$

b. An acid dissociation constant expression has the concentrations of the ions in the numerator and that of the un-ionized acid molecules in the denominator. The acid dissociation constant expression of ascorbic acid is

$$K_a = \frac{[H^+][C_6H_7O_6^-]}{[C_6H_8O_6]}$$

Problem 12.6   Citric acid, $C_6H_8O_7$, is found in the juice of lemons and other citrus fruits, hence the name. It is a weak electrolyte and ionizes in aqueous solution to form H^+ and the citrate ion, $C_6H_7O_7^-$.

a. Write the equilibrium equation for this ionization.

b. Write the expression for the acid dissociation constant of citric acid.

12.6   **Hydrogen Ion Concentration in Acid Solutions**

Acid dissociation constants are used to calculate the hydrogen ion concentration in the solution of an acid.

A. Hydrogen Ion Concentration in Solutions of Strong Acids

Strong acids with one ionizable hydrogen are completely ionized in aqueous solution; therefore, the hydrogen ion concentration of these solutions is equal to the molar concentration of the acid.

Example 12.7   What is the hydrogen ion concentration in 1.0 M HCl?

Solution   Hydrochloric acid is a strong acid that is completely ionized in water:

$$HCl \longrightarrow H^+ + Cl^-$$

Therefore, in a solution prepared by adding 1.0 mol HCl to enough water to make 1 L solution, the concentration of H^+ is 1.0 M, that of Cl^- is 1.0 M, and that of undissociated acid is 0.

Problem 12.7   What is the hydrogen ion concentration of 0.1 M HNO$_3$?

B. Hydrogen Ion Concentration in Solutions of Weak Acids

The hydrogen ion concentration of an aqueous solution of a weak acid depends on the value of its acid dissociation constant and is always less than the concentration of the weak acid. The hydrogen ion concentration can be calculated using the value of K_a and the molar concentration of the weak acid.

Acetic acid is a weak acid that ionizes according to the equation

$$HC_2H_3O_2 \rightleftharpoons H^+ + C_2H_3O_2^-$$

Its acid dissociation constant is

$$K_a = \frac{[H^+][C_2H_3O_2^-]}{[HC_2H_3O_2]} = 1.8 \times 10^{-5}$$

The hydrogen ion concentration of a 1.0 M acetic acid solution can be calculated as follows: The solution contains 1.0 mol acetic acid in 1.0 L solution. Because acetic acid is a weak electrolyte, only a small fraction of the molecules ionize to hydrogen and acetate ions; most remain as un-ionized acetic molecules. Let x stand for the number of moles of acetic acid that ionize. If x moles ionize, then $1.0 - x$ moles of acetic acid remain un-ionized. For x moles of acetic acid that ionize, x moles of H^+ and x moles of $C_2H_3O_2^-$ are formed. The resulting concentrations of acetic acid, hydrogen ion, and acetate ion at equilibrium are

$$\underset{1.0-x}{HC_2H_3O_2} \rightleftharpoons \underset{x}{H^+} + \underset{x}{C_2H_3O_2^-}$$

Substituting these values into the expression for the acid dissociation constant gives:

$$K_a = \frac{[H^+][C_2H_3O_2^-]}{[HC_2H_3O_2]} = \frac{(x)(x)}{1.0 - x} = 1.8 \times 10^{-5}$$

The tiny value of the acid dissociation constant suggests that the amount of acid dissociated is very small (less than 0.01 M). Using the rules for significant figures in addition and subtraction (Section 2.2C), we know that the quantity $1.0 - 0.01$ expressed to two significant figures is 1.0. If x has a value less than 0.01, it is appropriate to disregard x in the expression $1.0 - x$ and change the K_a expression to:

$$K_a = \frac{x^2}{1.0} = 1.8 \times 10^{-5}$$

Solving this equation gives:

$$x^2 = 1.8 \times 10^{-5} = 18 \times 10^{-6}$$
$$x = 4.2 \times 10^{-3} = [H^+] = [C_2H_3O_2^-]$$

These values are shown in Table 12.5.

TABLE 12.5   Concentrations of species in 1.0 M acetic acid solution

	In calculation	Calculated value
acetic acid molecules	$1.0 - x$	$1.0\ M$
hydrogen ions	x	$4.2 \times 10^{-3}\ M$
acetate ions	x	$4.2 \times 10^{-3}\ M$

The hydrogen ion concentration in 1.0 M acetic acid solution is, then, $4.2 \times 10^{-3}\ M$, or 0.0042 M. The number 0.0042 is not significant when subtracted from 1.0, so our simplification of the original equation was justified. If the acid is very dilute, for example $10^{-3}\ M$, or if it is one with a large acid dissociation constant, such as 10^{-2}, this simplification would not be valid.

These calculations emphasize the difference between strong and weak acids. The hydrogen ion concentration of a 1.0 M solution of a strong acid is 1.0 M (see Example 12.7). The hydrogen ion concentration of a 1.0 M solution of a weak acid can be calculated from the acid dissociation constant of the weak acid and is much less than 1.0 M.

■

Example 12.8

Calculate the hydrogen ion concentration in 0.10 M ascorbic acid, $C_6H_8O_6$, a weak acid. The K_a for ascorbic acid is 8.0×10^{-5}.

Solution

In 0.10 M ascorbic acid, the equilibrium equation is

$$\underset{0.10-x}{C_6H_8O_6} \rightleftharpoons \underset{x}{H^+} + \underset{x}{C_6H_7O_6^-}$$

The K_a for this equilibrium is

$$K_a = \frac{[H^+][C_6H_7O_6^-]}{[C_6H_8O_6]} = 8.0 \times 10^{-5}$$

Let $[H^+] = x$; then $[C_6H_7O_6^-]$ also equals x and

$$[C_6H_8O_6] = 0.10 - x$$

Substituting these values into the expression for the acid dissociation constant gives:

$$\frac{(x)(x)}{0.10 - x} = 8.0 \times 10^{-5}$$

Assuming, as before, that $[H^+]$ is so much less than 0.10 M as to be insignificant, we rewrite the equation as

$$\frac{x^2}{0.10} = 8.0 \times 10^{-5}$$

Solving for x, we get:

$$x^2 = 8.0 \times 10^{-6}$$
$$x = 2.8 \times 10^{-3}$$

Thus, in $0.10\ M$ ascorbic acid, $[H^+] = 2.8 \times 10^{-3}\ M$.

Problem 12.8 Calculate the hydrogen ion concentration of $0.1\ M$ formic acid. (See Table 12.4 for the molecular formula of formic acid and the value of its acid dissociation constant.)

C. Changing the Hydrogen Ion Concentration in Solutions of Weak Acids; The Common-Ion Effect

We know that the solution of a weak acid contains the equilibrium

$$HA \rightleftharpoons H^+ + A^-$$

and, as long as the equilibrium exists, everything is in balance so that

$$K_a = \frac{[H^+][A^-]}{[HA]}$$

If a substance is added to the solution that changes one of the concentrations, the system goes out of equilibrium. Then either more molecules dissociate or more ions combine until concentrations again fit the equilibrium constant expression. When this set of concentrations is established, an equilibrium is again present, the opposing rates are equal, and the concentrations become constant.

It is possible to calculate the new equilibrium concentrations after a change in one concentration. Suppose we have one liter of a solution containing one mole of acetic acid and one mole of sodium acetate. The acetic acid is present as an equilibrium mixture of acetic acid molecules, hydrogen ions, and acetate ions. The sodium acetate is present only as ions (recall from Chapter 7, Section 7.5C, that salts are completely ionized in solution). The acetate ions from both the acetic acid and the sodium acetate participate in the acetic acid equilibrium:

$$HC_2H_3O_2 \rightleftharpoons H^+ + C_2H_3O_2^- \qquad K_a = 1.8 \times 10^{-5}$$

Although sodium ions are also present in solution, they do not participate in the equilibrium, they do not appear in the equation for the equilibrium, and they play no role in determining the concentrations of those substances whose formulas do appear in the equilibrium expression. To calculate the hydrogen ion concentration in this solution, we begin as in previous problems when only the weak acid was present. If x equals the concentration of ionized acetic acid molecules, the concentrations at equilibrium are:

$1.0 - x =$ the concentration of un-ionized acetic acid molecules

$x =$ the concentration of hydrogen ions

$1.0 + x =$ the concentration of acetate ions (the concentration of sodium acetate in the solution plus the acetate ions from the ionization of acetic acid)

In the equilibrium equation, the concentrations are:

$$\underset{1.0-x}{HC_2H_3O_2} \rightleftharpoons \underset{x}{H^+} + \underset{1.0+x}{C_2H_3O_2^-}$$

As in the case of a solution containing only acetic acid, we predict that the concentration of ionized acetic acid, and therefore of hydrogen ions, is very small and not significant when added to or subtracted from 1.0 M. Given this assumption, $1.0 + x$ is approximately equal to 1.0, and $1.0 - x$ is also approximately equal to 1.0. Thus, the concentrations can be expressed as

$$\underset{1.0}{HC_2H_3O_2} \rightleftharpoons \underset{x}{H^+} + \underset{1.0}{C_2H_3O_2^-}$$

Substituting these values into the expression for the acid dissociation constant for acetic acid and solving gives:

$$K_a = \frac{[H^+][C_2H_3O_2^-]}{[HC_2H_3O_2]} = \frac{(x)(1.0)}{(1.0)} = 1.8 \times 10^{-5}$$

$$x = 1.8 \times 10^{-5}$$

$$[H^+] = 1.8 \times 10^{-5} \ M$$

The addition of sodium acetate has decreased the hydrogen ion concentration from $4.2 \times 10^{-3} \ M$ in 0.1 M acetic acid solution to $1.8 \times 10^{-5} \ M$ in a 1.0 M acetic acid solution that is also 1.0 M in acetate ion. This decrease is tremendous.

These calculations have shown that an ionic equilibrium such as the ionization of a weak acid can be shifted by the addition of another ionic substance (such as salt) that contains one of the ions present in the equilibrium. This effect is known as the **common-ion effect** because it is caused by the addition of an ion common to both substances.

Example 12.9   Calculate the hydrogen ion concentration of a solution that is 1.0 M in acetic acid and 0.2 M in sodium acetate.

Solution   If we let x equal the moles of acetic acid that ionize, the concentrations at equilibrium will be:

$x =$ hydrogen ion concentration

$1.0 - x =$ acetic acid concentration

$0.2 + x =$ acetate ion concentration

We can drop x from the concentration of acetic acid and acetate ions because, as has been shown before, x is not significant when added to or subtracted from a number as large as 1.0 M or 0.2 M. The equilibrium concentrations become:

$$HC_2H_3O_2 \Longleftrightarrow H^+ + C_2H_3O_2^-$$
$$\ 1.0 x 0.2$$

Substituting these values into the expression for the acid dissociation constant gives:

$$K_a = \frac{[H^+][C_2H_3O_2^-]}{[HC_2H_3O_2]} = \frac{(x)(0.2)}{1.0} = 1.8 \times 10^{-5}$$

or

$$0.2x = 1.8 \times 10^{-5}$$
$$x = 9.0 \times 10^{-5}\ M$$

A solution that is 1.0 M in acetic acid and 0.2 M in sodium acetate has a hydrogen ion concentration of $9.0 \times 10^{-5}\ M$.

Problem 12.9   Calculate the hydrogen ion concentration in one liter of solution that is 0.1 M in formic acid and 0.1 M in sodium formate.

12.7   pH

The **pH** of a solution describes its acidity and is the negative logarithm (log) of its hydrogen ion concentration. The term pH is used because the hydrogen ion concentration in solutions of weak acids and in many other fluids is frequently much less than 1. Therefore, when the concentration is expressed exponentially, it contains a negative exponent. Many people find numbers with negative exponents to be confusing, and they answer with some hesitation such questions as: Is 1.8×10^{-4} larger or smaller than 3.6×10^{-5}? (To answer the question, state both numbers with the same exponent of 10. This restatement changes 3.6×10^{-5} to 0.36×10^{-4}, a value that is clearly less than 1.8×10^{-4}.) To avoid confusion when dealing with small numbers, dissociation constants and ion concentrations are stated not in exponential form, but as the negative logarithms of the actual values. The letter p has been chosen to mean "negative logarithm of." Thus, pH means the negative log of the hydrogen ion concentration, and **pOH** means the negative log of the hydroxide ion concentration:

$$pH = -\log [H^+] \qquad pOH = -\log [OH^-]$$

A. Calculation of pH

On most hand-held calculators we can calculate pH with the push of a button. Nevertheless, it is wise to know how such a calculation is carried out.

To calculate the pH of a solution, the hydrogen ion concentration must be stated in exponential form. For a solution with a hydrogen ion concentration of 0.003 M, restate that concentration as 3.0×10^{-3} M. Next, determine the log of that number. The log of the product of two numbers is the sum of their logs; thus,

$$\log(3.0 \times 10^{-3}) = \log 3 + \log 10^{-3}$$

The log of the first term can be found in Table 12.6 or with a calculator; the log of the exponential term is its exponent.

$$\log 3.0 + \log 10^{-3} = 0.477 + (-3) = -2.52$$

The pH is the negative log of the hydrogen ion concentration; thus, for $[H^+] = 3.0 \times 10^{-3}$, pH $= 2.52$.

TABLE 12.6   Logarithms of small whole numbers

log 1.0 = 0.000	log 6.0 = 0.778
log 2.0 = 0.301	log 7.0 = 0.845
log 3.0 = 0.477	log 8.0 = 0.903
log 4.0 = 0.602	log 9.0 = 0.954
log 5.0 = 0.699	log 10.0 = 1.000

Example 12.10

a. Calculate the pH of a solution with a hydrogen ion concentration of 0.00040 M.

b. Calculate the hydrogen ion concentration of a solution of pH 8.52.

Solution

a. State $[H^+]$ in exponential form:

$$[H^+] = 0.00040 \ M = 4.0 \times 10^{-4} \ M$$

Determine the log of $[H^+]$ using Table 12.6 or your calculator:

$$pH = -\log(4.0 \times 10^{-4}) = -(\log 4.0 + \log 10^{-4})$$
$$= -[0.602 + (-4)] = -0.602 + 4 = 3.40$$

b. This calculation is performed by reversing the steps in part **a.**

$$pH = -\log[H^+] = 8.52 = 9 - 0.48$$
$$\log[H^+] = -9 + 0.48$$
$$[H^+] = (\text{antilog } 0.48) \times 10^{-9} \ M$$
$$= 3.0 \times 10^{-9} \ M$$

Problem 12.10   **a.** Calculate the pH of a solution with a hydrogen ion concentration of 5.0×10^{-4} M.

b. What is the hydrogen ion concentration of a solution of pH 3.16?

B. The Interpretation of pH Values

When the hydrogen ion concentration is stated in exponential notation, the smaller the exponent, the greater the acidity of the solution. Consequently, with pH values, the lower the pH, the more acidic the solution.

$$
\begin{array}{ll}
\text{In } 0.1 \ M \ \text{HCl} & \text{In } 0.0001 \ M \ \text{HCl} \\
[\text{H}^+] = 1 \times 10^{-1} \ M & [\text{H}^+] = 1 \times 10^{-4} \ M \\
\text{pH} = 1 & \text{pH} = 4
\end{array}
$$

Figure 12.2 shows the pH of several familiar fluids. Many of these values are the midpoint of a range. Human blood plasma normally varies only between pH 7.35 and pH 7.45. Human gastric fluid is much more acidic; its normal range is between pH 1.0 and pH 2.0.

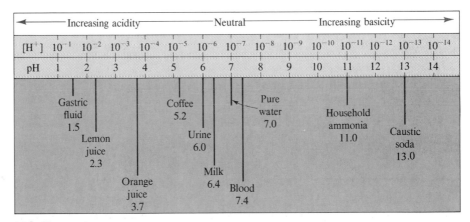

FIGURE 12.2   pH and hydrogen ion concentration.

12.8   pK_a

The **pK_a** of an acid is the negative logarithm of its acid dissociation constant. Just as pH can be used to describe the hydrogen ion concentration of a solution, pK_a can be used to describe the dissociation constant of a weak acid. The higher the pK_a of an acid, the weaker is the acid.

Table 12.7 repeats the weak acids listed in Table 12.4 and gives the pK_a of each. Notice that the acids with larger ionization constants have smaller pK_a's.

TABLE 12.7   The pK_a of some weak acids

Weak acid	K_a	pK_a
acetic acid	1.8×10^{-5}	4.74
formic acid	1.8×10^{-4}	3.74
nitrous acid	4.6×10^{-4}	3.34
hydrocyanic acid	4.9×10^{-10}	9.31
carbonic acid: K_{a_1}	4.3×10^{-7}	6.37
K_{a_2}	5.6×10^{-11}	10.25
phosphoric acid: K_{a_1}	7.5×10^{-3}	2.12
K_{a_2}	6.2×10^{-8}	7.21
K_{a_3}	2.2×10^{-13}	12.67
ammonium ion	5.5×10^{-10}	9.26

For example, formic acid ($K_a = 1.8 \times 10^{-4}$) is a stronger acid than acetic acid ($K_a = 1.8 \times 10^{-5}$). The pK_a of formic acid is 3.74, a smaller number than 4.74, the pK_a of acetic acid. Notice too that, for the polyprotic acids, the pK_a increases with each ionization. For example, the pK_a for the first ionization of phosphoric acid:

$$H_3PO_4 \rightleftharpoons H^+ + H_2PO_4^- \qquad pK_a = 2.12$$

is much smaller than that of the second ionization:

$$H_2PO_4^- \rightleftharpoons H^+ + HPO_4^{2-} \qquad pK_a = 7.21$$

Phosphoric acid is a much stronger acid and therefore much more completely ionized in solution than the dihydrogen phosphate ion.

12.9   Water as a Weak Acid

To a very small but very important extent, water is a weak acid that ionizes to hydrogen and hydroxide ions. This equilibrium reaction has the equation

$$H_2O \rightleftharpoons H^+ + OH^-$$

As with all other weak acids, this reaction has an acid dissociation constant expression:

$$K_a = \frac{[H^+][OH^-]}{[H_2O]}$$

In any amount of water, the concentration of water is so high (55.5 mol water molecules in 1000 mL water) and the number of ionized water molecules is so low (1.0×10^{-7} mol in 1000 mL water) that the concentration of the water molecules is a constant. Therefore we use a different expression when considering the ionization of water. This expression is an ion product called K_w, which has the value 1.0×10^{-14}.

$$K_w = K_a[H_2O] = [H^+][OH^-] = 1 \times 10^{-14}$$

The pK_w of water is the negative logarithm of this constant:

$$pK_w = 14$$

In pure water, the hydrogen ion concentration, $[H^+]$, equals the hydroxide ion concentration, $[OH^-]$. These concentrations can be calculated from the equation for the ionization of water,

$$H_2O \rightleftharpoons H^+ + OH^-$$

Let x equal the hydrogen ion concentration, $[H^+]$. Then x also equals the hydroxide ion concentration, $[OH^-]$. Substituting into the equilibrium expression we obtain

$$[H^+][OH^-] = (x)(x) = 1.0 \times 10^{-14}$$
$$x^2 = 1.0 \times 10^{-14}$$
$$x = 1.0 \times 10^{-7}$$
$$[H^+] = [OH^-] = 1.0 \times 10^{-7} \, M$$

The pH of pure water is 7, the negative logarithm of 1.0×10^{-7}.

A neutral solution is one that is neither acidic nor basic. The hydrogen ion concentration equals the hydroxide ion concentration, and both equal $1.0 \times 10^{-7} \, M$. In a neutral solution, then, pH = pOH = 7.

An acidic solution is one in which the hydrogen ion concentration is greater than the hydroxide ion concentration; in other words, the hydrogen ion concentration is greater than $1.0 \times 10^{-7} \, M$, and the hydroxide ion concentration is less than $1.0 \times 10^{-7} \, M$. In terms of pH, an acidic solution has a pH less than 7.

What is the hydroxide ion concentration in an acid solution? The following relationships exist whenever water is present:

$$[H^+][OH^-] = 1.0 \times 10^{-14} \quad \text{and} \quad pH + pOH = 14$$

If the hydrogen ion concentration is known, the first relationship can be used to calculate the hydroxide ion concentration. Suppose the hydrogen ion concentration is $0.10 \, M$. Substituting into and rearranging the first relationship gives

$$[OH^-] = \frac{1.0 \times 10^{-14}}{0.10} = 1.0 \times 10^{-13} \, M$$

Suppose the pH of an acidic solution is 1.0. The second relationship can be used to calculate the pOH.

$$1.0 + pOH = 14 \qquad pOH = 13$$

An alkaline or basic solution is one in which the hydrogen ion concentration is less than $1.0 \times 10^{-7} M$. In terms of pH, an alkaline solution is one in which pH is greater than 7.0. Table 12.8 shows the relationship between pH and pOH on a scale from 0 to 14.

TABLE 12.8   The relationship between pH and pOH

pH	0	1	2	3	4	5	6	7	8	9	10	11	12	13	14
pOH	14	13	12	11	10	9	8	7	6	5	4	3	2	1	0

increasingly acidic ⟵─────────────────── neutral ───────────⟶ increasingly alkaline (basic)

Example 12.11   Calculate the hydroxide ion concentration, the pH, and the pOH of 0.01 M nitric acid, HNO_3.

Solution   Nitric acid is a strong acid and is therefore completely ionized in solution. Thus, $[H^+] = 0.01 M$ and pH = 2. The hydroxide ion concentration can be calculated using the K_w constant:

$$K_w = [H^+][OH^-] = 1.0 \times 10^{-14}$$

Substituting and rearranging gives

$$0.01 \times [OH^-] = 1.0 \times 10^{-14}$$

$$[OH^-] = \frac{1.0 \times 10^{-14}}{0.01} = 1.0 \times 10^{-12} M$$

Because pH + pOH = 14, pOH = 12.

Problem 12.11   Calculate the hydrogen ion concentration of a solution having a hydroxide ion concentration of $1.0 \times 10^{-9} M$. What is the pH of this solution?

12.10   Hydrogen Ion Concentration in Solutions of the Salts of Weak Electrolytes

A. Equilibrium Considerations

The importance of the equilibria involved in the ionization of water and of weak electrolytes in water cannot be underestimated. Whenever water is present, whether in the ocean, in a raindrop, or in blood, hydrogen and hydroxide ions are present in amounts that satisfy the equation

$$K_w = [H^+][OH^-] = 1.0 \times 10^{-14}$$

Whenever an ion present in water is related through ionization to a weak acid, that equilibrium, too, must be satisfied. For example, whenever acetate ion is present in water, the equilibrium

$$\underset{\text{weak acid}}{HC_2H_3O_2} \rightleftharpoons H^+ + \underset{\text{ion}}{C_2H_3O_2^-}$$

is present. The concentrations of hydrogen ion, acetate ion, and acetic acid molecules must satisfy the acid dissociation constant for acetic acid:

$$K_a = \frac{[H^+][C_2H_3O_2^-]}{[HC_2H_3O_2]} = 1.8 \times 10^{-5}$$

In a solution of sodium acetate, the hydrogen ion concentration must satisfy two equilibrium constants: the ion product of water (K_w) and the acid dissociation constant of acetic acid (K_a). The hydrogen ion concentration of pure water is decreased by the amount of hydrogen ions that react with acetate ions to satisfy the acid dissociation constant of acetic acid. This decrease in the concentration of hydrogen ions means that the solution is no longer neutral (pH 7) but now contains an excess of hydroxide ions and is basic (pH > 7).

Similarly, if the cation of the salt is ammonium ion, NH_4^+, its equilibrium reaction in solution

$$NH_4^+ + H_2O \rightleftharpoons NH_3 + H_3O^+$$

produces hydronium ions and the solution is acidic (pH < 7).

B. Brønsted–Lowry Considerations

When a salt dissolves in water, it dissociates into ions. If the anion is a strong Brønsted–Lowry base, such as the acetate ion, or the cation is a weak Brønsted–Lowry acid, such as the ammonium ion, there is an acid–base reaction with water. The reaction is called **hydrolysis.**

The reaction of the anion base would produce hydroxide ions as the acetate ion does in the following equation:

$$C_2H_3O_2^- + H_2O \rightleftharpoons HC_2H_3O_2 + OH^-$$

If the cation of the salt is a weak Brønsted–Lowry acid, like ammonium ion, the reaction with water produces hydronium ions and the solution is acidic:

$$NH_4^+ + H_2O \rightleftharpoons NH_3 + H_3O^+$$

Example 12.12   Write equations for the hydrolysis of the appropriate ion of the following compounds:

      a. sodium carbonate   **b.** ammonium chloride
      c. potassium nitrate

Predict whether a solution of each is acidic, basic, or neutral. Justify your decision by considering the equilibrium involved.

Solution

a. Sodium ion is neither a weak acid nor a strong base. Carbonate ion is a strong base (the anion of a weak acid). The equation for its hydrolysis is

$$CO_3^{2-} + H_2O \rightleftharpoons HCO_3^- + OH^-$$

Solutions of sodium carbonate will be basic due to the presence of hydroxide ions.

b. A solution of ammonium chloride contains ammonium and chloride ions. Ammonium ion is a weak acid; chloride ion is a weak base. Weak acids will react with water to produce hydronium ion and an acidic solution:

$$NH_4^+ + H_2O \rightleftharpoons NH_3 + H_3O^+$$

Weak bases do not hydrolyze.

c. A solution of potassium nitrate contains potassium and nitrate ions. Potassium is not a weak acid; nitrate ion is a weak base. Neither ion will hydrolyze; the solution is neutral. The only equilibrium present is that of water with its ions:

$$H_2O \rightleftharpoons H^+ + OH^-$$

Problem 12.12   Write the equation for the hydrolysis of the ions of the following compounds:

a. sodium formate   **b.** ammonium bromide   **c.** lithium chloride

Predict whether a solution of each is acidic, basic, or neutral. Justify your decision by considering any equilibria present in the solutions.

12.11   **Summary**

In the traditional (Arrhenius) system, an acid is a substance whose aqueous solution contains more hydrogen ions than hydroxide ions, and a base is a substance whose aqueous solution contains more hydroxide ions than hydrogen ions. In the Brønsted–Lowry system, an acid is a proton donor, and a base is a proton acceptor. There is a system for the nomenclature of acids. Some names must be memorized.

The most common reactions of acids are: (1) neutralization, the reaction of an acid with a base, and (2) displacement of hydrogen by a metal.

Acids can be characterized as strong (completely ionized in solution) or weak (partially ionized in solution). The anion or conjugate base of a strong acid is a weak base; that of a weak acid is a strong base. Molecules of a weak acid exist in aqueous solution in equilibrium with ions. The concentrations of the weak acid molecules, hydrogen ion, and the anion of the acid are related by an equilibrium constant known as the acid dissociation constant, K_a, of the acid. This constant always has a value much less than 1.

Changing the concentration of one of the components of an equilibrium results in a change in the concentration of the other components. In the equilibrium of a weak acid with its ions, the addition of a salt of the acid profoundly decreases the concentration of the hydrogen ion. This effect is known as the common-ion effect.

The pH of a solution is the negative log of its hydrogen ion concentration. The pK_a of an acid is the negative log of its acid dissociation constant. Water is a weak acid with a dissociation constant K_w of 1×10^{-14}; p$K_w = 14$. A neutral solution has pH 7. An acidic solution has pH < 7; a basic solution has pH > 7.

Hydrolysis is the reaction of water with another substance. Salts whose anions are strong Brønsted–Lowry bases or whose cations are themselves weak acids hydrolyze, yielding either acidic or basic solutions.

Key Terms

acid dissociation constant (12.5)
Arrhenius definitions (12.1A)
Brønsted–Lowry definitions (12.1B)
common-ion effect (12.6C)
conjugate acid (12.1B)
conjugate base (12.1B)
equilibrium constant (12.5)
hydrolysis (12.10B)
hydronium ion (12.1B)
ionization constant (12.5)
K_a (12.5)

K_{eq} (12.5)
K_w (12.9)
nomenclature of acids (12.2)
pH (12.7)
pK_a (12.8)
pK_w (12.9)
pOH (12.7)
polyprotic acids (12.4B)
strong acids (12.4A)
weak acids (12.4A)

Multiple-Choice Questions

MC1. Which of the following statements are true of an acid?
   1. Its aqueous solution contains more hydroxide than hydrogen ions.
   2. Its aqueous solution turns litmus blue.
   3. It is a proton acceptor.
   4. It is more apt to contain a metal than a nonmetal.
   a. none   **b.** all   **c.** 1, 2, and 3   **d.** 2, 3, and 4
   e. 1 and 3

MC2. Which of the following can act as Brønsted–Lowry bases?

1. $H-\overset{\displaystyle\cdot\cdot}{\underset{\displaystyle |}{N}}-H$ **2.** Cl^- **3.** $\left[H-\overset{\displaystyle}{\underset{\displaystyle \|}{C}}-\overset{\displaystyle\cdot\cdot}{\underset{\displaystyle\cdot\cdot}{O}}\colon \right]^-$ **4.** Al^{3+}

a. none **b.** all **c.** 1, 2, and 3 **d.** 2, 3, and 4
e. 1 and 3

MC3. Which of the following formulas has been given an incorrect name? All are named as if in aqueous solution.
a. HBr, hydrobromic acid
b. H_2SO_3, sulfurous acid
c. HNO_3, nitrous acid
d. $Cu(CHO_2)_2$, copper(II) formate
e. H_3PO_4, phosphoric acid

MC4. Which of the following statements is *not* true of a weak acid?
a. Its solution contains fewer ions than molecules of the acid.
b. Its formula always shows two oxygen atoms.
c. Its solution contains a dynamic equilibrium.
d. The value of its equilibrium constant is less than 1.0.
e. All of these statements are true of a weak acid.

MC5. What is the molarity of a solution of hydrochloric acid that was prepared by diluting 5.0 mL of 12 M HCl to 100 mL?
a. 0.5 M **b.** 0.12 M **c.** 0.6 M **d.** 1.2 M **e.** 6.0 M

MC6. Which of the following statements are true of pH?
1. It is always less than 7 in the solution of an acid.
2. The pH of 0.1 M HCl is 1.
3. The pH of a 0.1 M solution of a weak acid is always less than 7 but more than 1.
a. none **b.** all **c.** 1 and 2 **d.** 2 and 3 **e.** 1 and 3

MC7. The addition of solid sodium nitrite to a solution of nitrous acid will
a. not change the pH of the solution.
b. lower the pH of the solution.
c. raise the pH of the solution.
d. increase the concentration of hydrogen ions in the solution.
e. lower the concentration of the NO_2^- ion.

MC8. Which of the following solutions contains the smallest number of hydrogen ions?
a. 1.0 L of 0.1 M HNO_3 **b.** 1.0 L of 0.1 M H_2SO_4
c. 1.0 L of 0.1 M $HC_2H_3O_2$ **d.** 1.0 L of 0.1 M HCl
e. They all contain the same concentration of hydrogen ions.

MC9. Magnesium metal is added in excess to 50 mL of the following solutions. From which would 0.10 g hydrogen be produced?
a. 0.05 M H_2SO_4 **b.** 0.10 M HCl **c.** 0.01 M $HC_2H_3O_2$
d. from none of them **e.** from all of them

MC10. What is the concentration of hydroxide ion in 0.010 M HCl?
a. 0.010 M **b.** 1.0×10^{-9} M **c.** 1.0×10^{-7} M
d. 1.0×10^{-12} M
e. The hydroxide ion concentration cannot be calculated unless the K_a of HCl is given.

Problems

12.1 Definitions of Acids and Bases

12.13. Name the following compounds and classify each as an acid or base in the Arrhenius system.
a. $Ca(OH)_2$ b. $HClO$
c. NH_3 d. H_2CO_3

12.14. Name the following and, using the equation for its reaction with water, state whether each is an acid or base in the Brønsted–Lowry system.
a. NH_3 b. OH^- c. $HC_2H_3O_2$
d. NH_4^+ e. SO_3^{2-} f. HBr

*12.15. The following are weak acids. For each, give its name and the formula and name of its conjugate base in the Brønsted–Lowry system.
a. HPO_4^{2-} b. HNO_2 c. HCN
d. HCO_3^- e. NH_4^+

12.16. Draw Lewis structures for the following acids and bases. Show how each acts as a Brønsted–Lowry acid or base.
a. $HC_2H_3O_2$ b. CN^-
c. formate ion, $HCOO^-$

12.17. Each of the following can act as either Brønsted–Lowry acid or base. Show this fact with appropriate equations.
a. HSO_4^- b. $H_2PO_4^-$

12.18. Give the formula and name of the conjugate base of the following acids. State whether each is a strong or weak base.
a. H_2SO_4 b. HCl c. HNO_3
d. HCO_3^- e. H_2SO_3

12.2 Nomenclature of Acids

*12.19. Given the $HBrO_3$ is bromic acid, name the following:
a. $KBrO_4$ b. $HBrO(aq)$
c. $LiBr$ d. $Ca(BrO_3)_2$
e. $Fe(BrO)_2$ f. $Al(BrO_3)_3$
g. $HBr(aq)$

12.20. Name the following:
a. $NaHSO_3$ b. Li_2S
c. $H_2SO_3(aq)$ d. $KHSO_4$
e. $H_2S(aq)$

12.3 Reactions of Acids

*12.21. Write both complete and net ionic equations for the following reactions. Name the products.
a. $H_2SO_4 + LiOH$
b. $HNO_2 + Ca(OH)_2$
c. $HC_2H_3O_2 + KOH$
d. $HCl + Zn$
e. $H_2SO_4 + Mg$
f. $Ba(OH)_2 + HNO_3$

12.22. Write both complete and net ionic equations for the following reactions. Name the products.
a. $HC_2H_3O_2 + Na$
b. $H_2SO_3 + KOH$ (one molecule)
c. $NaOH + NH_4^+$
d. $NH_3 + H_2SO_4$

12.4 Ionization of Acids

*12.23. Write equations for the equilibria present in aqueous solutions of the following weak acids. Name the acids and their conjugate bases.
a. $HClO$
b. HNO_2
c. H_2SO_3 (two equations)
d. HIO_2

12.24. Write equations for the equilibrium ionization of the weak acids in Problem 12.15. Give the expressions of their ionization constants.

12.6 Hydrogen Ion Concentration in Acid Solutions

12.25. Calculate the hydrogen ion concentration and pH in 0.1 M HNO_2.

12.26. Calculate the hydrogen ion concentration and the pH in 3.5 M formic acid.

12.27. Using the data in Table 12.4, calculate the $[H^+]$ in 1.0 M HCN. Calculate the pH of 1 L of this solution to which 0.1 mol solid sodium cyanide has been added.

12.28. Using the data in Table 12.4 for the ionization of NH_4^+ as a weak acid, calculate the pH of a solution of 0.10 M NH_4Cl.

12.7 pH

12.29. Calculate the pH of a solution of:
  a. 1×10^{-3} M HCl
  b. 0.10 M $Ca(OH)_2$
  c. 1 L of 0.1 M HCl containing 0.1 mol sodium chloride

12.30. Calculate the pH of 1 L of the following solutions:
  a. 0.5 M acetic acid that is also 1.0 M in sodium acetate

  b. 0.5 M formic acid that is also 0.5 M in sodium formate
  c. 0.1 M nitrous acid that is also 0.25 M in sodium nitrite

12.31. Vinegar is a 5% solution (wt/vol) of acetic acid in water.
  a. Calculate the molarity of the acetic acid.
  b. Calculate the hydrogen ion concentration and pH of this solution.
  c. What volume of 0.10 M sodium hydroxide would be necessary to react completely with 100 mL vinegar?

12.32. Calculate the pH of the solutions used in Examples 12.8 and 12.9 and Problems 12.8 and 12.9.

12.33. Calculate the hydrogen ion concentration and the pH of 1 L of 0.1 M acetic acid to which 31.6 g calcium acetate have been added.

Review Problems

12.34. The ionization constant for butyric acid is 1.5×10^{-5}. Calculate the pH of a solution that is 0.01 M in butyric acid and 0.02 M in sodium butyrate.

12.35. At body temperature, $pK_w = 13.6$. Calculate $[H^+]$, $[OH^-]$, pH, and pOH for pure water at body temperature.

12.36. Calculate the hydrogen ion concentration and the pH of a 0.1 M solution of nitrous acid ($K_a = 4.6 \times 10^{-4}$) that is also 0.1 M in nitrite ion.

12.37. Calculate the hydrogen ion concentration and the pH of a solution of acrylic acid, $K_a = 5.6 \times 10^{-5}$.

▪13▪

Reaction Rates and Chemical Equilibrium

Thus far we have written equations for reactions, calculated the amounts of reactants needed to form a given amount of product, and measured the enthalpy change. We have observed that some reactions are exothermic (releasing heat energy as they occur), and others are endothermic (requiring the input of heat energy). Throughout we have assumed that a reaction occurs as soon as the reactants are mixed.

Some reactions do begin immediately and are completed rapidly. Reactions between ions, such as neutralization reactions or precipitation reactions, take place as soon as the reactants are combined. Other reactions, like combustion, require added energy, such as a spark, to get started. Some reactions occur easily in one direction at some temperatures but equally easily in the opposite direction at other temperatures. We need to extend our study of reactions to explain these various phenomena. In this chapter we will consider:

1. The criteria for a reaction in terms of molecular orientation, free energy, and activation energy.
2. How to plot the progress of a reaction.
3. The rate of a reaction and how that rate is affected by changing concentrations, temperature, pressure, or surface area or by adding a catalyst.
4. Reversible reactions and their criteria, including especially the equilibrium constant.
5. The application of Le Chatelier's Principle in shifting equilibria and the effect of various stresses on different equilibria.
6. Equilibria involving ions: equilibria of sparingly soluble salts with their ions in solution and equilibria of weak electrolytes, especially as shown in the properties of buffers.

13.1   Requirements for a Reaction

A. Energy Requirements

Every reaction has an energy change associated with it. This energy change can be shown either as the enthalpy change (ΔH), discussed in Section 8.5, or as the **free energy change** (ΔG). Free energy G is the energy available to do useful work. These two energy changes are related by the equation

$$\Delta G = \Delta H - T\Delta S$$

in which T is the temperature (K) at which the reaction is taking place and ΔS is the associated **entropy change.**

1. Entropy

Entropy measures disorder. The entropy change accompanying a reaction measures how the reaction affects the orderliness of the system. Other factors being equal, it is the nature of matter to move toward a state of maximum disorder. When rocks are dumped from a truck, they do not spontaneously arrange themselves in a neat wall; rather, they land in an untidy pile. If you emptied a bag of red, white, and blue jelly beans onto a flat surface, they would not spontaneously settle into a replica of the flag but would become even more disordered than they were in the bag, for they would now be spread over a larger surface. In either case, energy must be added to decrease the entropy—to arrange the rocks into a neat fence or the jelly beans into a flag. See Figure 13.1 for another example of increase in entropy.

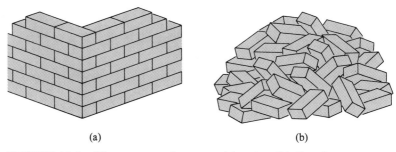

(a)   (b)

FIGURE 13.1   The concept of entropy: (a) order; (b) disorder.

In our study of chemistry we have seen changes in entropy. In Chapters 9 and 10 we saw that the structure of a solid is usually well ordered, the structure of a liquid is less ordered, and that of a gas is quite random. From these facts, we can deduce that the change from a solid to a liquid is accompanied by an increase in entropy (Figure 13.2); so is a change from a liquid to a gas. When molecules containing many atoms break apart to form smaller molecules, there is also an increase in entropy.

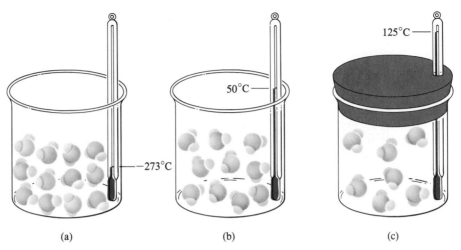

FIGURE 13.2 Entropy in the three physical states: (a) crystalline solid (order); (b) liquid (less order); and (c) gas (very little order).

2. Free energy

The free energy change ΔG associated with a reaction tells whether or not the reaction will occur under the specified conditions. For a reaction to occur spontaneously (that is, without the net addition of energy) and actually produce the products shown in its equation, the free energy change at the specified conditions must be negative. The three factors that play a role in determining the **spontaneity** of a reaction (that is, the sign of ΔG) are: (1) whether the reaction as written is exothermic ($\Delta H < 0$), (2) the temperature at which it is proposed to run the reaction, and (3) the entropy change ΔS associated with the reaction. These factors are related by the equation given earlier:

$$\Delta G = \Delta H - T\Delta S$$

Careful examination of this equation allows us to make the following predictions:

1. If a reaction is exothermic (ΔH is negative) and the entropy ΔS is positive (more disorder), the free energy change is always negative and the reaction is always spontaneous.

2. If a reaction is endothermic (ΔH is positive) and the entropy change ΔS is negative (less disorder), the free energy change is always positive and the reaction is never spontaneous.

3. If the enthalpy change ΔH and the entropy change ΔS are both positive or both negative, the spontaneity of the reaction depends on the temperature. These predictions are summarized in Table 13.1.

The phase changes of water illustrate the dependence of some free energy changes on temperature. When ice melts, the change is endothermic (ΔH is

TABLE 13.1   Enthalpy, entropy, and free energy

Enthalpy	Entropy	Free energy
exothermic, $\Delta H < 0$	increased disorder, $\Delta S > 0$	spontaneous, $\Delta G < 0$
exothermic, $\Delta H < 0$	decreased disorder, $\Delta S < 0$	spontaneity depends on temperature
endothermic, $\Delta H > 0$	increased disorder, $\Delta S > 0$	spontaneity depends on temperature
endothermic, $\Delta H > 0$	decreased disorder, $\Delta S < 0$	reaction is never spontaneous, $\Delta G > 0$

positive), and entropy increases (ΔS is positive) as the water molecules lose the ordered arrangement of ice crystals. The $T\Delta S$ factor for melting ice at 298 K is numerically larger than ΔH; the free energy change ΔG is then negative, so the melting is spontaneous at that temperature.

Similarly, the freezing of water at temperatures below 273 K is an example of a spontaneous change that is exothermic and accompanied by a decrease in entropy (an increase in order).

3. Activation energy

A reaction that is spontaneous is always accompanied by the net release of **free energy** (energy available to do useful work). However, some spontaneous reactions require added energy to get started. The energy they finally release includes both this added energy and the calculated free energy of the reaction. We all know that paper burns in air (the reaction has a negative free energy), but the reaction does not take place until extra energy is added (the heat from a burning match). That added energy is called **activation energy.**

B. Requirements at the Molecular Level

Consider now what happens at the molecular level when a reaction takes place. Bonds between the atoms in the reacting molecules break, and new bonds form to combine the atoms in a different way. For this reaction to occur, the reacting molecules must collide. Together they must have enough kinetic energy (energy of motion) to overcome the repulsion between the clouds of electrons that surround the molecules. As they collide, the two reacting molecules must be oriented so that those atoms that will be bonded together in the product are next to each other. Without this **molecular orientation,** the molecules will retreat from the collision without reacting (Figure 13.3). A collision that takes place with enough energy and with the correct molecular orientation is called an **effective collision.** Needless to say, not all collisions are effective.

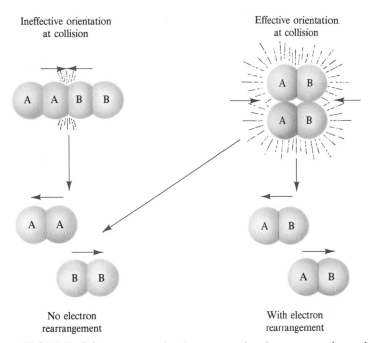

Ineffective orientation
at collision

Effective orientation
at collision

No electron
rearrangement

With electron
rearrangement

FIGURE 13.3   For a reaction between molecules to occur, the molecules must be correctly oriented when they collide.

13.2 The Course of a Reaction

Figure 13.4 plots the **course of a reaction.** The initial average energy of the reactants is indicated at the left side of each graph. If molecules are to collide effectively, they must have more than the average energy. They must have enough to overcome the repulsive forces between molecules. This added amount of energy, the activation energy, is the difference between the initial energy and the energy at the peak of each graph. Molecules having sufficient energy can collide, and, if they are correctly oriented at collision, their bonds may break and the bonds of the products form. As the new bonds form, energy is released, leaving the product molecules with the average energy shown at the right of the graph.

If the energy released is less than the activation energy—a net absorption of energy—the reaction is endothermic (Figure 13.4a). If the energy released is greater than the activation energy—a net release of energy—the reaction is exothermic (Figure 13.4b). Remember that, of all the molecules present, only some will collide; of those collisions, only some are effective and result in reactions.

This picture of a reaction is analogous to riding a bicycle over a mountain pass. The activation energy of the reaction is comparable to the energy needed

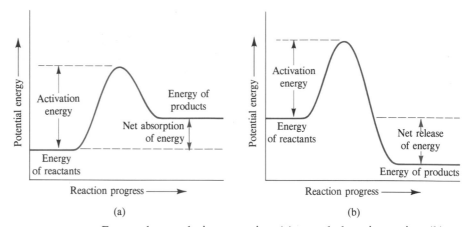

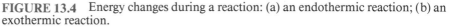

FIGURE 13.4 Energy changes during a reaction: (a) an endothermic reaction; (b) an exothermic reaction.

to pedal to the top of the pass. The energy released by the rearrangement of bonds is comparable to that gained in coasting down from the top of the pass to the floor of the next valley. If this second valley is higher than the one you started from, the energy gained in coasting down is less than the energy expended in pedaling up. This situation corresponds to an endothermic reaction, which has a net absorption of energy. If the second valley is lower than the one you started from, you gain more energy coasting down than was used pedaling up. This situation corresponds to an exothermic reaction, which results in a net release of energy.

13.3   The Rate of a Reaction

The **rate of a reaction** measures how fast the concentrations of the reactants decrease or how fast the concentrations of the products increase. The rate of a reaction is different from its spontaneity. In speaking of chemical reactions, spontaneity has no connotation of speed; it refers only to whether the reaction will occur with the release of free energy as the equation is written and at the specified temperature. The reaction of hydrogen with oxygen is a spontaneous reaction at 298 K:

$$H_2(g) + \tfrac{1}{2}O_2(g) \longrightarrow H_2O(g) \qquad \Delta G = -235 \text{ kJ/mol}$$

But a mixture of the two gases can be kept for years at room temperature (298 K) with only imperceptible amounts of water being formed. Only when more energy is added (see Section 13.1A3) does reaction take place at a measurable rate.

A. Changing the Rate of a Reaction

A reaction will form products more rapidly if the conditions under which the reaction occurs are changed so that more molecules have enough energy to reach the peak of either of the graphs in Figure 13.4. There are three ways to increase the size of this set of molecules.

1. Increase the concentration of reactant molecules present

The more molecules present in the reaction vessel, the more likely is a collision. We can increase the number of molecules by increasing the concentration of the reactants. If the reactants are both gases, an increase in pressure decreases the volume and brings the molecules closer together, thus increasing the likelihood of collision.

2. Increase the temperature of the reaction

The rate of a reaction will increase if the number of molecules with enough energy to provide the activation energy of the reaction increases. Figure 13.5 shows the distribution of energies in a collection of molecules at two different temperatures. (We considered the same distribution in Chapter 9.) In Figure 13.5, molecules with an energy greater than at point A are sufficiently energetic to provide the activation energy necessary for collision. The screened area under each curve represents the number of molecules at that temperature with an energy greater than A. The screened area is much larger under the higher-temperature curve. Therefore, at the higher temperature, more collisions occur and the reaction proceeds faster. At lower temperatures, these results are reversed and the reaction is slower.

We store food in a refrigerator because of this effect of temperature on reaction rates. The rates of the reactions that lead to food spoilage are decreased

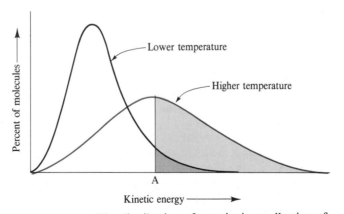

FIGURE 13.5 The distribution of energies in a collection of reacting molecules at two different temperatures.

considerably by cooling the food from room temperature to that in a refrigerator. The rates of these reactions are decreased even further by storing food in a freezer. Recent developments in low-temperature surgery have resulted from the application of this principle. By cooling the patient, metabolic reactions are slowed and the operation can be performed more deliberately, with less trauma to the patient.

3. Lower the activation energy required for reaction

We have said that a certain amount of energy, the activation energy, is necessary for reaction. If the activation energy could be lowered (in our bicycling analogy, if the pass were not quite so far above the first valley), more molecules would be able to react. In Figure 13.6, the color line represents a lower activation energy. How can the activation energy of a reaction be lowered? Just as another pass between the valleys might be lower than that originally used, another pathway for the reaction may have a lower activation energy.

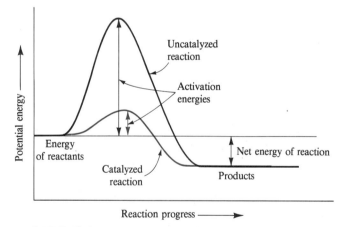

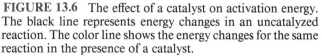

FIGURE 13.6 The effect of a catalyst on activation energy. The black line represents energy changes in an uncatalyzed reaction. The color line shows the energy changes for the same reaction in the presence of a catalyst.

A catalyst can provide such an alternative pathway. A **catalyst** is a substance that, when added to a reaction mixture, increases the rate of the overall reaction yet is recovered unchanged after the reaction is complete. Suppose a substance C is added to a reaction mixture. If the formation of the product occurs at a faster rate in the presence of C than in its absence and if C is recovered unchanged, then C is a catalyst for the reaction. The color line in Figure 13.6 shows the energy changes for the same reaction as shown by the black line but in the presence of a catalyst. Activation energy is still required, but it is less than that of the uncatalyzed reaction.

There are many examples of catalysts. Since the mid-1970s, many automobile exhaust systems have been manufactured with catalysts for the reaction

$$2\ CO(g) + O_2(g) \longrightarrow 2\ CO_2(g)$$

In the absence of a catalyst, this reaction requires a very high temperature and does not occur significantly at normal exhaust temperatures. The well-being of the public requires that cars stop spewing out large amounts of carbon monoxide. The introduction of a catalyst to the exhaust system of the car makes possible the oxidation of carbon monoxide to carbon dioxide at lower exhaust temperatures, with a considerable improvement in air quality.

The **enzymes** that trigger biological processes are catalysts. Enzymes have enormous power to change the rates of chemical reactions. In fact, most of the reactions that occur so readily in the living cell would, in the absence of enzymes, occur too slowly to support life.

For example, the enzyme carbonic anhydrase catalyzes the reaction of carbon dioxide and water to form carbonic acid:

$$CO + H_2O \underset{\text{anhydrase}}{\overset{\text{carbonic}}{\rightleftharpoons}} H_2CO_3$$

Carbonic anhydrase increases the rate of this reaction almost tenfold over that of the uncatalyzed reaction. Red blood cells are especially rich in this enzyme. For this reason, they are able to absorb carbon dioxide as it is produced in the body and transport it back to the lungs, where it is released as one of the waste products of the body.

Whether catalysts are inorganic like those in automotive emission-control systems or organic like the enzymes of living systems, they are remarkably effective because they provide an alternative pathway for a reaction, one that has a lower activation energy.

B. Other Factors That Affect the Rate of a Reaction

1. Surface area

When a reaction is to take place between reactants in two different physical states, the reaction rate is increased if we increase the **surface area** of the more-condensed reactant. Such reactions include a gas or a liquid with a solid or a gas with a liquid. Consider the reaction between oxygen in the air with cellulose, a reaction we call burning. Cellulose is the main component of wood and of flour. A match will ignite twigs but will not ignite a large log; the twigs have more surface area. If the cellulose is ground to a fine powder (enormous surface area) as in flour, a spark is sufficient to start a very rapid reaction — that is, an explosion. For this reason, flour mills operate under strict regulations designed to prevent static electricity.

Similarly, the reaction between a gas and a liquid will take place more rapidly if the liquid is sprayed in small drops through which the gas passes than if the gas is passed over the surface of a large body of the liquid.

2. Light

Some reactions, classified as **photochemical reactions,** are very sensitive to light. A mixture of the reactants in such a reaction will be stable in the dark indefinitely. When exposed to light of the correct wavelength, the reaction occurs, often at an explosive rate. The reaction of hydrogen with chlorine is one such reaction.

$$H_2(g) + Cl_2(g) \xrightarrow{\text{dark}} \text{no reaction}$$

$$H_2(g) + Cl_2(g) \xrightarrow{\text{light}} 2\ HCl(g)$$

The decomposition of nitrogen dioxide into nitrogen monoxide and atomic oxygen is another photochemical reaction. Small amounts of nitrogen dioxide are found in the exhaust gases from gasoline engines. Its decomposition on bright sunny days triggers the series of reactions that causes smog.

Example 13.1

The decomposition of hydrogen peroxide is exothermic. The reaction is catalyzed by iodide ion. The equation for the uncatalyzed reaction is

$$2\ H_2O_2(l) \longrightarrow 2\ H_2O(l) + O_2(g)$$

Sketch a possible graph for this reaction, first without a catalyst and then with a catalyst. On both graphs, label the energy of the products and the enthalpy of the reaction.

Solution

Because the uncatalyzed reaction is exothermic, the energy of the products will be less than the energy of the reactants. The shape of the graph will be like that of Figure 13.7a.

Because the catalyzed reaction is the same reaction as before, the energy of the reactants and the energy of the products will be the same as in Figure 13.7a. But, because it is catalyzed, the activation energy will be less in Figure 13.7b.

Problem 13.1

The formation of ammonia from nitrogen and hydrogen is an exothermic reaction. It can be catalyzed with heavy metal oxides. Sketch two possible graphs for this reaction: one uncatalyzed and one catalyzed. On both graphs label the energy of the reactants, the energy of the products, the activation energy, and the enthalpy of the reaction.

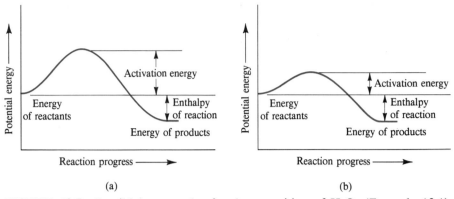

FIGURE 13.7 Possible energy plot for decomposition of H_2O_2 (Example 13.1): (a) uncatalyzed reaction; (b) catalyzed reaction.

13.4 Chemical Equilibrium

A. Definition of Chemical Equilibrium

Many chemical reactions are **reversible;** that is, the products of the reaction can combine to re-form the reactants. An example of a reversible reaction is that of hydrogen with iodine to form hydrogen iodide:

$$H_2(g) + I_2(g) \rightleftharpoons 2 \, HI \, (g)$$

We can study this reversible reaction by placing hydrogen and iodine in a reaction vessel and then measuring the concentrations of H_2, I_2, and HI at various times after the reactants are mixed. Figure 13.8 is a plot of the concentrations of reactants and products of this reaction versus time. The concentration of hydrogen iodide increases very rapidly at first, then more slowly, and finally, after the time indicated by the vertical line marked "Equilibrium," remains constant. Similarly, the concentrations of hydrogen and iodine are large at the start of the reaction but decrease, rapidly at first, and then more slowly. Finally, they, too, become constant.

If this reaction were not reversible, the concentrations of hydrogen and iodine would have continued to decrease and the concentration of hydrogen iodide to increase. This process does not happen. Instead, as soon as any molecules of hydrogen iodide are formed, some decompose into hydrogen and iodine. Two reactions are taking place simultaneously: the formation of hydrogen iodide and its decomposition. When the concentrations of all these components become constant (at the equilibrium point in Figure 13.8), the rate of the forward reaction ($H_2 + I_2 \longrightarrow 2 \, HI$) must be equal to the rate of the reverse reaction ($2 \, HI \longrightarrow H_2 + I_2$). A state of **dynamic chemical equilibrium** has then been reached, one in which two opposing reactions are proceeding at equal rates, with no net changes in concentration.

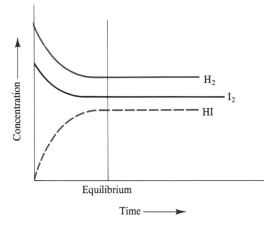

FIGURE 13.8   Concentration changes during the reversible reaction $H_2(g) + I_2(g) \rightleftharpoons 2\,HI(g)$ as it proceeds toward equilibrium.

We have encountered this criterion for equilibrium before. In the equilibrium between a liquid and its vapor, the rate of vaporization is equal to the rate of condensation. In the equilibrium of a saturated solution with undissolved solute, the rate of dissolution is equal to the rate of precipitation. In the equilibrium of a weak acid with its ions, the rate of dissociation is equal to the rate of recombination. Note that none of these reactions is static: Two opposing changes are occurring at equal rates.

B. The Characteristics of Chemical Equilibrium

1. Equal rates

At equilibrium, the rate of the forward reaction is equal to the rate of the reverse reaction.

2. Constant concentrations

At equilibrium, the concentrations of the substances participating in the equilibrium are constant. Although individual reactant molecules may be reacting to form product molecules and individual product molecules may be reacting to re-form the reactants, the concentrations of the reactants and the products remain constant.

3. No free energy change

At equilibrium, the free energy change is zero. Neither the forward nor the reverse reaction is spontaneous and neither is favored. Consider the ice–water change. Above 0°C, ice melts spontaneously to form liquid water; ΔG for this change is negative. Below $0°C$, the change from ice to water is not spontaneous; ΔG is positive. At $0°C$, the two states are in equilibrium. The rate of melting is equal to the rate of freezing: The amount of ice and the amount of liquid water present remain constant, and the free energy change is zero as long as no energy is added to or subtracted from the mixture.

C. The Equilibrium Constant

In Chapter 12, we introduced the mathematical relationship between the concentrations of the components of an equilibrium, known as the **equilibrium constant, K_{eq}**. We said that, for the general equation of a reversible reaction

$$aA + bB \rightleftharpoons cC + dD$$

the equilibrium constant expression is

$$K_{eq} = \frac{[C]^c[D]^d}{[A]^a[B]^b}$$

where the brackets mean concentration in mol/L. Therefore, for the reversible reaction

$$H_2 + I_2 \rightleftharpoons 2\,HI$$

the equilibrium constant expression is

$$K_{eq} = \frac{[HI]^2}{[H_2][I_2]}$$

In each of these expressions, the concentrations of the products of the reaction, each raised to a power equal to the coefficient of that product in the balanced equation for the reaction, are multiplied in the numerator. The concentrations of the reactants in the equation, each raised to a power equal to the coefficient of that reactant in the balanced equation, are multiplied in the denominator.

When we write an equilibrium constant expression, we omit the concentration of a substance that is a pure solid or a pure liquid. The concentration of a pure liquid or solid is so great that it remains essentially constant during a reaction. We tacitly admitted this fact in Chapter 12 by omitting water from the equation for the ionization of a weak acid and from the water ion product expression, K_w. The equation for ionization of acetic acid was given as

$$HC_2H_3O_2 \rightleftharpoons H^+ + C_2H_3O_2^- \qquad K_a = \frac{[H^+][C_2H_3O_2^-]}{[HC_2H_3O_2]}$$

even though water is required for the ionization to occur. Similarly, for the ionization of water

$$H_2O \rightleftharpoons H^+ + OH^- \qquad K_w = [H^+][OH^-]$$

the concentration of water molecules was included in the constant K_w. Applying this rule to an equilibrium involving solids:

$$CaCO_3(s) \rightleftharpoons CaO(s) + CO_2(g) \qquad K_{eq} = [CO_2]$$

The concentrations of the two solids, calcium carbonate and calcium oxide, do not appear in the equilibrium constant expression.

Example 13.2   Write the equilibrium constant expression for the reaction

$$N_2(g) + 3 H_2(g) \rightleftharpoons 2 NH_3(g)$$

Solution   1. The numerator of the constant contains the product NH_3 enclosed in brackets to represent concentration and raised to the second power, because 2 is the coefficient in the equation

$$[NH_3]^2$$

2. The denominator includes the reactants of the equation, N_2 and H_2, enclosed in brackets. The nitrogen term is to the first power; the hydrogen term is raised to the third power:

$$[N_2][H_2]^3$$

3. The complete expression is

$$K_{eq} = \frac{[NH_3]^2}{[N_2][H_2]^3}$$

Problem 13.2   Write the equilibrium constant expression for the reaction

$$2 NO(g) + O_2(g) \rightleftharpoons 2 NO_2(g)$$

Example 13.3   Write the equilibrium constant expression for the reaction

$$Fe_2O_3(s) + 3 H_2(g) \rightleftharpoons 3 H_2O(g) + 2 Fe(s)$$

Solution   Both Fe_2O_3 and Fe are solids, so their concentrations do not appear in the equilibrium constant expression. The numerator of the expression will be $[H_2O]^3$; the denominator will be $[H_2]^3$. The expression is

$$K_{eq} = \frac{[H_2O]^3}{[H_2]^3}$$

Problem 13.3   Write the equilibrium constant expression for the reaction

$$Mg(OH)_2(s) \rightleftharpoons MgO(s) + H_2O(g)$$

The value of an equilibrium constant does not depend on how equilibrium was reached. Table 13.2 presents data on the $H_2 + I_2 \rightleftharpoons 2\,HI$ equilibrium. It shows several different sets of initial concentrations and the accompanying concentrations at equilibrium. The value of the equilibrium constant is given for each experiment. Notice that the value of the equilibrium constant is the same, regardless of whether the initial materials were hydrogen and iodine or the hydrogen iodide molecule or a combination of all and whether the components were present in equal or different concentrations.

TABLE 13.2   The hydrogen iodide equilibrium constant at 490°C

Original concentrations (mol/L)			Final concentrations (mol/L)			Equilibrium constant $\dfrac{[HI]^2}{[H_2][I_2]}$
$[H_2]$	$[I_2]$	$[HI]$	$[H_2]$	$[I_2]$	$[HI]$	
1.0	1.0	0	0.228	0.228	1.544	45.9
0	0	2.0	0.228	0.228	1.544	45.9
1.0	2.0	3.0	0.316	1.316	4.368	45.9

The value of the equilibrium constant does depend on how the equation for the equilibrium is written. For example, the equilibrium constants given in Table 13.2 were calculated from the expression

$$K_{eq} = \frac{[HI]^2}{[H_2][I_2]} = 45.9$$

which is the equilibrium constant for the equation

$$H_2(g) + I_2(g) \rightleftharpoons 2\,HI(g)$$

If this equation is rewritten as

$$2\,HI(g) \rightleftharpoons H_2(g) + I_2(g)$$

the equilibrium constant becomes

$$K_{eq} = \frac{[H_2][I_2]}{[HI]^2} = 2.18 \times 10^{-2}$$

The value of an equilibrium constant does change with a change in temperature. The equilibrium constant for the $H_2 + I_2 \rightleftharpoons 2\,HI$ reaction is 45.9 only at 490°C. At 445°C, it is 64. At other temperatures, the equilibrium constant for this equation has other values, increasing as the temperature decreases and decreasing as the temperature increases.

The equilibrium constant is a very useful concept, for it allows the prediction and calculation of the concentrations of the various species present in a reaction mixture at equilibrium. This calculation is important in determining the pH of a solution, the solubility of a sparingly soluble salt, how far a reaction goes toward completion, and other similar data.

13.5   Shifting Equilibria; Le Chatelier's Principle

A system in equilibrium is a special situation where everything is in balance. Things rarely stay in balance; changes occur that shift the balance and the equilibrium that was present. We have discussed such changes in previous chapters. Recall the equilibrium that exists between a liquid and its vapor in a closed container:

$$\text{liquid} \rightleftharpoons \text{vapor}$$

At a given temperature, the vapor has a particular pressure; if the temperature is increased, it has a higher pressure. Increasing the temperature causes the equilibrium to shift to the right toward a higher concentration of vapor, but, if the system is maintained at that higher temperature, equilibrium will again be established.

In Chapter 12, we discussed the common-ion effect. We showed that, when acetate ion was added to a solution of acetic acid, the hydrogen ion concentration decreased. The addition of acetate ion caused the equilibrium

$$\text{HC}_2\text{H}_3\text{O}_2 \rightleftharpoons \text{H}^+ + \text{C}_2\text{H}_3\text{O}_2^-$$

to shift to the left. The original equilibrium was upset by the addition of the acetate ion. When a new equilibrium was established, there were more acetate ions, fewer hydrogen ions, and more acetic acid molecules. But remember, both before and after the addition of acetate ions, the concentrations of acetic acid, hydrogen ion, and acetate ion were related by the equilibrium constant

$$K_a = \frac{[\text{H}^+][\text{C}_2\text{H}_3\text{O}_2^-]}{[\text{HC}_2\text{H}_3\text{O}_2]}$$

It is possible to predict how a particular stress or change in conditions will affect an equilibrium. Such predictions are based on a principle first stated by the French chemist Henri Le Chatelier (1850–1936). In 1888, **Le Chatelier's Principle** was proposed as follows: When a stress or change in conditions is applied to a system in equilibrium, the system shifts to absorb the effect of that stress.

In considering this principle, it is important to realize that equilibrium is *not* present while the change is taking place. The sequence is: The system is in equilibrium, the stress is applied, the system changes to absorb the stress, and finally equilibrium is again reached. The following sections discuss the effect of various stresses on equilibria.

A. The Effect of Concentration Changes on Equilibria

If the stress applied to a system in equilibrium is a change in the concentration of a component of the equilibrium, the system shifts to counteract that change. If the concentration of a substance is increased, the reaction that consumes that substance is favored, and the equilibrium shifts away from that substance. If the concentration of a substance is decreased, the reaction that produces that substance is favored, and the equilibrium shifts toward that substance.

To elaborate, an equilibrium is a combination of two reactions, one forward and one reverse. Changing the concentration of a reactant in an equation changes the rate of that reaction (see Section 13.3A). When the concentration of a component in an equilibrium is increased, the rate of the reaction in which that substance is a reactant is increased; more of the product of that reaction is formed. When the concentration of a component is decreased, the rate of the reaction that uses that substance is decreased; less of its product is formed.

In studying the common-ion effect in Section 12.6C, we calculated the effect of a concentration change on the equilibrium between acetic acid and its ions:

$$HC_2H_3O_2 \rightleftharpoons H^+ + C_2H_3O_2^-$$

In 1.0 M acetic acid, the concentrations of both hydrogen and acetate ions were calculated to be 4.2×10^{-3} M. We then raised the concentration of acetate ions to 1.0 M by the addition of solid sodium acetate. How did this change affect the equilibrium? The reaction that consumes acetate ion is the one toward the left—the formation of acetic acid molecules. Every acetic acid molecule formed uses up a hydrogen ion, which decreases the concentration of hydrogen ions in the newly established equilibrium and increases the concentration of acetic acid molecules. We calculated the concentrations in the new equilibrium and found that the hydrogen ion concentration had decreased from 4.2×10^{-3} M to 1.8×10^{-5} M. The equilibrium had shifted to absorb the effect of the stress caused by the increase in the concentration of acetate ions.

Example 13.4   Consider the equilibrium

$$PCl_3(g) + Cl_2(g) \rightleftharpoons PCl_5(g)$$

Write the equilibrium constant expression for this reaction. Predict how the equilibrium position will be affected by

a. an increase in concentration of PCl_3

b. a decrease in concentration of Cl_2

Solution   The equilibrium constant expression for the reaction is

$$K_{eq} = \frac{[PCl_5]}{[PCl_3][Cl_2]}$$

a. The reaction that consumes phosphorus trichloride is the forward reaction. Increasing the concentration of PCl_3 will increase the rate of this reaction, decrease the concentration of chlorine, and increase the concentration of phosphorus pentachloride. The equilibrium will shift right.

b. The reaction that produces chlorine is the one toward the left (the reverse reaction). If the concentration of chlorine is reduced, the rate of the reverse reaction will increase. The equilibrium will shift to produce chlorine and phosphorus trichloride. When equilibrium is reestablished, there will be less phosphorus pentachloride and more phosphorus trichloride present. Thus, decreasing the concentration of chlorine will decrease the rate of the forward reaction, and the equilibrium will shift left.

Problem 13.4   Consider the equilibrium

$$H_2(g) + CO_2(g) \rightleftharpoons H_2O(g) + CO(g)$$

a. Write the equilibrium constant expression for this reaction.

b. Predict how the equilibrium position will be affected by (1) an increase in the concentration of carbon dioxide and (2) a decrease in the concentration of hydrogen.

B. The Effect of Pressure Changes on Equilibria Involving Gases

A **pressure change** is a stress to those equilibria that involve gases—that is, those equilibria that have different numbers of gaseous molecules on the left and right sides of the equilibrium equation. Increased pressure favors the reaction that decreases the number of gaseous molecules.

In the equilibrium

$$N_2(g) + 3 H_2(g) \rightleftharpoons 2 NH_3(g)$$

the forward reaction produces two molecules of gas, whereas the reverse reaction produces four molecules of gas. In terms of moles, there are four moles of gas on the left of this equilibrium and two on the right. Recall that the volume of a gas is independent of the composition of the gas; that is, at the same temperature and pressure, one mole of any gas behaving ideally occupies the same volume as one mole of any other gas. Thus, the gases on the left occupy a total of four volumes, and those on the right occupy two volumes. Increased pressure decreases the volume available to this gaseous equilibrium and favors the forward reaction, because the forward reaction decreases the number of gaseous molecules. An increase in pressure on this equilibrium will favor the formation of more ammonia (Figure 13.9).

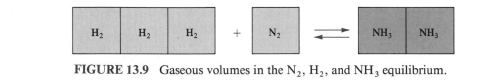

FIGURE 13.9 Gaseous volumes in the N_2, H_2, and NH_3 equilibrium.

Example 13.5 Consider the following equilibria and predict how they will shift in response to increasing pressure.

$$\textbf{a.} \quad PCl_3(g) + Cl_2(g) \rightleftharpoons PCl_5(g)$$
$$\textbf{b.} \quad CO_2(g) + H_2(g) \rightleftharpoons H_2O(g) + CO(g)$$

Solution **a.** This equilibrium has two gaseous molecules on the left and one on the right. An increase in pressure will shift the equilibrium to the right, forming more PCl_5.

b. This equilibrium has two molecules of gas on the left and the same number on the right. An increase in pressure will not shift the equilibrium in either direction.

Problem 13.5 Predict the effect of an increase in pressure on the following equilibria:

$$\textbf{a.} \quad C_2H_2(g) + H_2(g) \rightleftharpoons C_2H_4(g)$$
$$\textbf{b.} \quad N_2(g) + O_2(g) \rightleftharpoons 2\ NO(g)$$

C. The Effect of Temperature Changes on Equilibria

A change in temperature is a stress on a system in equilibrium. It changes the rate of both forward and reverse reactions and at the same time changes the value of the equilibrium constant.

In every equilibrium, two reactions proceed simultaneously, one forward and one reverse. One of these is endothermic ($\Delta H > 0$), and one is exothermic ($\Delta H < 0$). When the equilibrium is shown as an equation, the enthalpy term ΔH refers to the forward reaction. For example, in the hydrogen iodide equilibrium, the forward reaction is exothermic, as shown by the equation

$$H_2(g) + I_2(g) \rightleftharpoons 2\ HI(g) \qquad \Delta H = -51.0 \text{ kJ}$$

When the temperature of an equilibrium mixture is increased, the rate of both reactions increases (see Section 13.3A2), but the rate of the endothermic reaction (the reaction that absorbs the added energy) is increased more. For the hydrogen iodide equilibrium, an increase in temperature favors the endothermic reverse reaction. When the system returns to equilibrium, the hydrogen iodide concentration will be smaller and the concentrations of hydrogen and iodine larger. The equilibrium constant will be changed:

$$\text{at } 445°C, K_{eq} = \frac{[HI]^2}{[H_2][I_2]} = 64 \qquad \text{at } 490°C, K_{eq} = \frac{[HI]^2}{[H_2][I_2]} = 45.9$$

Example 13.6   How will an increase in temperature affect the following equilibrium?

$$PCl_3(g) + Cl_2(g) \rightleftharpoons PCl_5(g) \qquad \Delta H = -92.5 \text{ kJ}$$

Solution   In this equilibrium, the forward reaction (to form phosphorus pentachloride) is exothermic; the reverse reaction (to consume phosphorus pentachloride) is endothermic. An increase in temperature favors the endothermic reaction, and the equilibrium will shift to the left to absorb the added energy and produce more phosphorus trichloride.

Problem 13.6   How will an increase in temperature affect the following equilibrium?

$$H_2(g) + CO_2(g) \rightleftharpoons H_2O(g) + CO(g) \qquad \Delta H = -41.2 \text{ kJ}$$

D. The Effect of Catalysts on Equilibria

A catalyst changes the rate of a reaction by providing an alternative pathway with a lower activation energy. The lower-energy pathway is available to both the forward and the reverse reactions of the equilibrium. The addition of a catalyst to a system in equilibrium does not favor one reaction over the other. Instead, it increases equally the rates of both the forward and the reverse reactions. The rate at which equilibrium is reached is increased, but the relative concentrations of reactants and products at equilibrium, and hence the equilibrium constant, are unchanged.

Example 13.7 summarizes the information about equilibria and how they respond to changing conditions.

Example 13.7   Given the equilibrium

$$PBr_3(g) + Br_2(g) \rightleftharpoons PBr_5(g) \qquad \Delta H = -151.1 \text{ kJ}$$

a. Write the equilibrium constant expression for this reaction.

b. How will the equilibrium shift if the temperature is increased?

c. How will the equilibrium shift if more bromine, Br_2, is added to the reaction mixture? What will happen to the concentration of phosphorus tribromide, PBr_3?

d. How will an increase in pressure affect the relative concentrations of products and reactants?

e. How will the addition of a catalyst affect this equilibrium?

f. Is the value of K_{eq} increased, decreased, or unchanged by the change in conditions in parts **b, c,** and **d**?

Solution

a. The equilibrium constant expression is

$$K_{eq} = \frac{[PBr_5]}{[PBr_3][Br_2]}$$

b. The forward reaction is exothermic; the reverse reaction is endothermic. Increasing the temperature will increase the rate of both reactions but will increase more the rate of the reverse reaction, which is endothermic, thus changing the equilibrium constant. The concentration of phosphorus pentabromide will decrease; that of bromine and phosphorus tribromide will increase.

c. If more bromine is added to the reaction mixture, the rate of the forward reaction, the one that consumes the added bromine, will be increased. There will be more phosphorus pentabromide and less phosphorus tribromide.

d. In the forward reaction, two gaseous molecules combine to form one gaseous molecule. Increased pressure will favor the forward reaction. The concentrations of phosphorus tribromide and bromine will decrease; that of phosphorus pentabromide will increase.

e. The addition of a catalyst will change neither the concentrations nor the value of K_{eq}. The rate of both reactions will increase by the same amount.

f. K_{eq} will not be changed by the changes in conditions in parts **c** and **d**. K_{eq} will be decreased by the temperature increase in part **b**. The reverse reaction is endothermic and favored by the increase in temperature. The reverse reaction uses up phosphorus tribromide; thus, the numerator of the equilibrium constant is decreased, the denominator increased, and the value of K_{eq} decreased.

Problem 13.7   Given the equilibrium

$$2 \, H_2O(g) \rightleftharpoons 2 \, H_2(g) + O_2(g) \qquad \Delta H = +241 \text{ kJ}$$

a. Write the equilibrium constant expression for this reaction.

b. In what way will the addition of more O_2 affect this equilibrium?

c. How will a decrease in pressure affect this equilibrium?

d. In what way will an increase in temperature affect this equilibrium?

e. What effect will the addition of a catalyst have on the equilibrium?

f. Which of the four changes will affect the value of K_{eq}?

13.6   Equilibria Involving Ions

A. Equilibria of Sparingly Soluble Substances

When a **sparingly soluble salt** (the solubility of salts was discussed in Section 8.2D2) is added to water, a small amount dissolves until the solution is saturated at that temperature. An equilibrium is then present between the undissolved ions of the solid and those ions in solution. For example, when a saturated solution of calcium sulfate is prepared, the undissolved solute is in equilibrium with dissolved ions:

$$CaSO_4(s) \rightleftharpoons Ca^{2+} + SO_4^{2-}$$

The equilibrium constant for this equation is

$$K_{sp} = [Ca^{2+}][SO_4^{2-}] = 2.45 \times 10^{-5}$$

Notice that solid calcium sulfate is not included in the equilibrium constant expression. The reason for this omission was given in Section 13.4C. To emphasize this omission, as well as the fact that this constant is for an ion product rather than a true equilibrium constant expression, the constant is known as the **solubility product constant, K_{sp}**.

If the solubility of a substance is known, its solubility product constant can be calculated.

Example 13.8   The solubility of silver chloride is 6.56×10^{-4} g/L. Calculate the solubility product constant for silver chloride.

Solution   For silver chloride, $K_{sp} = [Ag^+][Cl^-]$, where the brackets mean concentration in mol/L. For every mole of silver chloride dissolved, one mole of silver ion and one mole of chloride ion are found in solution. Therefore, the first step in solving the problem is to convert the solubility from g/L to mol/L. The formula weight of silver chloride is $107.8 + 35.4 = 143.3$. The solubility of AgCl is

$$\frac{6.56 \times 10^{-4} \text{ g}}{1 \text{ L}} \times \frac{1 \text{ mol}}{143.3 \text{ g}} = 4.58 \times 10^{-6} \text{ mol/L}$$

Therefore, $[Ag^+] = [Cl^-] = 4.58 \times 10^{-6}$ mol/L. Substituting this value into the solubility product expression gives:

$$K_{sp} = [Ag^+][Cl^-] = (4.58 \times 10^{-6})(4.58 \times 10^{-6})$$
$$= 20.9 \times 10^{-12} = 2.11 \times 10^{-11}$$

Problem 13.8   The solubility of barium sulfate is 2.43×10^{-3} g/L. Calculate the solubility product constant of barium sulfate.

If the solubility product constant of a sparingly soluble salt is known, its solubility can be calculated. Table 13.3 lists the solubility product constants of several sparingly soluble salts.

TABLE 13.3   Solubility product constants

Sparingly soluble salt	Equilibrium	K_{sp}	pK_{sp}
silver bromide	$AgBr(s) \rightleftharpoons Ag^+ + Br^-$	5.2×10^{-13}	12.3
calcium carbonate	$CaCO_3(s) \rightleftharpoons Ca^{2+} + CO_3^{2-}$	4.7×10^{-9}	8.3
lead(II) chloride	$PbCl_2(s) \rightleftharpoons Pb^{2+} + 2\,Cl^-$	1.6×10^{-5}	4.8
aluminum hydroxide	$Al(OH)_3(s) \rightleftharpoons Al^{3+} + 3\,OH^-$	5.0×10^{-33}	32.3
calcium fluoride	$CaF_2(s) \rightleftharpoons Ca^{2+} + 2\,F^-$	1.7×10^{-10}	9.7
strontium sulfate	$SrSO_4(s) \rightleftharpoons Sr^{2+} + SO_4^{2-}$	7.6×10^{-7}	6.1
copper(II) sulfide	$CuS(s) \rightleftharpoons Cu^{2+} + S^{2-}$	8.0×10^{-45}	44.1

Example 13.9   Using the data in Table 13.3, calculate the solubility of strontium sulfate.

Solution   From Table 13.3, the solubility product constant of strontium sulfate is

$$K_{sp} = 7.6 \times 10^{-7} = [Sr^{2+}][SO_4^{2-}]$$

Because $[Sr^{2+}] = [SO_4^{2-}]$,

$$K_{sp} = [Sr^{2+}]^2 = 7.6 \times 10^{-7} = 76 \times 10^{-8}$$
$$[Sr^{2+}] = \sqrt{76 \times 10^{-8}} = 8.7 \times 10^{-4}\ mol/L$$

Because solubility should be given in g/L, we must convert this concentration to g/L. The formula weight of $SrSO_4$ is $87.6 + 32.0 + 4(16.0) = 183.6$. Thus the solubility of $SrSO_4$ in units of g/L is

$$\text{Solubility} = \frac{8.7 \times 10^{-4}\ mol}{1\ L} \times \frac{183.6\ g}{1\ mol} = 0.16\ g/L$$

Problem 13.9   Using the data in Table 13.3, calculate the solubility of copper(II) sulfide.

In the preceding exercises, the anion and cation of the salt were in a $1:1$ ratio. If the ions of the salt are in some other ratio, such as the $1:2$ ratio in lead(II) chloride, the concentrations of the ions are not equal. In a solution that contains only lead(II) chloride, the concentration of chloride ion is twice that of the lead ion. The equilibrium is

$$PbCl_2(s) \rightleftharpoons Pb^{2+} + 2\,Cl^-$$

and the solubility product constant expression is

$$K_{sp} = [Pb^{2+}][Cl^-]^2$$

The value of the solubility product constant of a salt depends on the solubility of the salt. Its usefulness is shown by the fact that any sparingly soluble salt will precipitate from a solution in which the product of the concentration of its ions, each raised to the power shown in the solubility product constant expression, is greater than the solubility product constant. For example, the solubility product constant of silver bromide is 5.2×10^{-13}. Whenever $[Ag^+][Br^-]$ is greater than 5.2×10^{-13} in a solution, silver bromide will precipitate. The two concentrations need not be equal. Silver bromide will precipitate when

$$[Ag^+] = 0.1 \ M, [Br^-] = 10^{-11} \ M \quad \text{because } [Ag^+][Br^-] = 10^{-12}$$
$$[Ag^+] = [Br^-] = 7.4 \times 10^{-7} \ M \quad \text{because } [Ag^+][Br^-] = 5.4 \times 10^{-13}$$
$$[Ag^+] = 10^{-10} \ M, [Br^-] = 6 \times 10^{-2} \ M \text{ because } [Ag^+][Br^-] = 6 \times 10^{-12}$$

In each case, $[Ag^+][Br^-]$ is greater than 5.2×10^{-13}. In a saturated solution of silver bromide,

$$[Ag^+][Br^-] = 5.2 \times 10^{-13}$$

and

$$AgBr(s) \rightleftharpoons Ag^+ + Br^-$$

If more bromide ions are added to the solution, the equilibrium will shift to the left, forming more precipitate and decreasing the amount of silver ion in solution. Conversely, if bromide ions are removed from solution, the precipitate will redissolve until the product of $[Ag^+]$ and $[Br^-]$ again equals 5.2×10^{-13}.

Example 13.10   A 1-L sample of solution contains 3.6×10^{-6} mol calcium ion. The equilibrium is

$$CaC_2O_4(s) \rightleftharpoons Ca^{2+} + C_2O_4^{2-} \qquad K_{sp} = 1.8 \times 10^{-9}$$

Will calcium oxalate precipitate if 0.1 mol oxalate ion is added to the solution?

Solution   For this equilibrium

$$K_{sp} = 1.8 \times 10^{-9}$$

In this solution, $[Ca^{2+}] = 3.6 \times 10^{-6} \ M$ and $[C_2O_4^{2-}] = 0.1 \ M$. Therefore

$$[Ca^{2+}][C_2O_4^{2-}] = (3.6 \times 10^{-6})(0.1) = 3.7 \times 10^{-7}$$

Because $3.7 \times 10^{-7} > 1.8 \times 10^{-9}$, calcium oxalate will precipitate.

Problem 13.10   A solution is 0.01 M in carbonate ion. Calcium ion (as solid calcium chloride) is added to the solution. At what concentration of calcium ion will calcium carbonate begin to precipitate? The solubility product constant expression of calcium carbonate is $K_{sp} = [Ca^{2+}][CO_3^{2-}] = 4.7 \times 10^{-9}$.

Example 13.11   Solid lead(II) nitrate is added to 0.1 M sodium chloride. At what concentration of lead ion will lead chloride start to precipitate?

Solution   From Table 13.3, we know that the solubility product constant of lead chloride is

$$K_{sp} = [Pb^{2+}][Cl^-]^2 = 1.6 \times 10^{-5}$$

The $[Cl^-]$ is 0.1 M; the $[Pb^{2+}]$ is unknown. Substituting and rearranging,

$$[Pb^{2+}](0.1\ M)^2 = 1.6 \times 10^{-5}$$

$$[Pb^{2+}] = \frac{1.6 \times 10^{-5}}{(0.1)^2} = 1.6 \times 10^{-3}\ M$$

As soon as $[Pb^{2+}] = 1.6 \times 10^{-3}\ M$, lead(II) chloride will start to precipitate.

Problem 13.11   If solid magnesium nitrate is added to 0.1 M sodium hydroxide, at what concentration of magnesium ion will magnesium hydroxide start to precipitate? The solubility product constant of magnesium hydroxide is

$$K_{sp} = 2.0 \times 10^{-13}$$

B. Equilibria of Weak Acids; Buffer Solutions

In Chapter 12 we discussed the equilibria between the molecules and ions of weak acids and showed how such equilibria respond to the stress of changing concentrations. Another interesting and important example of how such equilibria respond to concentration changes comes from the study of buffers. A buffer is a system that resists change. A chemical **buffer** system is one that resists change in hydrogen ion concentration (pH). Buffers are important because many chemical reactions, particularly those in biological systems, proceed best at a particular pH. If the reaction takes place in a solution that remains at that pH throughout the reaction, the most satisfactory results will be obtained.

A chemical buffer system that is designed to resist changes in hydrogen ion concentration will contain a **weak acid** and its conjugate base, both in such concentrations as will give the solution the desired pH. For example, if a buffer of pH 4.74 is needed, a solution that is 0.1 M in both acetic acid and acetate ion would be suitable (see Section 12.7A for the calculations that assign this pH to this solution). This buffer of pH 4.74 contains the following equilibrium:

$$HC_2H_3O_2 \rightleftharpoons H^+ + C_2H_3O_2^-$$
$$\underset{0.1\ M}{} \quad \underset{1.8 \times 10^{-5}\ M}{} \quad \underset{0.1\ M}{}$$

Notice that the concentrations of both the acid and its conjugate base are much larger than that of the hydrogen ion. From our study of equilibria, we know that, if an acid (H^+) is added to this solution, the equilibrium will shift to the left because the reverse reaction consumes the added hydrogen ions. If, on the other hand, hydroxide ion or some other strong base is added to the solution, that base will react with the hydrogen ions present; more acid will ionize to replenish the supply and keep the system in equilibrium. In the buffer we describe, if these added amounts are small, the concentrations of the acetic acid molecules and of acetate ion are sufficiently large that the pH of the solution will remain essentially constant. The calculations in the following example illustrate this point.

Example 13.12   Calculate the following:

   a. The pH of a buffer solution, 1 L of which contains 0.10 mol acetic acid and 0.10 mol acetate ion. The K_a of acetic acid is 1.8×10^{-5}.

   b. The pH of 1.0 L of the buffer solution in part **a** after the addition of 0.010 mol hydrochloric acid to the solution.

   c. The pH of 1.0 L of the buffer solution in part **a** after 0.010 mol hydroxide ion as NaOH has been added.

Solution   The equilibrium involved is

$$HC_2H_3O_2 \rightleftharpoons H^+ + C_2H_3O_2^-$$

The original concentrations are

$$[HC_2H_3O_2] = 0.10\ M \quad \text{and} \quad [C_2H_3O_2^-] = 0.10\ M$$

   a. We can calculate the pH of this solution using the K_a for acetic acid:

$$K_a = \frac{[H^+][C_2H_3O_2^-]}{[HC_2H_3O_2]} = 1.8 \times 10^{-5}$$

Rearranging this equation gives

$$[H^+] = \frac{K_a[HC_2H_3O_2]}{[C_2H_3O_2^-]}$$

Solving for $[H^+]$ and substituting:

$$[H^+] = \frac{(1.8 \times 10^{-5})(0.10\ M)}{0.10\ M} = 1.8 \times 10^{-5}$$

For this concentration of hydrogen ions, pH = 4.74.

   b. Before the addition of HCl, 1.0 L of the solution contains 0.10 mol each of acetate ion and acetic acid:

$$[C_2H_3O_2^-] = 0.10\ M \quad \text{and} \quad [HC_2H_3O_2] = 0.10\ M$$

When 0.010 mol hydrochloric acid is added to the solution, 0.010 mol H^+ is added to the solution. The added hydrogen ions combine with acetate ions to form more un-ionized molecules of acetic acid. The concentration of acetate ion decreases, and the concentration of acetic acid in the solution increases:

$$[C_2H_3O_2^-] = 0.10\ M - 0.010\ M = 0.09\ M$$
$$[HC_2H_3O_2] = 0.10\ M + 0.010\ M = 0.11\ M$$

To determine the hydrogen ion concentration in this solution, the same equality is used as was used for the original solution. Substituting these values into that equality:

$$[H^+] = \frac{(1.8 \times 10^{-5})[HC_2H_3O_2]}{[C_2H_3O_2^-]}$$

$$[H^+] = \frac{(1.8 \times 10^{-5})(0.11\ M)}{0.09\ M} = 2.2 \times 10^{-5}\ M$$

The addition to the buffer of 0.010 mol hydrogen ions as hydrochloric acid changed the pH to

$$pH = -\log[H^+] = -\log(2.2 \times 10^{-5}) = 4.66$$

a change of 0.08 pH units.

c. Before addition of NaOH, 1 L of the solution contained 0.10 mol acetic acid and 0.10 mol acetate ion. The added NaOH (0.010 mol) reacts with the acetic acid to form 0.010 mol sodium acetate. This reaction increases the concentration of acetate ion by 0.010 mol and decreases the concentration of acetic acid by the same amount.

$$[C_2H_3O_2^-] = 0.10\ M + 0.010\ M = 0.11\ M$$
$$[HC_2H_3O_2] = 0.10\ M - 0.010\ M = 0.09\ M$$

For this solution

$$[H^+] = \frac{(1.8 \times 10^{-5})(0.09\ M)}{0.11\ M} = 1.47 \times 10^{-5}$$

$$pH = -\log[H^+] = -\log(1.47 \times 10^{-5}) = 4.83$$

a change of 0.09 pH units from the original solution.

Problem 13.12   One liter of a buffer solution contains 0.10 mol formic acid and 0.20 mol sodium formate.

    a. Calculate the pH of this solution.

    b. Calculate the pH after 0.010 mol hydrogen ion is added to 1 L of the buffer.

    c. Calculate the pH after 0.010 mol hydroxide ion is added to 1 L of the buffer.

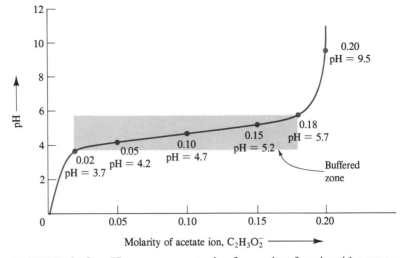

FIGURE 13.10   pH versus concentration for a series of acetic acid–acetate ion solutions.

Figure 13.10 shows a plot of pH versus concentration of acetate ion for the buffer solution whose pH was calculated in Example 13.12. The graph shows a change of only two pH units as the acetate ion concentration changes from 0.02 M to 0.18 M. Within this range the solution acts as a buffer.

The addition of small amounts of either hydrogen or hydroxide ions to a buffer does not appreciably change the pH. Example 13.13 illustrates how different the results would be if, instead of using a buffer based on acetic acid, we used a solution of a strong acid with the same pH.

Example 13.13   A solution contains the same number of moles of HCl and NaCl and has a pH of 4.74. Calculate the final pH of 1 L of this solution after the addition of 0.01 mol hydrogen ion.

Solution   Because HCl is a strong acid, it is completely ionized. Cl^- has no tendency to combine with H^+ to form HCl molecules. At pH 4.74, the concentration of hydrogen ion is 1.8×10^{-5} M. Any addition of H^+ will simply increase this concentration. The final $[H^+]$ of the solution will be

$$[H^+] = 0.01 \ M + 0.000018 \ M = 0.010018 \ M \quad \text{and} \quad pH = 2$$

Thus, the addition of a small amount of H^+ to a very dilute solution of a strong acid causes a large change in pH.

Problem 13.13   Calculate the change in pH when 0.01 mol of hydroxide ion is added to 1 L 0.000018 M HCl.

In choosing the buffer for a particular reaction, the following criteria should be considered.

1. ***pH.*** Buffering is most effective if the pH required for the reaction is close to the pK_a of the weak acid of the buffer. An effective buffer has a pH that equals the pK_a of the weak acid ±1 unit.

   For acetic acid, the pK_a is 4.74; therefore, a solution of acetic acid and sodium acetate will function as a buffer within a pH range of approximately 3.74–5.74.

2. ***Concentration.*** The concentration of a buffer refers to the total concentration of both the weak acid and the anion of the weak acid. A 0.1 M acetic acid–acetate ion buffer may be made up of 0.025 mol acetic acid and 0.075 mol sodium acetate in a liter of solution, or any other combination in which $[HC_2H_3O_2] + [C_2H_3O_2^-] = 0.1$ M. The buffer used in Figure 13.10 has a concentration of 0.2 M. At each point on the graph, the sum of the concentrations of acetic acid and acetate ion equals 0.2 M. The concentration of the buffer should be greater than the amount of H^+ or OH^- that will be produced by the reaction being buffered.

3. ***Capacity.*** The capacity of a buffer, or its effectiveness, is the amount of hydrogen or hydroxide ion that the buffer can absorb without a significant change in pH. Capacity depends on concentration and pH. The most effective buffer is one with equal concentrations of a weak acid and its salt — that is, one in which the pH of the solution is equal to the pK_a of the weak acid. Although a solution of acetic acid and sodium acetate will buffer within the pH range 3.74 5.74, it is most effective as a buffer at, or very near, pH 4.74. Table 13.4 lists some common buffers and the pH at which each is most effective.

TABLE 13.4   Some common buffers and the pH at which each is most effective

Name of acid	Formula	K_a	Anion	pH of buffer when anion/acid = 1
phosphoric	H_3PO_4	7.5×10^{-3}	$H_2PO_4^-$	2.12
formic	HCO_2H	1.8×10^{-4}	HCO_2^-	3.74
acetic	$HC_2H_3O_2$	1.8×10^{-5}	$C_2H_3O_2^-$	4.74
carbonic	$CO_2 + H_2O$	4.3×10^{-7}	HCO_3^-	6.37
dihydrogen phosphate ion	$H_2PO_4^-$	6.2×10^{-8}	HPO_4^{2-}	7.21
bicarbonate ion	HCO_3^-	5.6×10^{-11}	CO_3^{2-}	10.25
monohydrogen phosphate ion	HPO_4^{2-}	2.2×10^{-13}	PO_4^{3-}	12.66

13.7   Summary

For two molecules to react, they must collide and the electrons rearrange so that old bonds are broken and new bonds formed. For a reaction to occur spontaneously, the enthalpy of the reaction, the temperature at which it is to occur, and the entropy change must be such that the free energy of the reaction as calculated by the equation $\Delta G = \Delta H - T\Delta S$ is less than zero. Often activation energy must be added to initiate the reaction. The course of a chemical reaction can be plotted versus the energy of the reactants and products. This plot shows the activation energy necessary for an effective collision between reactants, as well as the exothermic or endothermic nature of the reaction. The number of collisions between reactants can be increased by increasing their concentrations. Increasing the temperature of a reaction will increase the number of molecules that are sufficiently energetic to collide effectively. The addition of a catalyst provides a reaction pathway that requires a lower activation energy. Catalysts in biological systems (enzymes) are particularly important as they allow complex reactions to occur at body temperature.

Many chemical reactions are reversible. Allowed to proceed spontaneously, they reach an equilibrium state in which the rate of the forward reaction equals the rate of the reverse reaction. At equilibrium, a relationship exists between the concentrations of the components of the reaction known as the equilibrium constant. For the reaction

$$a\text{A} + b\text{B} \rightleftharpoons c\text{C} + d\text{D}$$

the equilibrium constant expression is

$$K_{eq} = \frac{[\text{C}]^c[\text{D}]^d}{[\text{A}]^a[\text{B}]^b}$$

Le Chatelier's Principle predicts how an equilibrium will adjust to stresses imposed by changing concentrations, pressure, or temperature. A catalyst does not shift an equilibrium.

TABLE 13.5   Equilibrium constants

Name of constant	Symbol	Typical equation	Expression of constant
equilibrium constant	K_{eq}	$A_2 + B_2 \rightleftharpoons 2\,AB$	$\dfrac{[AB]^2}{[A_2][B_2]}$
acid dissociation constant	K_a	$HA \rightleftharpoons H^+ + A^-$	$\dfrac{[H^+][A^-]}{[HA]}$
ionization constant of water	K_w	$H_2O(l) \rightleftharpoons H^+ + OH^-$	$[H^+][OH^-]$
solubility product constant	K_{sp}	$M_aN_b(s) \rightleftharpoons aM^{b+} + bN^{a-}$	$[M^{b+}]^a[N^{a-}]^b$

Ionic equilibria may involve sparingly soluble salts. The equilibrium constant for the equilibrium between a sparingly soluble salt and its ions in solution is known as the solubility product constant, K_{sp}. The concentration of the undissolved salt does not appear in this constant. Ionic equilibria also involve weak electrolytes. These equilibria were discussed in Chapter 12 and repeated here in a discussion of buffers.

A buffer solution resists change. An acid–base buffer resists changes in pH. Buffers, prepared from weak acids or bases and their salts, are particularly effective within one pH unit of the pK_a of the acid. Table 13.5 lists the types of equilibrium constants discussed.

Key Terms

activation energy (13.1A3)
buffers (13.6B)
catalyst (13.3A3)
chemical equilibrium (13.4A)
course of a reaction (13.2)
dynamic chemical equilibrium
  (13.4A)
effective collision (13.1B)
entropy (13.1A1)
entropy change (13.1A)
enzyme (13.3A3)
equilibrium constant, K_{eq} (13.4C)
free energy (13.1A3)
free energy change (13.1A)

Le Chatelier's Principle (13.5)
molecular orientation (13.1B)
photochemical reaction (13.3B2)
pressure change (13.5B)
rate of a reaction (13.3)
reversible chemical reactions (13.4A)
solubility product constant, K_{sp}
  (13.6A)
sparingly soluble salt (13.6A)
spontaneity (13.1A2)
surface area (and reaction rate)
  (13.3B1)
weak acid (13.6B)

Multiple-Choice Questions

MC1. Which of the following terms must have a negative sign for a reaction to be spontaneous?
  a. the free energy change   **b.** the entropy change
  c. the enthalpy change   **d.** They must all have negative signs.
  e. The reaction will be spontaneous only if all these terms are positive.

MC2. Which of the following statements are always true of a spontaneous reaction?
  1. It always occurs rapidly.
  2. It is always accompanied by an increase in temperature.
  3. It is always exothermic.
  a. none   **b.** 1 and 2   **c.** 1 and 3   **d.** 2 and 3
  e. All are true.

MC3. Which of the following statements are correct in describing the reaction whose course is shown in the figure?

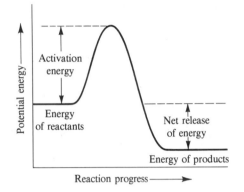

1. The reaction is endothermic.
2. The addition of a catalyst will raise the activation energy.
3. The reactants have less energy than the products.
4. All the molecules have enough energy to react.
   a. none   **b.** 2, 3, and 4   **c.** 1 and 4   **d.** 1, 2, and 3
   e. all

MC4. Which of the following changes will increase the rate of the reaction shown in this equation?

$$4\,NH_3(g) + 5\,O_2(g) \longrightarrow 4\,NO(g) + 6\,H_2O(g)$$

1. increasing the concentration of oxygen
2. addition of a catalyst
3. increasing the temperature of the reaction
   a. all   **b.** none   **c.** 1   **d.** 2 and 3   **e.** 1 and 3

MC5. Which of the following statements is/are true of a catalyst?
1. It increases the rate of a reaction.
2. It can be recovered unchanged after the reaction ceases.
3. It prevents the reaction from reaching equilibrium.
   a. all   **b.** 1   **c.** 2   **d.** 2 and 3   **e.** 1 and 2

MC6. For the equilibrium

$$2\,P(s) + 3\,H_2(g) \rightleftharpoons 2\,PH_3(g) \qquad \Delta H = -9.6\ kJ$$

The equilibrium constant expression is:

a. $\dfrac{[PH_3]}{[H_2]}$ **b.** $\dfrac{[PH_3]^2}{[P]^2[H_2]^3}$ **c.** $\dfrac{[PH_3]^2}{[H_2]^3}$ **d.** $\dfrac{[P][H_2]^2}{[PH_3]^3}$

e. none

MC7. Which of the following will not change the position of the equilibrium shown in Question 6?
a. increasing the temperature
b. increasing the pressure
c. adding a catalyst
d. increasing the concentration of hydrogen
e. decreasing the temperature

MC8. Calculate the concentration of copper(II) ion in a saturated solution of copper(II) sulfide (K_{sp} of CuS $= 8.5 \times 10^{-45}$).
a. 2.9×10^{-23} mol/L   b. 9.2×10^{-23} mol/L
c. 8.5×10^{-23} mol/L   d. 2.9×10^{-23} g/L
e. not enough data to calculate

MC9. Which of the following silver salts is the most soluble?
a. AgCl, $K_{sp} = 1.5 \times 10^{-10}$ b. AgIO$_3$, $K_{sp} = 9.2 \times 10^{-9}$
c. AgBr, $K_{sp} = 7.7 \times 10^{-13}$ d. AgBrO$_3$, $K_{sp} = 4.0 \times 10^{-5}$
e. AgI, $K_{sp} = 1.0 \times 10^{-16}$

MC10. An acid–base buffer may:
a. contain a strong acid and the salt of that acid.
b. resist pH change no matter how much strong acid is added.
c. contain a weak acid and a salt of that weak acid.
d. always have a pH of 7.
e. never be found in biological systems.

Problems

13.1 Requirements for a Reaction

13.14. Define entropy. Is the entropy of the universe increasing or decreasing?

13.15. Which member of the following pairs has the higher entropy?
a. H$_2$O(l) or H$_2$O(g)
b. 6 HCHO or C$_6$H$_{12}$O$_6$ both in solution
c. ten students in class or ten students after class

13.16. Define free energy. Explain why an exothermic reaction is not always spontaneous.

13.2 The Course of a Reaction

13.17. For the reaction

$$\tfrac{1}{2}H_2(g) + \tfrac{1}{2}Cl_2(g) \longrightarrow HCl(g)$$
$$\Delta H = -92.0 \text{ kJ}$$

a. Show how the molecules must be oriented at collision for the reaction to occur.
b. Plot the course of the reaction.

*13.18. Plot the course of a reaction for:
a. an exothermic reaction.
b. an endothermic reaction.
c. an exothermic reaction in the presence of a catalyst.

*13.19. Label the activation energy and the enthalpy in each part of Problem 13.18.

13.3 The Rate of a Reaction

*13.20. Explain why increasing the concentration of the reactants may increase the rate of a reaction. Explain why increasing the pressure of a reaction involving gases increases the rate of the reaction.

13.21. Describe how a catalyst can change the rate of a reaction.

13.22. Draw a curve showing the normal distribution of energies in a collection of molecules. Draw another curve showing how this distribution of energies changes as the temperature is increased. Why does the rate of a reaction increase with increased temperature?

13.23. What role do enzymes play in biological reactions?

13.24. Consider the reaction

$$4 \text{ NH}_3(g) + 5 \text{ O}_2(g) \longrightarrow$$
$$4 \text{ NO}(g) + 6 \text{ H}_2\text{O}(g)$$
$$\Delta H = -1.61 \times 10^3 \text{ kJ}$$

What changes would increase the rate of the reaction?

13.4 Chemical Equilibrium

13.25. a. Calculate the equilibrium constant for the hypothetical reaction

$$\text{AB} \rightleftharpoons \text{A} + \text{B}$$

if the concentrations at equilibrium are [AB] = 2, [A] = 2, [B] = 2.

   b. After a stress has been absorbed by the above reaction, [A] = 8 and [AB] = 16. What is the new concentration of B?

13.26. a. Write the equilibrium constant expression, K_{a_1} for the reaction

$$\text{NH}_3 + \text{H}_2\text{O} \rightleftharpoons \text{NH}_4^+ + \text{OH}^-$$

   b. Write the equilibrium constant expression, K_{a_2} for the reaction

$$\text{NH}_4^+ \rightleftharpoons \text{NH}_3 + \text{H}^+$$

   c. Show for these reactions that $K_{a_1} \times K_{a_2} = K_w$.

13.27. At body temperature, $pK_a = 13.6$. Calculate [H$^+$], [OH$^-$], pH, and pOH for pure water at body temperature.

13.28. Write the equilibrium constant expression for each of the following equilibria.
   a. the ionization of propanoic acid, $\text{HC}_3\text{H}_5\text{O}_2$
   b. the reaction of nitric oxide with oxygen to form nitrogen dioxide
   c. a precipitate of aluminum hydroxide with its ions in solution

13.29. For the reaction

$$2 \text{ Sb}(s) + 3 \text{ Cl}_2(g) \rightleftharpoons 2 \text{ SbCl}_3(g)$$
$$\Delta H = -315 \text{ kJ/mol SbCl}_3$$
$$\Delta G = -302 \text{ kJ/mol SbCl}_3$$

   a. Draw a graph showing the progress of the forward reaction.
   b. Draw a graph showing the progress of the reverse reaction.
   c. Write the equilibrium constant expression.
   d. Indicate what is the value of ΔG for (1) the forward reaction, (2) the reverse reaction, and (3) the equilibrium.

13.5 Shifting Equilibria; Le Chatelier's Principle

*13.30. The Haber process for the industrial preparation of ammonia uses the reaction

$$\text{N}_2(g) + 3 \text{ H}_2(g) \rightleftharpoons 2 \text{ NH}_3(g)$$
$$\Delta H = -92 \text{ kJ/mol}$$

   a. Write the equilibrium constant expression for this reaction.
   b. How will increased concentration of nitrogen change the concentration of ammonia?
   c. How will increased pressure change the concentration of ammonia?
   d. How will the addition of a catalyst change the equilibrium?
   e. How will increased temperature change the concentration of ammonia?

13.31. One of the components of smog is the brown gas nitrogen dioxide, NO_2. It participates in the equilibrium

$$2 NO_2 \rightleftharpoons N_2O_4$$
$$\Delta H = -61.5 \text{ kJ/mol}$$

Dinitrogen tetroxide (N_2O_4) is colorless. Is smog apt to be darker in winter or in summer? If the nitrogen dioxide were in a closed cylinder, would the gas get lighter or darker as the pressure is increased?

13.32. **a.** Write an equation for the equilibrium that exists between a precipitate of chromium(III) hydroxide and its ions in solution.
b. What is the ratio between $[Cr^{3+}]$ and $[OH^-]$ in this equilibrium situation?
c. What will happen to the concentration of Cr^{3+} if a solution of sodium hydroxide is added?
d. What will happen to the concentration of Cr^{3+} if hydrochloric acid is added?

*13.33. Consider the equilibrium

$$2 H_2O(g) + 2 Cl_2(g) \rightleftharpoons$$
$$4 HCl(g) + O_2(g) \qquad \Delta H = -120 \text{ kJ}$$

What will happen to the concentration of chlorine if:
a. the temperature is raised?
b. the pressure is increased?
c. the concentration of HCl is increased?
d. the concentration of oxygen is decreased?
e. a catalyst is added to the system?

13.6 Equilibria Involving Ions

13.34. A solution contains 0.1 M strontium nitrate. How many moles of sulfate ion can be added to one liter of this solution before strontium sulfate ($K_{sp} = 7.6 \times 10^{-7}$) will begin to precipitate?

*13.35. A saturated solution of silver bromide ($K_{sp} = 5.2 \times 10^{-13}$) is 0.1 M in bromide ion (from sodium bromide). What is the concentration of dissolved silver ion?

13.36. Calculate the concentration of calcium ion in a saturated solution of calcium oxalate ($K_{sp} = 1.3 \times 10^{-9}$).

13.37. Calculate the concentration of fluoride ion in a saturated solution of barium fluoride ($K_{sp} = 2.4 \times 10^{-5}$).

13.38. Calculate the concentration of lead ion in a 0.1 M solution of potassium iodide that is saturated with lead iodide ($K_{sp} = 8.3 \times 10^{-9}$).

13.39. Given pure acetic acid and pure sodium acetate, what weight of each would you mix in water to prepare one liter of a 0.1 M buffer solution of pH 4.74? The total concentration of acetic acid and sodium acetate should equal 0.1 M.

13.40. Calculate the pH change on adding 0.04 g solid NaOH to 100 mL of a solution containing 0.05 mol acetic acid and 0.05 mol sodium acetate. Assume the volume remains constant.

Review Problems

13.41. A sample of silver chloride ($K_{sp} = 2.1 \times 10^{-11}$) is to be prepared by adding 0.10 M silver nitrate solution to 0.100 L of a solution of sodium chloride that contains 35.0 g sodium chloride per liter of solution. At what silver concentration will a precipitate first appear? What volume of silver nitrate solution must be added to form 5.0 g silver chloride?

13.42. What is the molarity of a sulfuric acid solution if 10.00 mL of this solution reacts exactly with 15.67 mL of 0.15 M $BaCl_2$? Will barium sulfate ($K_{sp} = 1.1 \times 10^{-10}$) have precipitated at the endpoint of this reaction?

13.43. What volume of 0.15 M nitrous acid will react exactly with 0.15 g magne-

sium? How will the rate of this reaction compare with the rate of the reaction of 0.15 M HCl with a similar sample of magnesium?

13.44. What mass of calcium oxide must be added to 100 mL water to prepare a 0.01 M solution of calcium hydroxide? What is the concentration of hydroxide ion in this solution? Could you prepare a 0.1 M solution of $Ca(OH)_2$ the same way?

K_{sp} of $Ca(OH)_2 = 1.3 \times 10^{-6}$

■14■

Oxidation–Reduction

The original meaning of the term *oxidation* was "combination with oxygen." At that same time, *reduction* referred to the reduction of a metallic ore, usually an oxide, to a free metal. Modern chemists have extended those definitions. Oxidation now means any change that results in an increase in oxidation number. Reduction now means any change that results in a decrease in oxidation number. In this chapter, we consider the various aspects of oxidation–reduction reactions, such as:

1. The terms used in describing oxidation and reduction.
2. Two methods of balancing the equations of complex oxidation–reduction reactions.
3. Reduction potentials and their implications in determining the relative ease with which various substances are oxidized or reduced and the relative activities of metals and of nonmetals.
4. How to determine the spontaneity of an oxidation–reduction reaction.
5. The electrochemical application of oxidation–reduction reactions.

14.1   Review of Terms

In Section 8.3, those reactions that involved a transfer of electrons were classified as oxidation–reduction reactions. Many definitions were given. In this chapter on oxidation–reduction, we will remind you of those definitions. For a more complete discussion of the material, reread Section 8.3.

When an iron nail is immersed in a solution of copper(II) ion, a thin coating of copper is deposited on the nail and iron(II) ion is found in the solution. The equation for this reaction is:

$$\underset{0}{Fe(s)} + \underset{+2}{Cu^{2+}(aq)} \longrightarrow \underset{0}{Cu(s)} + \underset{+2}{Fe^{2+}(aq)}$$

Note that we have used a net ionic equation to describe the reaction. The essential components of an oxidation–reduction reaction are much more easily identified in a net ionic equation than in a formula equation. We will use almost exclusively net ionic equations in this chapter.

Below each component of the equation we have written its **oxidation number.** We have done so to stress that a reaction is classified as oxidation–reduction if in its course two elements change oxidation number; one must increase in oxidation number, and the other must decrease in oxidation number. Table 14.1 reviews the rules for assigning oxidation numbers. A complete discussion of the subject is in Section 6.2.

TABLE 14.1   Rules for assigning oxidation numbers

Substance	Oxidation number
any uncombined element	0
a monatomic ion	the charge on the ion
oxygen (except in peroxides)	−2
in a compound	the sum equals zero
in a polyatomic ion	the sum equals the charge on the ion

In the reaction whose equation is shown above, iron has been oxidized: its oxidation number has increased from 0 to +2. If an element's oxidation number increases, the element must have lost electrons, which is the criterion for oxidation. Therefore, iron was oxidized in the reaction. Table 14.2 lists the criteria of **oxidation** and **reduction** and states that the substance oxidized is the reducing agent in the reaction. Therefore, iron is the **reducing agent** in this reaction.

Copper(II) ion is reduced in this reaction. It meets the requirements of having decreased in oxidation number and gained electrons. Copper(II) ion is the **oxidizing agent** in this reaction.

We will see later that, in the balanced equation for this or any other oxidation–reduction reaction, the number of electrons lost by the substance

TABLE 14.2 Criteria for oxidation and reduction

Substance oxidized	*Substance reduced*
loses electrons	gains electrons
attains a more positive oxidation number	attains a more negative oxidation number
is the reducing agent	is the oxidizing agent

oxidized equals the number of electrons gained by the substance reduced. Electrons can neither be plucked from the air to use in an equation nor left over after a reaction.

Example 14.1

When hydrogen sulfide is bubbled through an acidic solution of iron(III) chloride, iron(II) ion and uncombined (free) sulfur are obtained. The net ionic equation for this reaction is:

$$2 \text{ Fe}^{3+} + \text{HS}^- \longrightarrow 2 \text{ Fe}^{2+} + \text{S}(s) + \text{H}^+$$

For this reaction identify the substance oxidized, the substance reduced, the oxidizing agent, and the reducing agent.

Solution

1. Assign an oxidation number to each element each time it occurs in the equation.

$$\underset{+3}{2 \text{ Fe}^{3+}} + \underset{+1, -2}{\text{HS}^-} \longrightarrow \underset{+2}{2 \text{ Fe}^{2+}} + \underset{0}{\text{S}} + \underset{+1}{\text{H}^+}$$

2. Iron has changed from $+3$ to $+2$. Iron has decreased in oxidation number, therefore it has gained electrons and has been reduced. The iron(III) ion is the oxidizing agent.

3. Sulfur has changed from -2 in the bisulfide ion, HS^-, to free sulfur with an oxidation number of 0. The sulfur has lost electrons; it has been oxidized. The bisulfide ion is the reducing agent.

Problem 14.1

When manganese dioxide is reacted with hydrochloric acid, the manganese(II) ion, free chlorine, and water are the products. The net ionic equation for this reaction is:

$$\text{MnO}_2 + 2 \text{ Cl}^- + 4 \text{ H}^+ \longrightarrow \text{Mn}^{2+} + \text{Cl}_2(g) + 2 \text{ H}_2\text{O}$$

Identify what has been oxidized and what has been reduced in this reaction. Name the oxidizing and the reducing agents.

14.2   Half-Reactions

Although oxidation and reduction proceed simultaneously and an oxidation–reduction reaction can be shown in a single equation, the processes of oxidation and reduction are often shown as separate equations known as **half-reactions.** We encountered several examples of half-reactions in Section 8.3 during the introduction to oxidation, but we did not use the term itself. The half-reactions for the oxidation of sodium and magnesium are:

$$Na \longrightarrow Na^+ + e^-$$
$$Mg \longrightarrow Mg^{2+} + 2\ e^-$$

In these oxidation half-reactions, electrons are found as products. Similarly, we have already encountered reduction half-reactions for chlorine and oxygen:

$$Cl_2 + 2\ e^- \longrightarrow 2\ Cl^-$$
$$O_2 + 4\ e^- \longrightarrow 2\ O^{2-}$$

In a reduction half-reaction, the electrons are reactants.

An oxidation–reduction reaction results from the combination of an oxidation half-reaction with a reduction half-reaction. The reaction of iron with copper(II) ion (discussed in Section 14.1) combines the two half-reactions:

oxidation: $Fe(s) \longrightarrow Fe^{2+}(aq) + 2\ e^-$
reduction: $Cu^{2+}(aq) + 2\ e^- \longrightarrow Cu(s)$

The equations have been labeled oxidation and reduction. Note that electrons are a product (have been lost) in the oxidation half-reaction and are a reactant (have been gained) in the reduction half-reaction.

By adding together an oxidation half-reaction and a reduction half-reaction, the net ionic equation for an oxidation–reduction reaction is obtained. This equation will be balanced if the numbers of electrons in the two half-reactions are equal. Both half-reactions shown above involve two electrons. Adding them together gives the balanced net ionic equation for the overall reaction.

$$Fe(s) + Cu^{2+}(aq) + \cancel{2\ e^-} \longrightarrow Fe^{2+}(aq) + \cancel{2\ e^-} + Cu(s)$$

If the number of electrons in the two half-reactions is not the same, as, for example, in the reaction of aluminum with hydrogen ion, each equation must be multiplied by an appropriate factor. The unbalanced equation for this reaction is:

$$Al(s) + H^+ \xrightarrow{\ \ /\ \ } Al^{3+} + H_2(g)$$

The half-reactions for this equation are:

oxidation: $Al(s) \longrightarrow Al^{3+}(aq) + 3\ e^-$
reduction: $2\ H^+(aq) + 2\ e^- \longrightarrow H_2(g)$

Three electrons are lost in the oxidation half-reaction, but only two are gained in the reduction half-reaction. Before we can add them to obtain the net ionic equation for the overall reaction, we must multiply the oxidation half-reaction by 2 and the reduction half-reaction by 3. We then obtain

oxidation: $2 \text{ Al}(s) \longrightarrow 2 \text{ Al}^{3+}(aq) + 6 \text{ e}^-$

reduction: $6 \text{ H}^+(aq) + 6 \text{ e}^- \longrightarrow 3 \text{ H}_2(g)$

which can be added to give the balanced equation:

$2 \text{ Al}(s) + 6 \text{ H}^+(aq) + \cancel{6 \text{ e}^-} \longrightarrow 2 \text{ Al}^{3+}(aq) + \cancel{6 \text{ e}^-} + 3 \text{ H}_2(g)$

Example 14.2

Write the equation for the reaction of zinc with hydrochloric acid. Isolate the half-reactions involved. Show that the balanced net ionic equation can be obtained from these half-reactions.

Solution

The equation is:

$\text{Zn} + 2 \text{ HCl} \longrightarrow \text{H}_2 + \text{ZnCl}_2$

Zinc's oxidation number changes from 0 to +2; it is oxidized.

$\text{Zn} \longrightarrow \text{Zn}^{2+} + 2 \text{ e}^-$

Hydrogen's oxidation number changes from +1 to 0; it is reduced. Because the product hydrogen is a diatomic gas, we must use two hydrogen ions as reactants.

$2 \text{ H}^+ + 2 \text{ e}^- \longrightarrow \text{H}_2$

Both half-reactions use two electrons; therefore, they can be added as is to give:

$\text{Zn} + 2 \text{ H}^+ + \cancel{2 \text{ e}^-} \longrightarrow \text{H}_2 + \text{Zn}^{2+} + \cancel{2 \text{ e}^-}$

Problem 14.2

Write the net ionic equation for the reaction of bromine with sodium iodide to form sodium bromide and free iodine. Isolate the half-reactions involved. Show that the balanced net ionic equation can be obtained by the addition of these half-reactions.

14.3   Balancing More Complex Oxidation–Reduction Equations

The equations for some redox (a term meaning oxidation–reduction) reactions can be balanced by inspection as we have balanced all equations thus far in this text. However, many redox reactions have equations much too complex to be balanced by inspection without a great deal of frustrating trial and error. Such complex equations can more easily be balanced in a series of steps that utilize either (1) the changes in oxidation numbers or (2) the half-reactions involved. In both methods we will use net ionic equations rather than complete formulas.

A. Balancing Redox Equations Using Oxidation Numbers

This method of writing balanced equations for redox reactions depends on knowing how oxidation numbers change during the reaction. Each step will be illustrated by developing the equation for the reaction of copper metal with nitric acid to form copper(II) ion and nitrogen dioxide.

Step 1. Write a skeletal equation for the reaction using the formulas of all ions, molecules, and atoms participating in the reaction. Example:

$$Cu + H^+ + NO_3^- \not\longrightarrow Cu^{2+} + NO_2$$

Step 2. If the reaction occurs in acid, add hydrogen ion and water as reactants or products. If the reaction occurs in base, add hydroxide ion and water as reactants or products. Example: This reaction takes place in acid. We will leave H^+ as a reactant and add water as a product, knowing we might need H^+ as a product or water as a reactant.

$$Cu + H^+ + NO_3^- \not\longrightarrow Cu^{2+} + NO_2 + H_2O$$

Step 3. Assign an oxidation number to each element in the skeletal equation. Example:

$$Cu + H^+ + NO_3^- \not\longrightarrow Cu^{2+} + NO_2 + H_2O$$
$$\;\;0 \quad\; +1 \quad\; +5,-2 \qquad +2 \qquad +4,-2 \quad +1,-2$$

Step 4. For those elements that change oxidation number, use a line above or below the equation to connect the reactant form with the product form. Write above or below the line the loss or gain of electrons that accompanies the change. Example:

$$\overset{-2\,e^-}{Cu + H^+ + NO_3^- \not\longrightarrow Cu^{2+} + NO_2 + H_2O}$$
$$+1\,e^-$$

Step 5. Using multipliers, equate the loss and gain of electrons. Example:

$$\overset{-2\,e^-}{Cu + H^+ + NO_3^- \not\longrightarrow Cu^{2+} + NO_2 + H_2O}$$
$$2(+1\,e^-)$$

Step 6. Multiply the coefficient of the substance oxidized or reduced by the multiplier used to equate the electron changes. Example:

$$\overset{\displaystyle -2\ e^-}{Cu + H^+ + 2\ NO_3^- \nrightarrow Cu^{2+} + 2\ NO_2 + H_2O}$$

$$2(+1\ e^-)$$

Balance the oxygen by changing the number of water molecules present. Balance the hydrogen ions to match the water molecules. Example: Each nitrate ion yields one NO_2 and one O^{2-}. Two nitrate ions yield two O^{2-}. These require four hydrogens to form two water molecules, so the coefficient of H^+ is changed to 4. The equation should now be balanced.

$$Cu + 4\ H^+ + 2\ NO_3^- \longrightarrow Cu^{2+} + 2\ NO_2 + 2\ H_2O$$

Step 7. Check that the charge is balanced. Example: The charge on the left is $+2$; the charge on the right is also $+2$. The equation is balanced.

Example 14.3

Potassium permanganate reacts with hydrochloric acid to produce manganese(II) ion and free chlorine. Write the balanced equation for the reaction using oxidation numbers.

Solution

Step 1. The skeletal equation for the reaction is written below. Because the product chlorine is diatomic, a multiple of two chlorine atoms must be used in the equation. We prepare for this requirement by using 2 Cl^- as a reactant.

$$MnO_4^- + H^+ + 2\ Cl^- \nrightarrow Mn^{2+} + Cl_2$$

Step 2. The reaction takes place in acid; add H^+ as a reactant and water as a product.

$$MnO_4^- + H^+ + 2\ Cl^- \nrightarrow Mn^{2+} + Cl_2 + H_2O$$

Step 3. The oxidation numbers are:

$$\underset{\substack{+7,-2 \quad\;\; +1 \qquad\; -1 \qquad\qquad +2 \qquad 0 \quad\;\; +1,-2}}{MnO_4^- + H^+ + 2\ Cl^- \nrightarrow Mn^{2+} + Cl_2 + H_2O}$$

Step 4. Connect the two forms of the substances that change oxidation number and show electron change.

$$\overset{\displaystyle +5\ e^-}{MnO_4^- + H^+ + 2\ Cl^- \nrightarrow Mn^{2+} + Cl_2 + H_2O}$$

$$-2\ e^-$$

Note that we have included the electron change for both chlorine atoms.

Step 5. Equate the loss and gain of electrons.

$$2(+5\ e^-)$$

$$MnO_4^- + H^+ + 2\ Cl^- \not\longrightarrow Mn^{2+} + Cl_2 + H_2O$$

$$5(-2\ e^-)$$

Step 6. Multiply the coefficients of the substances oxidized or reduced by the multipliers used to equate the electron change.

$$2\ MnO_4^- + H^+ + 10\ Cl^- \not\longrightarrow 2\ Mn^{2+} + 5\ Cl_2 + H_2O$$

Balance the oxygen and hydrogen using water and H^+. We have 8 oxygens on the left; we need 8 water molecules on the right. This change will require 16 H^+ on the left. Inserting these numbers gives the balanced equation

$$2\ MnO_4^- + 16\ H^+ + 10\ Cl^- \not\longrightarrow 2\ Mn^{2+} + 5\ Cl_2 + 8\ H_2O$$

Step 7. Check that the charges are balanced. If they are, the equation is probably balanced too. On the left, $-2 + 16 - 10 = 4$; on the right, $+4$. The equation is balanced.

Problem 14.3 Lead dioxide reacts with aqueous hydrogen iodide to yield lead(II) ion and free iodine. Using oxidation numbers, write the balanced ionic equation for this reaction.

Example 14.4 Using oxidation numbers, write the balanced ionic equation for the reaction of metallic zinc with acidified potassium dichromate solution to yield zinc(II) and chromium(III) ions.

Solution **Step 1.** Write the skeletal equation for the reaction

$$Zn + Cr_2O_7^{2-} + H^+ \not\longrightarrow Zn^{2+} + Cr^{3+}$$

Step 2. The reaction takes place in acid, so add H^+ and H_2O as needed. At the same time, note that the dichromate ion has two chromium atoms, so we will need a multiple of two chromium atoms in the product. To meet this requirement, put the coefficient 2 in front of the chromium(III) ion.

$$Zn + Cr_2O_7^{2-} + H^+ \not\longrightarrow Zn^{2+} + 2\ Cr^{3+} + H_2O$$

Step 3. Assign oxidation numbers.

$$Zn + Cr_2O_7^{2-} + H^+ \not\longrightarrow Zn^{2+} + 2\ Cr^{3+} + H_2O$$
$$0 \quad +6,-2 \quad +1 \qquad +2 \qquad +3 \qquad +1,-2$$

Step 4. Connect the two forms of those substances that change oxidation number and show electron change.

$$\overset{-2\,e^-}{Zn + Cr_2O_7^{2-} + H^+ \longrightarrow Zn^{2+} + 2\,Cr^{3+} + H_2O}$$
$$+6\,e^-$$

Step 5. Using multipliers, equate the loss and gain of electrons.

$$\overset{3(-2\,e^-)}{Zn + Cr_2O_7^{2-} + H^+ \longrightarrow Zn^{2+} + 2\,Cr^{3+} + H_2O}$$
$$+6\,e^-$$

Step 6. Multiply the coefficients of the oxidized and reduced substances in the equation by the same number used to balance the electrons. Balance the oxygen and hydrogen using water and hydrogen ions. There are 7 oxygen on the left; we will need 7 water molecules on the right, which will require 14 hydrogen ions on the left.

$$3\,Zn + Cr_2O_7^{2-} + 14\,H^+ \longrightarrow 3\,Zn^{2+} + 2\,Cr^{3+} + 7\,H_2O$$

Step 7. Check that the charge is balanced. On the left, $0 - 2 + 14 = 12$; on the right, $3(+2) + 2(+3) = 12$. The charges are balanced; the equation is balanced.

Problem 14.4 Using oxidation numbers, write the ionic equation for the reaction of potassium iodide with an acidified solution of potassium dichromate. The products of the reaction are chromium(III) ion and free iodine.

B. Balancing Redox Equations Using Half-Reactions

The equations for oxidation–reduction reactions can also be balanced using half-reactions. Again we will illustrate this method using the reaction of nitric acid with copper metal to form nitrogen dioxide and copper(II) ion. The steps using this method are as follows:

Step 1. Write the skeletal equation for the reaction using the ions, atoms, and molecules that participate in the reaction. Example:

$$Cu + H^+ + NO_3^- \longrightarrow Cu^{2+} + NO_2$$

Step 2. Isolate the product and reactant form of each substance except H^+ that occurs in the skeletal equation. Example:

$$Cu \longrightarrow Cu^{2+}$$
$$NO_3^- \longrightarrow NO_2$$

Step 3. Balance these half-reactions by mass using hydrogen ions and water in acid solutions and hydroxide ions and base in basic solutions. Example:

$$Cu \nrightarrow Cu^{2+}$$
$$2\,H^+ + NO_3^- \nrightarrow NO_2 + H_2O$$

Step 4. Balance the charges in these half-reactions using electrons as reactants or products. Remember that electrons are negative; they must be added to the more positive side of the equation. Example:

$$Cu \longrightarrow Cu^{2+} + 2\,e^-$$
$$2\,H^+ + NO_3^- + e^- \longrightarrow NO_2 + H_2O$$

Step 5. Using multipliers, equate the numbers of electrons in the half-reactions. Example:

$$Cu \longrightarrow Cu^{2+} + 2\,e^-$$
$$2(2\,H^+ + NO_3^- + e^- \longrightarrow NO_2 + H_2O)$$

Step 6. Add the multiplied half-reactions. Delete the electrons from both sides of the equation. Example:

$$Cu + 4\,H^+ + 2\,NO_3^- + 2\,e^- \longrightarrow$$
$$Cu^{2+} + 2\,e^- + 2\,NO_2 + 2\,H_2O$$

Step 7. Check that the equation is balanced by mass and by charge. This equation has on the left $4(+1) + 2(-1) = +2$; on the right, $+2$. The equation is balanced. Notice that, in this method, it is not necessary to determine the oxidation numbers of the various elements present.

Example 14.5

Using half-reactions, write the balanced equation for the reaction of sulfuric acid with iodide ion to form free iodine and hydrogen sulfide gas.

Solution

Step 1. Write the skeletal ionic equation for the reaction.

$$H^+ + SO_4^{2-} + I^- \nrightarrow H_2S + I_2$$

Note that hydrogen sulfide is in the gaseous state; it is therefore not ionic and is written as a molecule.

Step 2. Isolate the half-reactions.

$$H^+ + SO_4^{2-} \nrightarrow H_2S$$
$$I^- \nrightarrow I_2$$

Step 3. Balance these half-reactions by mass using hydrogen ions and water.

$$10\,H^+ + SO_4^{2-} \nrightarrow H_2S + 4\,H_2O$$

We have 4 oxygen atoms on the left, so we add 4 H_2O on the right. They will require 8 hydrogen atoms. The hydrogen sulfide requires 2 hydrogen atoms. Together we need 10 hydrogen ions on the left. The second equation requires only a change of coefficient on the left.

$$2\ I^- \not\longrightarrow I_2$$

Step 4. Balance these half-reactions by charge by adding electrons to the more positive side.

$$10\ H^+ + SO_4^{2-} + 8\ e^- \longrightarrow H_2S + 4\ H_2O$$
$$2\ I^- \longrightarrow I_2 + 2\ e^-$$

Step 5. Equate the electrons in these half-reactions using multipliers.

$$10\ H^+ + SO_4^{2-} + 8\ e^- \longrightarrow H_2S + 4\ H_2O$$
$$4(2\ I^- \longrightarrow I_2 + 2\ e^-)$$

Step 6. Add the multiplied half-reactions. Delete the electrons.

$$10\ H^+ + SO_4^{2-} + \cancel{8\ e^-} + 8\ I^- \longrightarrow H_2S + 4\ H_2O + 4\ I_2 + \cancel{8\ e^-}$$

Step 7. Check that the equation is balanced by equating charges. On the left this equation has $10 + (-2) + 8(-1) = 0$; on the right, 0. The charges balance; the equation is balanced.

Problem 14.5   When bromide ion reacts with sulfuric acid, the products are sulfur dioxide, water, and free bromine. Using half-reactions, write the balanced ionic equation for this redox reaction.

Example 14.6   The reaction of permanganate ion, MnO_4^-, with oxalate ion, $C_2O_4^{2-}$, in acid solution forms manganese(II) ion and carbon dioxide. Using half-reactions, write the balanced equation for this reaction.

Solution   **Step 1.** Write the skeletal equation for the reaction.

$$MnO_4^- + H^+ + C_2O_4^{2-} \not\longrightarrow Mn^{2+} + CO_2$$

Step 2. Isolate the half-reactions.

$$MnO_4^- \not\longrightarrow Mn^{2+}$$
$$C_2O_4^{2-} \not\longrightarrow CO_2$$

Step 3. Balance the half-reactions by mass.

$$MnO_4^- + 8\ H^+ \not\longrightarrow Mn^{2+} + 4\ H_2O$$
$$C_2O_4^{2-} \not\longrightarrow 2\ CO_2$$

Step 4. Balance these half-reactions by charge by adding electrons to the more positive side.

$$MnO_4^- + 8\ H^+ + 5\ e^- \longrightarrow Mn^{2+} + 4\ H_2O$$
$$C_2O_4^{2-} \longrightarrow 2\ CO_2 + 2\ e^-$$

Step 5. Equate the electrons by using multipliers.

$$2(MnO_4^- + 8\ H^+ + 5\ e^- \longrightarrow Mn^{2+} + 4\ H_2O)$$
$$5(C_2O_4^{2-} \longrightarrow 2\ CO_2 + 2\ e^-)$$

Step 6. Add the multiplied half-reactions. Delete the electrons.

$$2\ MnO_4^- + 16\ H^+ + \cancel{10\ e^-} + 5\ C_2O_4^{2-} \longrightarrow$$
$$2\ Mn^{2+} + 8\ H_2O + 10\ CO_2 + \cancel{10\ e^-}$$

Step 7. Check that the equation is balanced by checking the total charge on each side of the equation: on the left, $2(-1) + 16 + 5(-2) = +4$; on the right, $2(+2) = +4$.

Problem 14.6 Acidified dichromate solution will oxidize iron(II) ion to iron(III) ion. The dichromate ion is reduced to chromium(III) ion. The reaction is accompanied by a dramatic color change as the orange dichromate ion changes to green chromium(III). Using half-reactions, write the balanced ionic equation for the reaction.

14.4 Reduction Potentials

Elements and ions differ in the ease with which they are reduced or oxidized. Reduction half-reactions for several ions and elements are listed in Table 14.3. After each equation is a number called the **reduction potential** (symbol, E^0), which compares the tendency for the listed reaction to occur with the tendency of aqueous hydrogen ion to be reduced:

$$2\ H^+ + 2\ e^- \longrightarrow H_2 \qquad E^0 = 0.0\ V$$

The reduction potential for the reduction of hydrogen ion to hydrogen has been arbitrarily assigned a value of 0.0 volts (V). A substance whose reduction potential is less than that of hydrogen ion (has a negative value) is reduced less easily than is hydrogen ion. A substance whose reduction potential is positive (is greater than that of a hydrogen ion) is more easily reduced than is hydrogen.

The reduction potentials listed were measured at 25 °C and 1 atm pressure, using solutions in which one mole of the reactant was dissolved in 1 L solution. At different concentrations, temperatures, or pressures, the values of the reduction potentials are slightly different.

Polyatomic ions (MnO_4^-, $Cr_2O_7^{2-}$, NO_3^-, and so on) are often involved in electron-transfer reactions. Reduction potentials for these ions are also shown

TABLE 14.3   A partial list of reduction potentials at 25°C

Oxidized form	Reduced form	E^0 (volts)
$Li^+ + e^-$	$\rightleftharpoons Li$	-3.05
$K^+ + e^-$	$\rightleftharpoons K$	-2.93
$Mg^{2+} + 2\,e^-$	$\rightleftharpoons Mg$	-2.38
$Zn^{2+} + 2\,e^-$	$\rightleftharpoons Zn$	-0.76
$Fe^{2+} + 2\,e^-$	$\rightleftharpoons Fe$	-0.44
$2\,H^+ + 2\,e^-$	$\rightleftharpoons H_2$	0.00
$SO_4^{2-} + 4\,H^+ + 2\,e^-$	$\rightleftharpoons H_2SO_3 + H_2O$	0.20
$Cu^{2+} + 2\,e^-$	$\rightleftharpoons Cu$	0.34
$I_2 + 2\,e^-$	$\rightleftharpoons 2\,I^-$	0.54
$Fe^{3+} + e^-$	$\rightleftharpoons Fe^{2+}$	0.77
$Hg_2^{2+} + 2\,e^-$	$\rightleftharpoons 2\,Hg$	0.79
$Ag^+ + e^-$	$\rightleftharpoons Ag$	0.80
$NO_3^- + 4\,H^+ + 3\,e^-$	$\rightleftharpoons NO + 2\,H_2O$	0.96
$Br_2 + 2\,e^-$	$\rightleftharpoons 2\,Br^-$	1.09
$O_2 + 4\,H^+ + 4\,e^-$	$\rightleftharpoons 2\,H_2O$	1.23
$Cr_2O_7^{2-} + 14\,H^+ + 6\,e^-$	$\rightleftharpoons 2\,Cr^{3+} + 7\,H_2O$	1.33
$Cl_2 + 2\,e^-$	$\rightleftharpoons 2\,Cl^-$	1.36
$MnO_4^- + 8\,H^+ + 5\,e^-$	$\rightleftharpoons Mn^{2+} + 4\,H_2O$	1.49

in Table 14.3. Why do we show the whole ion instead of just the element? The element is not alone in solution but is part of the polyatomic ion. For example, in the permanganate ion, MnO_4^-, manganese has an oxidation number of $+7$. Manganese in this state is not an ion but part of a covalently bonded polyatomic ion. The half-reactions show only substances that actually exist in solution.

A. The Use of Reduction Potential Tables in Writing Redox Equations

We have seen that an oxidation–reduction equation is the sum of an oxidation half-reaction and a reduction half-reaction. Reduction half-reactions can be obtained from a table of reduction potentials, but what about oxidation half-reactions? These equations can be obtained in a balanced form by reversing the reduction equations of the table. Let us see how equations such as those in Table 14.3 can be used in balancing redox equations.

Example 14.7   When copper reacts with dilute nitric acid, copper(II) ion and nitrogen oxide (NO) are formed:

$$Cu + HNO_3 \longrightarrow\!\!\!/ \, Cu^{2+} + NO$$

Write the balanced ionic equation for this reaction.

Solution   Copper has changed oxidation number from 0 to $+2$; it has been oxidized. We can find the equation for the reduction of copper ion in the table and reverse it to represent oxidation.

$$Cu \longrightarrow Cu^{2+} + 2\ e^-$$

The nitrogen of nitric acid was reduced in this reaction. We can find the equation for this reduction of nitric acid to nitrogen oxide in the table.

$$NO_3^- + 4\ H^+ + 3\ e^- \longrightarrow NO + 2\ H_2O$$

We can balance the loss and gain of electrons by multiplying the oxidation half-reaction by 3 and the reduction half-reaction by 2. Adding the resulting equations gives us the equation

$$3\ Cu + 2\ NO_3^- + 8\ H^+ + \cancel{6\ e^-} \longrightarrow 4\ H_2O + 2\ NO + 3\ Cu^{2+} + \cancel{6\ e^-}$$

After canceling electrons, we have the balanced equation for this reaction.

Problem 14.7   Using equations from Table 14.3, write the balanced ionic equation for the reaction of chloride ions with permanganate ion to form chlorine and manganese(II) ions.

B. The Spontaneity of Redox Reactions

A table of reduction potentials allows prediction of the spontaneity of a redox reaction. We saw earlier that a spontaneous reaction has a negative free energy change (see Section 13.1A). In a spontaneous reaction, the sum of the potentials of the half-reactions is positive.

   The potential of the reduction half-reaction is obtained from a table of reduction potentials such as Table 14.3; the potential of the oxidation half-reaction is equal in magnitude to the reduction potential of the reversed reduction half-reaction but is opposite in sign. For example, we can use the reduction half-reaction

$$Mg^{2+} + 2\ e^- \longrightarrow MgO \qquad E^0 = -2.38\ V$$

to write the oxidation half-reaction

$$MgO \longrightarrow Mg^{2+} + 2\ e^- \qquad E^0 = +2.38\ V$$

   The sum of the potentials of a redox reaction can be calculated as follows. For the redox reaction

$$Fe + Cu^{2+} \longrightarrow Fe^{2+} + Cu$$

the half-reactions are:

oxidation: $Fe \longrightarrow Fe^{2+} + 2\ e^-$

reduction: $Cu^{2+} + 2\ e^- \longrightarrow Cu$

The E^0 of the reduction half-reaction is obtained from Table 14.3:

$$Cu^{2+} + 2\,e^- \longrightarrow Cu \qquad E^0 = +0.34\ V$$

Table 14.3 also gives the potential for the reaction

$$Fe^{2+} + 2\,e^- \longrightarrow Fe \qquad E^0 = -0.44\ V$$

If the equation is reversed, we must also reverse the sign of the E^0 for the oxidation half-reaction

$$Fe \longrightarrow Fe^{2+} + 2\,e^- \qquad E^0 = +0.44\ V$$

The sum of the potentials for the overall reaction is:

$$E^0 = +0.34\ V + 0.44\ V = +0.78\ V$$

The reaction is spontaneous and will occur without the net input of energy.

Consider another situation. Suppose a student proposed preparation of hydrogen gas by adding copper metal to hydrochloric acid. Is this proposal sensible? The equation for the proposed reaction is:

$$Cu + 2\ HCl \longrightarrow Cu^{2+} + 2\ Cl^- + H_2(g)$$

The half-reactions for this equation and their potentials are:

oxidation: $Cu \longrightarrow Cu^{2+} + 2\,e^- \qquad E^0 = -0.34\ V$

reduction: $2\ H^+ + 2\,e^- \longrightarrow H_2(g) \qquad E^0 = 0.0\ V$

The potential for the reduction half-reaction was obtained from Table 14.3. The potential for the oxidation half-reaction was obtained by reversing the equation and sign of the corresponding reduction half-reaction in Table 14.3. The sum of the potentials is negative.

$$E^0 = -0.34\ V + 0.0\ V = -0.34\ V$$

The reaction is not spontaneous. The student will not be able to prepare hydrogen gas by this reaction.

In summary, a redox reaction is spontaneous if the sum of the potentials of its half-reactions is positive. If the sum of the potentials is negative, the reaction will not occur without the net input of energy.

Example 14.8

Will acidified potassium dichromate solution oxidize chloride ion? The equation for the proposed reaction is:

$$2\ Cl^- + 14\ H^+ + Cr_2O_7^{2-} \longrightarrow 2\ Cr^{3+} + Cl_2 + 7\ H_2O$$

Solution

The half reactions for the equation are:

oxidation: $2\ Cl^- \longrightarrow Cl_2 + 2\,e^- \qquad E^0 = -1.36\ V$

reduction: $Cr_2O_7^{2-} + 14\ H^+ \longrightarrow 2\ Cr^{3+} + 7\ H_2O \qquad E^0 = +1.33\ V$

The potential of the reduction half-reaction was obtained from Table 14.3. The potential for the oxidation of chloride ion was obtained by reversing the sign of the potential for the half-reaction

$$Cl_2 + 2\ e^- \longrightarrow 2\ Cl^-$$

The sum of these potentials is negative.

$$E^0 = -1.36\ V + 1.33\ V = -0.03\ V$$

Acidified potassium dichromate will not, under standard conditions, oxidize chloride ion.

Problem 14.8   Will acidified potassium dichromate solution oxidize bromide ion to free bromine?

C. Activity Series

The activity of an element determines how easily the free element becomes an ion. For metals to become ions, they must be oxidized:

$$Li \longrightarrow Li^+ + e^-$$

This equation is the reverse of the equation shown in Table 14.3. A metal is said to be very active (easily oxidized) if the reduction potential of its ion is near the top of the table (has a very large negative value). For example, lithium has a reduction potential of -3.05 V. Lithium is a very active metal and is very easily oxidized to lithium ion.

An **activity series** of the metals lists the metals in order of decreasing ease of oxidation. In terms of reduction potentials, this listing would be in order of increasing reduction potentials. Table 14.4 shows an activity series for metals. This activity series can be used to predict displacement reactions (see Section 8.2C). A metal will displace any metal below it in the series. For example, copper is above silver in the activity series. Copper will displace silver, as shown in the equation

$$Cu + 2\ Ag^+ \longrightarrow Cu^{2+} + 2\ Ag$$

The activity of a nonmetal also depends on the relative ease with which it forms ions. For a nonmetal, the formation of ions involves reduction. The reduction potential of chlorine measures the tendency of the following reaction to occur:

$$Cl_2 + 2\ e^- \longrightarrow 2\ Cl^-$$

The most active nonmetal has the lowest reduction potential. Table 14.5 is an activity series for the halogens. Any nonmetal in the series will displace a

TABLE 14.4   Activity series
for metals

Metals
lithium
potassium
calcium
sodium
magnesium
aluminum
zinc
chromium
iron
nickel
tin
lead
hydrogen
copper
mercury
silver
platinum
gold

TABLE 14.5   Activity series for nonmetals

Increasing ease of reducing element $\longrightarrow$

iodine	bromine	chlorine	fluorine
I$^-$	Br$^-$	Cl$^-$	F$^-$

$\longleftarrow$ Increasing ease of oxidizing ion

nonmetal to its left in the series. Thus, Table 14.5 predicts that chlorine will displace bromine as shown in the following equation:

$$2 \text{ NaBr} + \text{Cl}_2 \longrightarrow 2 \text{ NaCl} + \text{Br}_2$$

Example 14.9   Use Tables 14.3, 14.4, and 14.5 to predict the spontaneity of the following reactions. If the reaction goes as written, balance the equation. If the reaction will not go as written but requires the input of energy, cross out the products and write "no reaction."

　　　a. $\text{Ni} + \text{Zn}^{2+} \longrightarrow \text{Ni}^{2+} + \text{Zn}$
　　　b. $\text{I}_2 + 2 \text{ Br}^- \longrightarrow \text{Br}_2 + 2 \text{ I}^-$
　　　c. $\text{Fe} + 2 \text{ Ag}^+ \longrightarrow \text{Fe}^{2+} + 2 \text{ Ag}$

Solution

a. No reaction. Nickel is below zinc in the activity series of metals (Table 14.4). It will not displace zinc.

b. No reaction. Iodine is to the left of bromine in the nonmetal activity series (Table 14.5). Iodine will not displace bromine.

c. $Fe + 2 Ag^+ \longrightarrow Fe^{2+} + 2 Ag$. Iron is above silver in the activity series (Table 14.4). It will displace silver.

Problem 14.9   Use Tables 14.3, 14.4, and 14.5 to predict the spontaneity of the following reactions. If the reaction is not spontaneous as written, cross out the products and write "no reaction."

> **a.** $Zn + Pb^{2+} \longrightarrow Zn^{2+} + Pb$
>
> **b.** $Hg + 2 H^+ \longrightarrow H_2 + Hg^{2+}$
>
> **c.** $Br_2 + 2 I^- \longrightarrow 2 Br^- + I_2$

14.5   Electrochemical Cells

We have assumed that the oxidation–reduction reactions discussed so far have taken place in a single container and that electrons are transferred directly from the substance oxidized to the substance reduced. Although a redox reaction may take place in this manner, it is also possible to transfer the electrons through a wire or electrical circuit. Such an arrangement is called an **electrochemical cell.**

An electrochemical cell that utilizes a spontaneous redox reaction to produce a flow of electrons is called either a galvanic cell, after the Italian chemist Luigi Galvani (1737–1798), or a voltaic cell, after Galvani's friend Alessandro Volta (1745–1827). Galvani and Volta were pioneers in the study of the production of electricity by chemical means. In a **voltaic cell,** the half-reactions of the spontaneous redox reaction are kept separated and the electrons produced flow through an external wire. Figure 14.1 shows one such cell.

The overall reaction of the cell shown is:

$$Zn(s) + CuSO_4(aq) \longrightarrow Cu(s) + ZnSO_4(aq)$$

In this apparatus, the two half-reactions are separated from each other by a porous membrane. On one side of the membrane is a solution of zinc sulfate and a piece of copper. A wire connects the zinc electrode and the copper electrode. At the zinc electrode the reaction is:

$$Zn \longrightarrow Zn^{2+} + 2 e^-$$

The electrons produced in this half-reaction move through the wire to the copper-containing solution, where they combine with the copper ions to form copper atoms:

$$Cu^{2+} + 2 e^- \longrightarrow Cu$$

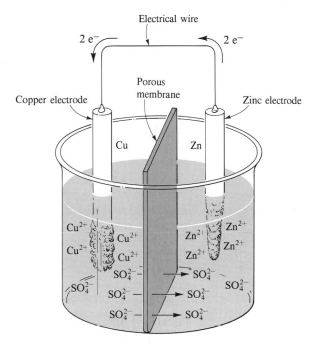

FIGURE 14.1 An electrochemical cell that produces electricity.

The excess sulfate ions left behind as the copper ions are reduced pass through the porous membrane to balance the charge of the newly formed zinc ions.

The electrons flowing through the wire are an electric current, and the arrangement or cell we have just described is a battery. The voltage of the cell can be calculated from the reduction potentials of the two half-reactions. In the cell of Figure 14.1, the half-reactions are:

$$Cu^{2+} + 2\ e^- \longrightarrow Cu \qquad E^0 = 0.34\ V$$
$$Zn \longrightarrow Zn^{2+} + 2\ e^- \qquad E^0 = 0.76\ V$$

The voltage of the cell (if the concentrations are 1 M) is 1.10 V.

Theoretically, any oxidation–reduction reaction can be used in a cell that produces an electrical current. The batteries that we use in flashlights, calculators, radios, and so on, produce electricity by various oxidation–reduction reactions.

14.6 Summary

Reactions in which electrons are transferred are oxidation–reduction or redox reactions. In a redox reaction, the substance oxidized increases in oxidation number and thus loses electrons; the substance reduced decreases in oxidation number and thus gains electrons. The substance oxidized is the reducing agent;

the substance reduced is the oxidizing agent. Simple redox reactions can be balanced by inspection but, for more complex reactions, either oxidation numbers or half-reactions can be used to aid in balancing. The reduction potential of an element measures its tendency to gain electrons in comparison with the tendency of hydrogen ion to gain electrons. The potential of an oxidation half-reaction is equal in magnitude but is opposite in sign to the potential of the reduction half-reaction found in tables such as Table 14.3.

A redox reaction is spontaneous if the sum of its potentials is positive. An active metal is easily oxidized; an active nonmetal is easily reduced. A spontaneous redox reaction can be used to produce an electric current by physically separating the half-reactions and transferring the electrons through a wire. Nonspontaneous redox reactions are often used in industry to produce chemicals. These processes require the input of large amounts of electrical energy.

Key Terms

activity series (14.4C)
electrochemical cell (14.5)
half-reaction (14.2)
oxidation (14.1)
oxidation number (14.1)

oxidizing agent (14.1)
reducing agent (14.1)
reduction (14.1)
reduction potential, E^0 (14.4)
voltaic cell (14.5)

Multiple-Choice Questions

MC1. In which of the following compounds does nitrogen have an oxidation number of -2?
  a. NO   **b.** KNO_3 **c.** NH_4Cl **d.** N_2O_3 **e.** none of these

MC2. Which of the following is/are oxidation–reduction reactions?
  1. $NaOH + HCl \longrightarrow NaCl + H_2O$
  2. $Cu + 2\ AgNO_3 \longrightarrow Cu(NO_3)_2 + 2\ Ag$
  3. $Mg(OH)_2 \longrightarrow MgO + H_2O$
  a. 1   **b.** 2   **c.** 3   **d.** 1 and 2   **e.** 2 and 3

MC3. What is the oxidizing agent in the following equation?

$$SnCl_2 + FeCl_3 \longrightarrow\!\!\!\!/\ \ SnCl_4 + FeCl_2$$

  a. Sn^{2+} **b.** Cl^- **c.** Fe^{3+} **d.** Sn^{4+} **e.** Fe^{2+}

MC4. When the equation in Question 3 is balanced, the coefficient of $FeCl_3$ is
  a. 1   **b.** 2   **c.** 3   **d.** 4
  e. The equation is incomplete and cannot be balanced as it stands.

MC5. What is oxidized in the reaction shown in Question 3?
  a. Sn^{2+} **b.** Cl^- **c.** Fe^{3+} **d.** Sn^{4+} **e.** Fe^{2+}

MC6. Which of the following is a balanced half-reaction for the reduction of nitric acid to nitrogen dioxide?

  a. $3 e^- + 4 H^+ + NO_3^- \longrightarrow NO + 2 H_2O$
  b. $e^- + 2 H^+ + NO_3^- \longrightarrow NO_2 + H_2O$
  c. $4 H^+ + NO_3^- \longrightarrow NO_2 + 2 H_2O + 3 e^-$
  d. $3 e^- + 4 HNO_3 \longrightarrow 4 NO_2 + 2 H_2O$
  e. $e^- + 2 HNO_3 \longrightarrow 2 NO_2 + H_2O$

MC7. Use Table 14.3 to determine whether the following equation represents a spontaneous reaction.

$$3 Cl_2 + 2 HNO_3 \longrightarrow 6 Cl^- + 2 NO + 4 H_2O$$

  a. yes   **b.** no

MC8. When a piece of iron is added to a solution of copper(II) sulfate,

  a. nothing happens.   **b.** the iron is oxidized.
  c. the copper(II) ion is oxidized.   **d.** the iron is reduced.
  e. hydrogen is evolved.

MC9. Which is the most easily reduced halogen?

  a. fluorine   **b.** chlorine   **c.** bromine   **d.** iodine
  e. None of the halogens can be reduced.

MC10. Which of the following metallic oxides is most easily reduced to a free metal? (Use the information in Table 14.3 for help in answering this question.)

  a. copper(II) oxide   **b.** iron(II) oxide   **c.** lithium oxide
  d. magnesium oxide   **e.** zinc(II) oxide

Problems

14.1 Review of Terms

*14.10. Balance the following equations. Identify each as neutralization, precipitation, oxidation–reduction, or other. For any oxidation–reduction reaction, identify the substance oxidized, the substance reduced, the oxidizing agent, and the reducing agent.

  a. $CO_2 + Ca(OH)_2 \longrightarrow\!\!\!\!/ \; Ca(HCO_3)_2$
  b. $Zn + H_2SO_4 \longrightarrow\!\!\!\!/ \; ZnSO_4 + H_2$
  c. $CaO + HCl \longrightarrow\!\!\!\!/ \; CaCl_2 + H_2O$
  d. $Pb + Cu(NO_3)_2 \longrightarrow\!\!\!\!/$
   $\qquad\qquad\qquad Cu + Pb(NO_3)_2$
  e. $Br_2 + KI \longrightarrow\!\!\!\!/ \; I_2 + KBr$

14.11. Write balanced equations for the following reactions:

  a. 3 mol potassium hydroxide added to 1 mol phosphoric acid
  b. zinc added to acetic acid
  c. sodium iodide added to lead(II) nitrate
  d. magnesium hydroxide decomposed by heat to water and magnesium oxide
  e. chlorine added to sodium bromide solution
  f. copper added to nitric acid, forming copper(II) nitrate and nitrogen dioxide

14.12. For those equations in Problem 14.11 that are oxidation–reduction, identify the substance oxidized, the substance reduced, the oxidizing agent, and the reducing agent.

14.2 Half-Reactions

14.13. Write half-reactions for those equations in Problem 14.10 that are oxidation–reduction. Label the half-reactions as oxidation or reduction.

14.14. Write half-reactions for the redox reactions in Problem 14.11. Identify which half-reaction is oxidation and which is reduction.

*14.15. Complete and balance the following half-reactions, using hydrogen ions, hydroxide ions, and water as needed.
a. $MnO_4^- + H^+ \longrightarrow MnO_2$
b. $MnO_4^- + H^+ \longrightarrow Mn^{2+}$
c. $MnO_4^- + OH^- \longrightarrow Mn^{3+}$

14.16. Complete and balance the following half-reactions using hydrogen ions and water as needed.
a. $IO_3^- \longrightarrow I_2$ b. $H_2SO_4 \longrightarrow SO_2$
c. $Sn^{2+} \longrightarrow Sn^{4+}$

14.17. Complete and balance the following half-reactions, using H_2O, H^+, and electrons as needed. State whether each reaction is oxidation or reduction.
a. $Fe^{2+} \longrightarrow Fe^{3+}$
b. $NO_3^- \longrightarrow NO_2$
c. $Cl_2 \longrightarrow Cl^-$
d. $Cr \longrightarrow Cr^{3+}$
e. $S \longrightarrow S^{2-}$

14.3 Balancing More Complex Oxidation–Reduction Equations

*14.18. Balance the following equations using oxidation numbers.
a. $Pb + HNO_3 \longrightarrow Pb(NO_3)_2 + NO + H_2O$
b. $Fe^{3+} + Br^- \longrightarrow Fe^{2+} + Br_2$
c. $SO_2 + HNO_3 \longrightarrow H_2SO_4 + NO$

14.19. Using oxidation numbers, write balanced ionic equations for the following reactions. Identify the substance oxidized and the substance reduced.
a. $KMnO_4 + H_2SO_3 \longrightarrow MnSO_4 + K_2SO_4 + H_2SO_4 + H_2O$ (acidic)
b. $Cu + HNO_3 \longrightarrow Cu(NO_3)_2 + NO + H_2O$ (acidic)
c. $MnO_4^- + Cl^- \longrightarrow Cl_2 + Mn^{2+} + H_2O$ (acidic)

14.20. The lead storage battery uses the following half-reactions during use (discharge):

$$Pb + SO_4^{2-} \longrightarrow PbSO_4 + 2\ e^{2-}$$
$$E^0 = 0.356\ V$$
$$PbO_2 + 4\ H^+ + SO_4^{2-} + 2\ e^- \longrightarrow PbSO_4 + 2\ H_2O$$
$$E^0 = 1.685\ V$$

a. What is the potential energy difference of this cell?
b. What are the half-reactions and the potential of the cell while it is being charged? Remember that nothing but electrical energy is added when a lead storage battery is being charged.
c. The density of sulfuric acid solution in a new battery is 1.25 g/mL. Why does the density of this solution decrease with use?

14.21. a. What is a reduction potential?
b. Using Table 14.3, arrange the following in order of increasing ease of reduction:

$$Cu^{2+}\quad MnO_4^-\quad Br_2\quad Li^+\quad Fe^{2+}$$

*14.22. Using half-reactions, balance the following ionic equation.

$$Mo_{24}O_{37} + MnO_4^- + H^+ \longrightarrow MoO_3 + Mn^{2+} + H_2O$$

14.23. Predict whether the following reaction is spontaneous. If it is, write the half-reactions for the cell and calculate E^0. If the reaction is not spontaneous, write "no reaction."

$$Ag^+ + Pb \longrightarrow Ag + Pb^{2+}$$

14.24. Balance the following equation using half-reactions. Identify the substance oxidized and the substance reduced.

$$Cr_2O_7^- + C_2O_4^{2-} + H^+ \longrightarrow$$
$$Cr^{3+} + CO_2 + H_2O$$

14.25. Name five metals that are more active than hydrogen. Name four that are less active than hydrogen.

Review Problems

14.26. Mercury is prepared by heating its sulfide HgS with oxygen to form free mercury and sulfur dioxide. Write the balanced equation for the reaction. If 4.6 g mercury(II) sulfide is heated with 0.224 L oxygen at STP, what mass of pure mercury is obtained?

14.27. Pure phosphorus is prepared by the reduction of the oxide

$$P_4O_{10} + 10\ C \longrightarrow 10\ CO + 4\ P$$

What mass of phosphorus would be formed by the reaction of 71 g P_4O_{10} with 72 g carbon?

14.28. Nitric acid acts not only as an oxidizing agent but as a strong acid. What volume of 0.200 M nitric acid will neutralize a 1.48-g sample of calcium hydroxide?

14.29. Fluorine is prepared by the electrolysis of hydrogen fluoride. Is the reaction spontaneous? What weight of fluorine would be obtained by the electrolysis of 33.5 g hydrogen fluoride? What volume of hydrogen would be obtained as a by-product (measured at STP)?

14.30. Magnesium is obtained by the electrolysis of magnesium chloride prepared from seawater. Of seawater, 0.13% by mass is magnesium ion. What mass of seawater would have to be processed to obtain 1.65 kg magnesium?

▪15▪

A Brief Look
at Organic Chemistry

Organic chemistry is one of the larger areas within the field of chemistry. Its original definition as "the chemistry of compounds found in living matter" has long been discarded in favor of "the chemistry of carbon and the covalent bond." If you look back through Chapter 7, in which covalent bonding was introduced, you will see that many of the examples contain carbon. Of all the elements, carbon is by far the most versatile in forming covalent bonds with itself and with other nonmetals, particularly hydrogen, oxygen, nitrogen, and the halogens.

In your daily life, you encounter many of these compounds. All fibers, whether natural or synthetic, are organic compounds. So are the sugars, fats, and proteins of the food you eat and of your own body. So are the vitamins, antibiotics, and cosmetics that you use.

As you read this chapter and encounter many familiar compounds, you may wonder why a study of chemistry does not begin with the familiar compounds of organic chemistry. The reason is that the principles of chemistry you have already learned—electrons and the structure of atoms, bonding, formulas, and so on—are a necessary foundation to understanding organic chemistry.

Remember also that a true introduction to organic chemistry takes at least a semester of study. We give here only the barest essentials of the field and in doing so will:

1. Introduce the characteristics of organic compounds.
2. Introduce the nomenclature of organic compounds.
3. Define the composition of the functional groups of organic compounds and describe the characteristic properties of compounds containing each of these functional groups.
4. Describe the nature of polymers.

15.1   The General Characteristics of Organic Compounds

Covalent bonding predominates over ionic bonding in organic chemistry. The most important element in organic chemistry is carbon. Carbon holds this place because of its ability to bond with itself, thus forming long arrays of carbon atoms. Carbon also bonds covalently with other nonmetals. In order of decreasing importance in organic chemistry, these nonmetals are: hydrogen, oxygen, nitrogen, the halogens, sulfur, and phosphorus.

Because organic bonding is predominantly covalent, organic molecules are usually lower-melting and frequently slower to react than the ionic compounds we studied earlier.

An organic compound usually consists of an array of carbon atoms attached to a functional group. The array of carbon atoms may be a long, single chain of atoms; it may be a chain of atoms with several branches; or it may be a ring of atoms. A long, single chain of carbon atoms is referred to as a **straight chain,** although, because of the tetrahedral nature of a carbon atom, the chain is anything but straight. In a **branched chain,** one or more short chains of carbon atoms have replaced one or more hydrogens on the carbons of a straight chain. Examples of each of these structures are shown in Figure 15.1.

A compound might also combine two of these structures as, for example, a ring attached to a branched chain of atoms. The bonds between the carbon atoms may be single, double, or triple bonds. A **functional group** is a group of atoms attached to one of the carbon atoms of the compound. A functional group imparts characteristic chemical properties to the compound. Examples of functional groups are the —OH group that is present in alcohols, the —NH_2 present in amines, and the —COOH group characteristic of organic acids. The reactions of organic chemistry tend to occur at the sites of functional groups.

Because of the varieties of carbon arrays possible, the number of organic compounds is enormous. Luckily, the chemistry of these compounds depends primarily on the functional groups they contain and the arrangement of carbon atoms in the neighborhood of the functional group. It is possible, then, to predict the properties of a compound if the properties of another compound containing the same functional group are known. A compound may contain more than one functional group. If so, the properties of the compound approximate the properties associated with each of these functional groups.

15.2   Hydrocarbons

A **hydrocarbon** is a compound that, as might be predicted from its name, contains only carbon and hydrogen. The fossil fuels, petroleum and coal, are naturally occurring sources of many hydrocarbons. Other hydrocarbons, and in fact most other organic compounds, are prepared from the hydrocarbons isolated from these sources. Both coal and petroleum are complex mixtures. The hydrocarbons present in petroleum can be separated from one another by the differences in their boiling points. Hydrocarbons that boil below room

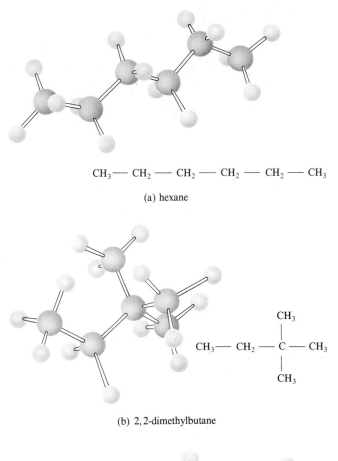

$$CH_3 \!-\! CH_2 \!-\! CH_2 \!-\! CH_2 \!-\! CH_2 \!-\! CH_3$$

(a) hexane

$$CH_3 \!-\! CH_2 \!-\! \overset{\displaystyle CH_3}{\underset{\displaystyle CH_3}{\overset{\mid}{\underset{\mid}{C}}}} \!-\! CH_3$$

(b) 2,2-dimethylbutane

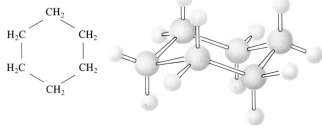

(c) cyclohexane

FIGURE 15.1 (a) A straight chain of carbon atoms in hexane; (b) a branched chain of carbon atoms in 2,2-dimethylbutane; (c) a ring of carbon atoms in cyclohexane.

temperature are found in natural gas. These hydrocarbons contain between 1 and 5 carbon atoms per molecule. The next-higher-boiling fraction of petroleum contains the hydrocarbons we find in gasoline; they have between 6 and 10 carbon atoms per molecule. The next-higher-boiling fractions are, respectively, kerosene (11 to 12 carbon atoms per molecule), fuel oils (13 to 25 carbon atoms per molecule), lubricants (26 to 37 carbon atoms per molecule), and asphalt. The molecules of asphalt contain more than 37 carbon atoms per molecule.

There are several classes of hydrocarbons. The first three—alkanes, alkenes, and alkynes—are grouped together as **aliphatic** hydrocarbons. Aromatic hydrocarbons constitute another class. We will discuss each of these classes separately.

A. Alkanes

In an **alkane,** all carbon–carbon linkages are single bonds; all other bonds are to hydrogen. Alkanes are also called **saturated** hydrocarbons, saturation meaning that they contain the largest number of hydrogen atoms possible for that number of carbon atoms. An alkane containing n carbon atoms will contain $2n + 2$ hydrogen atoms. Thus, the alkane containing 5 carbon atoms will contain 12 hydrogen atoms and have the molecular formula C_5H_{12}. This formula does not show the arrangement of those 5 carbon atoms. They could be in either a straight chain or a branched chain. The possible arrangements are:

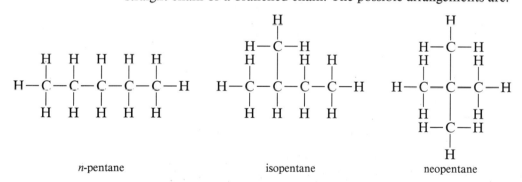

n-pentane   isopentane   neopentane

These formulas are called **structural formulas** because they show the structure of the molecules, how the atoms are bonded together. These same formulas are often displayed in a formula, intermediate between molecular and structural formulas, but one that clearly shows the bonding:

$$CH_3CH_2CH_2CH_2CH_3 \qquad (CH_3)_2CHCH_2CH_3 \qquad C(CH_3)_4$$
$$\text{n-pentane} \qquad\qquad \text{isopentane} \qquad\qquad \text{neopentane}$$

Compounds that have the same molecular formula but different structural formulas are called **isomers.** As the number of carbon atoms in a compound increases, the number of isomers that it can have increases dramatically. We have shown the 3 possible isomers of C_5H_{12}. There are 18 possible isomers for the molecular formula C_8H_{18}, and more than 62 billion possibilities for an alkane with the molecular formula $C_{40}H_{82}$.

The formulas and names of the lowest-molecular-weight straight-chain alkanes are shown in Table 15.1. Also shown in this table are the names and formulas of the groups (called **alkyl groups**) that remain when a terminal hydrogen is removed from each alkane.

TABLE 15.1   The names and formulas of some low-molecular-weight straight-chain alkanes, the boiling point of each, and the name and formula of the related alkyl group

Alkane	Formula	bp, °C	Alkyl group	Formula
*meth*ane	CH_4	−164	methyl	CH_3-
*eth*ane	CH_3CH_3	−88	ethyl	CH_3CH_2-
*prop*ane	$CH_3CH_2CH_3$	−42	propyl	$CH_3CH_2CH_2-$
*but*ane	$CH_3CH_2CH_2CH_3$	0	butyl	$CH_3CH_2CH_2CH_2-$
*pent*ane	$CH_3CH_2CH_2CH_2CH_3$	36	pentyl	$CH_3CH_2CH_2CH_2CH_2-$

Notice that each name has a prefix that denotes the number of carbon atoms it contains. In Table 15.1 we have italicized this prefix. The ending *ane* means the compound is a saturated hydrocarbon. In the name of the alkyl group derived from the alkane, the ending *ane* has been replaced with *yl*. If the compound is straight-chained, the name may carry the adjective **normal** (abbreviated as in *n*-pentane). Naming hydrocarbons that are not normal or are unsaturated is a complex procedure that need not be included in a brief introduction to organic chemistry.

The alkanes shown in Table 15.1 represent a homologous series. In a **homologous series,** each new member differs by a $-CH_2-$ group from the member next lower in molecular weight. The next member of the homologous series of alkanes in Table 15.1 would be hexane, $CH_3CH_2CH_2CH_2CH_2CH_3$. Hexane contains one more $-CH_2-$ group than pentane. In a homologous series, there is usually a regular increase in boiling point and in other physical properties

The alkanes react with only a few substances. They react with chlorine in the presence of light as in the following equation for the reaction of ethane with chlorine:

$$C_2H_6 + Cl_2 \xrightarrow{hv} C_2H_5Cl + HCl$$

In an excess of chlorine, more than one hydrogen can be replaced by chlorine.

By far the most familiar reaction of alkanes is their reaction with oxygen in combustion (see Section 8.3B), as in the combustion of gasoline or diesel fuel. If gasoline were pure octane, the equation for its complete combustion would be

$$2\,C_8H_{18} + 25\,O_2 \longrightarrow 16\,CO_2 + 18\,H_2O$$

B. Alkenes

An **alkene** is a hydrocarbon containing a carbon–carbon double bond. The simplest alkene is ethene, $CH_2{=}CH_2$ (more frequently called by its common name, ethylene). Table 15.2 shows the lowest-molecular-weight straight-chain alkenes.

TABLE 15.2 Low-molecular-weight alkenes		
Alkene	*Molecular formula*	*Structural formula*
ethene	C_2H_4	$CH_2{=}CH_2$
propene	C_3H_6	$CH_2{=}CHCH_3$
1-butene*⎫ 2-butene ⎬	C_4H_8	$\{CH_2{=}CHCH_2CH_3$ $CH_3CH{=}CHCH_3$
1-pentene⎫ 2-pentene⎬	C_5H_{10}	$\{CH_2{-}CHCH_2CH_2CH_3$ $CH_3CH{=}CHCH_2CH_3$ (This structure is the same as $CH_3CH_2CH{=}CHCH_3$.)

* The carbon atoms of a molecule can be numbered and the location of a functional group or unsaturation can be shown by including in the name of the compound the number of the carbon atom to which the functional group is attached.

Note that this series is like that of the alkanes, in that each new member has one more $-CH_2-$ group than the compound next lower in molecular weight. The general formula of the alkenes is C_nH_{2n}. Notice also that, starting with butene, the molecular formula does not give a clear indication of structure because of the existence of isomers with different locations of the double bond. As the molecular weight of the alkene increases, other isomers are possible due to branching of the chains. For example, there is another 1-pentene with the structural formula $CH_2{=}CHCH(CH_3)_2$ and another 2-pentene with the structural formula $CH_3CH{=}C(CH_3)_2$.

Alkenes are much more reactive than alkanes, owing to the presence of the double bond. Many reagents can add across the bond, as shown, for example, by the reaction of ethene with hydrogen chloride to form ethyl chloride:

$$CH_2{=}CH_2 + HCl \longrightarrow CH_3CH_2Cl$$
$$\text{ethene} \qquad\qquad\qquad \text{ethyl chloride}$$

An important application of this reactivity is in the formation of **polymers,** in which alkene molecules react with one another to form a long chain of like groups. The following equation shows the polymerization of propene to form polypropylene:

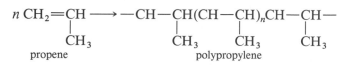

Other addition polymers formed from substituted ethenes are discussed in Section 15.7.

C. Alkynes

An **alkyne** contains a carbon–carbon triple bond. The simplest alkyne is ethyne, CH≡CH, more commonly known as acetylene. The homologous series of alkynes resembles that of the alkenes. The general formula of the alkynes is C_nH_{2n-2}. Structural isomers of the alkynes are known. The chemistry of the alkynes resembles that of the alkenes in that reactions take place across the triple bond, as illustrated in the following equation for the reaction of acetylene with hydrogen chloride:

$$CH{\equiv}CH + HCl \longrightarrow CH_2{=}CHCl$$
vinyl chloride

$$CH_2{=}CHCl + HCl \longrightarrow CH_3CHCl_2$$
1,1-dichloroethane

Notice the products of this reaction. The intermediate product is vinyl chloride, the monomer from which the familiar polymer polyvinyl chloride is prepared. To name the final product, the carbon atoms are numbered, and the atoms to which the chlorine atoms are attached are specified. This numbering is necessary because the compound has an isomer, 1,2-dichloroethane, CH_2ClCH_2Cl.

D. Aromatic Hydrocarbons

The nature of an **aromatic** hydrocarbon is best described by discussing the structural characteristics of benzene, the best-known aromatic hydrocarbon. Benzene is a ring compound with the formula C_6H_6. Its structure can be shown in any of the following ways. The carbon atoms of one ring have been numbered.

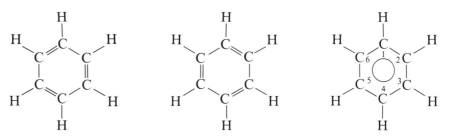

The first two of these structures show alternating single and double bonds. The difficulty with these two structures is that this notation implies a different bond length between carbon-1 and carbon-2 than between carbon-2 and carbon-3, because a carbon–carbon single bond is longer than a carbon–carbon double bond. Careful measurements of the carbon–carbon bond lengths in benzene have shown that they are all the same. Therefore, benzene cannot contain alternating single and double bonds. We conclude then that benzene has a resonating structure. (For an earlier description of resonance see Section 7.2D.)

This structure is illustrated in the third diagram, in which the benzene ring is shown with its six carbon atoms joined by single bonds. The circle inside the ring shows the other six electrons, which are not located between two specific atoms but are spread around the whole ring.

Aromatic structures are very stable. Their reactions are unique to them, and are *not* like those of the carbon–carbon double bond in alkenes. The reaction of benzene with chlorine is an example of this uniqueness. Instead of adding across the double bond as it would with cyclohexene, chlorine substitutes on benzene.

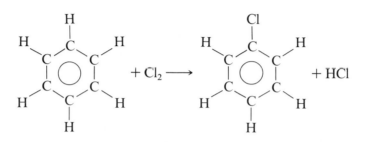

Naphthalene, $C_{10}H_8$, and anthracene, $C_{14}H_{10}$, are examples of more complex aromatic hydrocarbons. Their ring structures are shown in Figure 15.2.

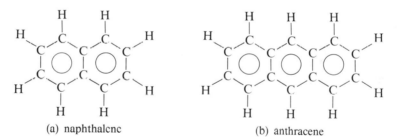

(a) naphthalene   (b) anthracene

FIGURE 15.2 Structural representations of (a) naphthalene, $C_{10}H_8$, and (b) anthracene, $C_{14}H_{10}$.

15.3 Halogens in Organic Compounds

In organic halides, a halogen atom — chlorine, bromine, or iodine — has replaced one or more hydrogen atoms of a hydrocarbon. These atoms are not ions but are covalently bonded to carbon. We have shown in the previous section how some organic halides can be prepared. The reactivity of the halogen atom in an organic molecule depends on the bonding around the carbon atom to which it is attached. This dependence is illustrated by the difference in reactivity of the two butyl chlorides shown. The chlorine of *n*-butyl chloride is very difficult to remove (as is the chlorine of chlorobenzene). The chlorine of *t*-butyl chloride is very easy to remove.

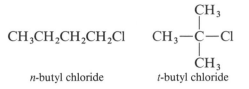

$$CH_3CH_2CH_2CH_2Cl$$

n-butyl chloride   *t*-butyl chloride

Until recently, organic halides were widely used. Carbon tetrachloride, CCl_4, was used in fire extinguishers and as a solvent, especially in the dry-cleaning industry. Chloroform, $CHCl_3$, was used as an anesthetic; iodoform, CHI_3, as a disinfectant. One of the most powerful pesticides known, DDT, is an organic halide. So are the PCBs that found widespread use in electrical insulators. However, studies have shown conclusively that most, if not all, organic halides are carcinogenic (cancer-causing agents), and their use has been sharply curtailed.

15.4   Oxygen in Organic Compounds

An oxygen atom has six valence electrons. To gain a complete octet, it can form a double bond with another atom or it can form two single bonds. Both of these bonding situations are found in organic molecules.

A. Alcohols

In an **alcohol,** oxygen is singly bonded to carbon and singly bonded to hydrogen. An alcohol, then, contains an —OH group attached to a carbon atom. The lowest-molecular-weight alcohols that contain a single —OH group attached to an alkyl group are shown in Table 15.3. Alcohols can be considered to be formed by replacing a hydrogen in an alkane with an —OH group. The IUPAC name of these alcohols replaces the final *e* of the name of the alkane with *ol.* Thus, the alcohol with one carbon atom is methanol, with two carbons is ethanol, and so on. Table 15.3 shows the more common nomenclature that

TABLE 15.3   Low-molecular-weight alcohols

Name	Formula	Properties
methyl alcohol (wood alcohol)	CH_3OH	solvent, toxic, causes blindness
ethyl alcohol (grain alcohol)	CH_3CH_2OH	beverage, causes drunkenness
propyl alcohol	$CH_3CH_2CH_2OH$	solvent
isopropyl alcohol	$(CH_3)_2 CHOH$	used in rubbing alcohol

combines the name of the alkyl group with the word *alcohol.* The low-molecu-lar-weight alcohols are completely miscible with water and have widespread use as solvents.

All sales of ethyl alcohol are under strict government regulation. Absolute alcohol is 100% ethyl alcohol; exact records of its sale must be kept for government inspection. Its use is restricted to applications in science and medicine. Denatured alcohol is 95% by volume ethyl alcohol in water; a toxic substance is added to this solution to prevent its use as a beverage. Frequently this toxic substance is methyl alcohol. All ethyl alcohol used in beverages is heavily taxed. Its common name, *grain alcohol,* comes from the fact that beverage alcohol is produced by fermentation of sugars, frequently those found in grain. The "proof" of an alcoholic beverage is twice the percent by volume of ethyl alcohol in the liquid. Thus 100 proof means 50% alcohol. Table 15.4 lists several types of alcoholic beverages, the percent ethyl alcohol they contain, and their source.

TABLE 15.4 Alcoholic beverages

Type	Alcohol content	Derivation
beer	3–6%	malt sugars
wine	5–15%	fruit sugars
whisky	approx. 50%	grain
vodka	approx. 50%	potatoes

A compound containing two —OH groups is a dihydric alcohol. Ethylene glycol, $HOCH_2CH_2OH$, the simplest of these compounds, is the most frequently used antifreeze compound. It is marketed under various trade names. Glycerine is a trihydric alcohol. It is a viscous water-soluble liquid used in many food, cosmetic, and drug formulations.

Aromatic alcohols are known as **phenols.** They are all weak acids. The ionization of the simplest phenol is shown:

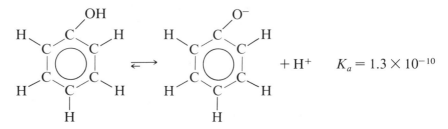

$K_a = 1.3 \times 10^{-10}$

B. Ethers

In an **ether,** an oxygen atom is bonded to two carbon atoms. Ethers are excellent solvents for other organic compounds but unfortunately are quite flammable; their use is therefore not widespread. Diethyl ether was formerly used as an anesthetic. Its flammability is responsible for the precautions in operating rooms against sparks from static electricity. Other anesthetics have been developed that are less flammable and have fewer unpleasant side effects than does diethyl ether.

C. Carbonyl Compounds

In a carbonyl compound, an oxygen atom is doubly bonded to a carbon atom. This group is quite polar. Many of the chemical reactions and the physical properties of compounds containing a carbonyl group can be attributed to the presence of this polarity. When the carbon atom of the carbonyl group is also attached to a hydrogen atom, the compound is known as an **aldehyde.** The simplest aldehyde is formaldehyde, HCHO. When the carbon of the carbonyl is attached to two carbon atoms, the compound is called a **ketone.** The simplest ketone is acetone. Structures of formaldehyde, of acetaldehyde (the next-higher aldehyde), and of acetone are shown:

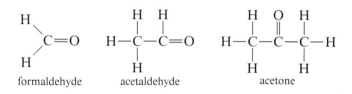

Formaldehyde is a preservative widely used as a 40% solution known as formalin in biological laboratories. It is also used in the preparation of a polymer known as Bakelite. The chief use of acetone is as a solvent and a cleaning agent.

D. Carboxylic Acids

Carboxylic acids contain the group —COOH. Structures of the first three members of the homologous series are shown:

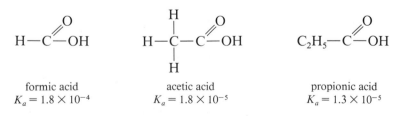

Carboxylic acids are only partially ionized. Their ionization constants depend on the nature of the group attached to the —COOH group. Under each acid is shown its acid dissociation constant.

Acetic acid is the most common of the organic acids. Vinegar is a 5% solution of acetic acid. The flavor of the vinegar depends on the substance from which it was produced.

High-molecular-weight nonaromatic organic acids, known as **fatty acids,** are found in animal and vegetable fats and oils. Several are listed in Table 15.5. A healthful diet should contain the unsaturated fatty acids linoleic, linolenic, and arachidonic acids.

Name	Structural formula
stearic	$CH_3(CH_2)_{16}COOH$
oleic	$CH_3(CH_2)_7CH{=}CH(CH_2)_7COOH$
linoleic	$CH_3(CH_2)_4CH{=}CHCH_2CH{=}CH(CH_2)_7COOH$
linolenic	$CH_3CH_2(CH{=}CHCH_2)_3(CH_2)_6COOH$

TABLE 15.5 Some naturally occurring fatty acids

E. Esters

An **ester** is formed by the reaction of an alcohol with an organic acid. The following equation shows the reaction of acetic acid with ethyl alcohol to form the ester ethyl acetate. The alkyl group of the alcohol replaces the acidic hydrogen of the acid.

$$CH_3COOH + C_2H_5OH \longrightarrow CH_3COOC_2H_5 + H_2O$$

Fats and oils are esters in which the alcohol groups come from the trihydric alcohol glycerol and the acids are high-molecular-weight straight-chain acids. Lower-molecular-weight esters are good solvents of all types of organic compounds. They also find widespread use as flavoring agents because of their pleasant and often fruity odor. Table 15.6 lists some esters, as well as the characteristic odor and the natural source of each.

TABLE 15.6 Some frequently encountered esters

Name	Odor	Source
ethyl butyrate	fruity	strawberries
ethyl hexanoate	pineapple	pineapple
ethyl isobutyrate	apple	honey
methyl salicylate	wintergreen	wintergreen, birch
methyl butyrate	banana	wood oil

15.5   Nitrogen in Organic Compounds

Although the nitro group, $-NO_2$, and the nitrile group, $-CN$, are sometimes encountered in organic compounds, the compounds of most interest to the average person are those organic compounds that can be considered to be derivatives of ammonia. These compounds are amines and their related compounds, amides. The formulas of several amines are shown here.

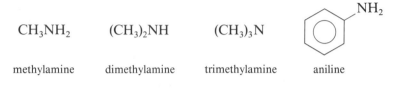

CH_3NH_2	$(CH_3)_2NH$	$(CH_3)_3N$	
methylamine	dimethylamine	trimethylamine	aniline

Amines are basic compounds like ammonia. They frequently have a very unpleasant fishy odor. The low-molecular-weight amines are soluble in water. Aniline, the simplest aromatic amine, was until recently widely used in dye manufacture, but its characterization as a carcinogen has greatly reduced its use.

An **amide** results from the reaction of ammonia with an acid. The formation of acetamide by the reaction of acetic acid with ammonia is typical:

$$CH_3COOH + NH_3 \longrightarrow CH_3\overset{O}{\overset{\|}{C}}-NH_2 + H_2O$$

acetic acid   acetamide

15.6   Polyfunctional Compounds

A **polyfunctional compound** contains more than one functional group. Several polyfunctional groups of compounds are important in biological chemistry. Sugars are polyhydric aldehydes or ketones; that is, they contain several $-OH$ groups and a carbonyl group. The formulas of glucose and fructose are shown in Figure 15.3. Glucose molecules can and do react with themselves to form starch, which is polymerized glucose. A glucose unit is also the fundamental unit of cellulose. In cellulose, the glucose molecules have combined in a slightly different way.

Amino acids comprise another polyfunctional group of compounds. An amino acid has an amino group attached to the alkyl group of an organic acid. Amino acids that have an amino group attached to the first carbon of the carbon chain of the acid are the building units of proteins. Some of these compounds are included in the group of amino acids essential to the human diet. Figure 15.4 shows several amino acids as well as a section of a protein.

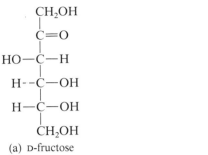

(a) D-fructose   (b) D-glucose

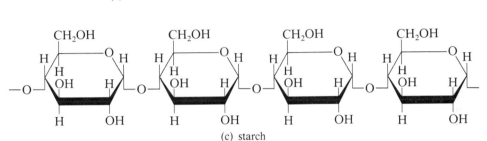

(c) starch

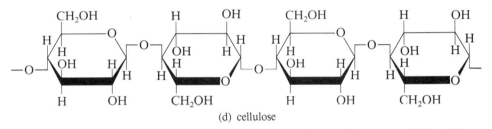

(d) cellulose

FIGURE 15.3 Structures of (a) D-fructose, (b) D-glucose, (c) starch, and (d) cellulose.

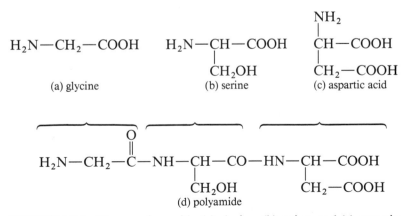

FIGURE 15.4 Three amino acids: (a) glycine, (b) serine, and (c) aspartic acid and the polyamide or tripeptide formed by their combination.

15.7   Polymers

Since World War II, synthetic materials are being used in almost every endeavor. These materials have been fabricated by chemists to have specific properties. For example, there are fabrics that dry quickly and are wrinkle-free, and other materials that are excellent insulators. Some polymers are made from small molecules containing a carbon–carbon double bond; these molecules add to one another forming what is known as an addition polymer. Table 15.7 lists some monomers and the trade names of several addition polymers.

TABLE 15.7   Polymers derived from substituted ethylene monomers

Monomer	Monomer name	Polymer name or trade name
$CH_2{=}CH_2$	ethylene	polyethylene, Polythene, for unbreakable containers and tubing
$CH_2{=}CHCH_3$	propylene	polypropylene, Herculon, fibers for carpeting and clothes
$CH_2{=}CHCl$	vinyl chloride	polyvinyl chloride, PVC, Koroseal
$CH_2{=}CCl_2$	1,1-dichloroethylene	Saran, food wrappings
$CH_2{=}CHCN$	acrylonitrile	polyacrylonitrile, Orlon
$CF_2{=}CF_2$	tetrafluoroethylene	polytetrafluoroethylene, Teflon
$CH_2{=}CHC_6H_5$	styrene	polystyrene, Styrofoam, for insulation
$CH_2{=}CCO_2CH_3$ $\quad\mid$ $\quad CH_3$	methyl methacrylate	polymethyl methacrylate, Lucite, Plexiglas, for glass substitutes
$CH_2{=}CHCO_2CH_3$	methyl acrylate	polymethyl acrylate, for latex paints

Other polymers are made by the reaction of a polycarboxylic acid with a polyhydric alcohol to form a polyester (Dacron) or the reaction of a polyamine with a polyacid to form a polyamide (nylon). Figure 15.5 shows reactions to form these polymers.

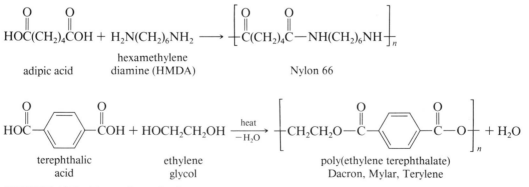

adipic acid + hexamethylene diamine (HMDA) → Nylon 66

terephthalic acid + ethylene glycol → poly(ethylene terephthalate) Dacron, Mylar, Terylene + H_2O

FIGURE 15.5 Formation of polymers.

15.8   ## Summary

Organic chemistry is the study of carbon and covalent bonds. The elements most often encountered in organic compounds are carbon, hydrogen, halogens, oxygen, and nitrogen. There are several kinds of hydrocarbons: alkanes, or saturated hydrocarbons; alkenes and alkynes, which are unsaturated hydrocarbons; and aromatic hydrocarbons, which usually contain a ring of carbon atoms joined by alternating single and double bonds producing a resonating structure that is more stable than that of aliphatic hydrocarbons. There are many organic halides. The oxygen-containing compounds are: alcohols, ethers, carbonyl compounds (aldehydes and ketones), carboxylic acids, and esters. Some of the nitrogen-containing compounds are amines and amides. Many compounds will contain more than one functional group.

Key Terms

alcohol (15.4A)	*fatty acids (15.4D)*
aldehyde (15.4C)	*functional group (15.1)*
aliphatic (15.2)	*homologous series (15.2A)*
alkane (15.2A)	*hydrocarbon (15.2)*
alkene (15.2B)	*isomer (15.2A)*
alkyl group (15.2A)	*ketone (15.4C)*
alkyne (15.2C)	*normal (15.2A)*
amide (15.5)	*phenols (15.4A)*
amine (15.5)	*polyfunctional compounds (15.6)*
aromatic (15.2D)	*polymer (15.2B)*
branched chain (15.1)	*saturated (15.2A)*
carboxylic acid (15.4D)	*straight chain (15.1)*
ester (15.4E)	*structural formula (15.2A)*
ether (15.4B)	

Multiple-Choice Questions

The possible answers to Questions 1–5 are:

$$\overset{\displaystyle O}{\overset{\displaystyle \|}{}}$$

1. $C_2H_5C—CH_3$ **2.** $(CH_3)_2CHOH$
3. C_3H_7COOH **4.** C_4H_9CHO **5.** C_3H_7CHO

MC1. Which is a carboxylic acid?
    a. 1   **b.** 2   **c.** 3   **d.** 4   **e.** 5

MC2. Which contains a butyl group?
    a. 1   **b.** 2   **c.** 3   **d.** 4   **e.** 5

MC3. Which is isomeric with $C_2H_5OCH_3$?
    a. 1   **b.** 2   **c.** 3   **d.** 4   **e.** 5

MC4. Which is a ketone?

    a. 1   **b.** 2   **c.** 3   **d.** 4   **e.** 5

MC5. Which two are isomers?

    a. 1 and 2   **b.** 1 and 4   **c.** 1 and 5   **d.** 2 and 4

    e. 2 and 5

MC6. $CH_3CH=CH_2$ is an

    a. alkane.   **b.** alkene.   **c.** alkyne.

    d. aromatic hydrocarbon.   **e.** none of these

MC7. Which of the following is a saturated hydrocarbon?

    a. $CH_3CH_2CH_2CH_3$ **b.** $CH_3CH=CHCH_3$

    c. $CH_3CH=CHCH=CH_2$ **d.** $CH_3C\equiv CH$

    e. $(CH_3)_2C=C(CH_3)_2$

MC8. What linkage joins the monomers of proteins?

    a. amine   **b.** amide   **c.** ester   **d.** ether   **e.** alcohol

MC9. Which of the following is ethyl acetate?

    a. $C_2H_5COOC_2H_5$ **b.** $CH_3COOC_2H_5$ **c.** $HCOOC_2H_5$

    d. $C_2H_5COOCH_3$

MC10. What is the molecular formula of the next member of the homologous series: methyl chloride, ethyl chloride, propyl chloride?

    a. $C_5H_{10}Cl$ **b.** C_4H_9Cl **c.** $C_5H_{11}Cl$ **d.** C_4H_8Cl

    e. $C_6H_{11}Cl$

Problems

15.2 Hydrocarbons

15.1. Why are fossil fuels important to organic chemists?

*__15.2.__ Draw the structure of the five isomers of hexane.

*__15.3.__ Write balanced equations for the complete combustion of ethane, propane, and butane to carbon dioxide and water.

15.4. Show the resonance structures of naphthalene.

15.3 Halogens in Organic Compounds

15.5. Draw the structures of five hexyl bromides.

15.4 Oxygen in Organic Compounds

15.6. Give the molecular formulas and names of the first six members of the homologous series of straight-chain alcohols.

*__15.7.__ Draw the structural formulas for all alcohols that have the molecular formula C_4H_9OH.

15.8. Draw structural formulas for all ketones that contain five carbon atoms.

15.9. Draw structural formulas for the five aldehydes that have molecular weight less than 70 g/mol.

*__15.10.__ Draw the structural formula for propyl alcohol. Show that it is isomeric with methyl ethyl ether.

*15.11. Write the balanced equation for the reaction of propyl alcohol with acetic acid to form propyl acetate.

15.5 Nitrogen in Organic Compounds

15.12. Draw the structural formula of ethyl amine.

15.6 Polyfunctional Compounds

15.13. Salicylic acid has the structure

What functional groups does it contain? Is it an aromatic or aliphatic compound?

Appendix

This appendix has been compiled to serve as a quick review of the material covered in more depth in the text itself. It contains most of the definitions, physical constants, equations, laws, and problem-solving algorithms that are in the text. Major topics are arranged in alphabetical order, not in the order presented in the text. We expect this appendix to serve as a quick reference for use in problem solving and in review. If you find the material here does not seem familiar, use the references given to locate the appropriate section of the text for a more extensive discussion.

Acids and Bases (Chapter 12)

Definitions, reaction with indicators, and composition

	Acid	*Base*
Definitions (Table 12.2)		
Arrhenius	excess of H^+ in aqueous solution	excess of OH^- in aqueous solution
Brønsted-Lowry	proton donor	proton acceptor
Reaction with indicators		
litmus	red	blue
phenolphthalein	colorless	red
methyl orange	red	yellow
Composition	always hydrogen, frequently oxygen, usually nonmetal	often OH^-

Extent of ionization

Strong acids are completely ionized in aqueous solution.
Weak acids are partially ionized in aqueous solution; their solutions contain both molecules and ions.

Common acids

Name (aq)	Formula	Anion name	Anion formula
Common strong acids			
hydrochloric	HCl	chloride	Cl^-
nitric	HNO_3	nitrate	NO_3^-
sulfuric	H_2SO_4	sulfate	SO_4^{2-}
Common weak acids			
acetic	$HC_2H_3O_2$	acetate	$C_2H_3O_2^-$
carbonic	H_2CO_3	carbonate	CO_3^{2-}

Reactions

Neutralization: acid + base $\longrightarrow$ salt + water
Displacement: metal + acid $\longrightarrow$ hydrogen + salt

Atoms (Chapter 4)

Subatomic particles (Table 4.1)

Particle	Actual mass (g)	Relative mass	Relative charge
proton	1.6726×10^{-24}	1.007	$+1$
neutron	1.6749×10^{-24}	1.008	0
electron	9.108×10^{-28}	5.45×10^{-4}	-1

Composition of atoms (Section 4.2)

Mass number = number of protons + number of neutrons
Atomic number = number of protons = number of electrons
Atomic weight (g/mol) = mass (g) of one mole of naturally occurring
atoms

Designation

$_\text{atomic number}^\text{mass number}$symbol of element   or $_z^A X$

Compounds: Bonding and Geometry (Chapter 7)

Bond angles and molecular shapes (Section 7.3)

Number of regions of high electron density around central atom	Arrangement of regions of high electron density in space	Lone pairs of electrons on central atom	Predicted bond angles	Example	Geometry of molecule
4	tetrahedral	0	109.5°	CH_4 methane	tetrahedral
		1		NH_3 ammonia	pyramidal
		2		H_2O water	bent
3	trigonal planar	0	120°	H_2CO formaldehyde	trigonal planar
		0		$H_2C=CH_2$ ethylene	planar
		1		SO_2 sulfur dioxide	bent
2	linear	0	180°	CO_2 carbon dioxide	linear
		0		$HC\equiv CH$ acetylene	linear

Concentration of Solutions (Section 11.3)

Molarity = mol solute/L solution

Density (Section 2.4)

$$\text{Density} = \frac{\text{mass}}{\text{volume}}$$

Units: for solids and liquids, g/mL; for gases, g/L

For an ideal gas, density at STP = molecular weight/22.4 L

Electron-Dot Structures, *see* Lewis Structures

Equilibrium Constants (Sections 12.5, 13.4C; Table 13.5)

Name of constant	Symbol	Typical equation	Expression of constant
equilibrium constant	K_{eq}	$A_2 + B_2 \rightleftharpoons 2\,AB$	$\dfrac{[AB]^2}{[A_2][B_2]}$
acid dissociation constant	K_a	$HA \rightleftharpoons H^+ + A^-$	$\dfrac{[H^+][A^-]}{[HA]}$
ionization constant of water	K_w	$H_2O \rightleftharpoons H^+ + OH^-$	$[H^+][OH^-]$
solubility product constant	K_{sp}	$M_aN_b \rightleftharpoons aM^{b+} + bN^{a-}$	$[M^{b+}]^a[N^{a-}]^b$

Electrons (Chapter 5)

Electron configuration (Section 5.4; Figure 5.9)

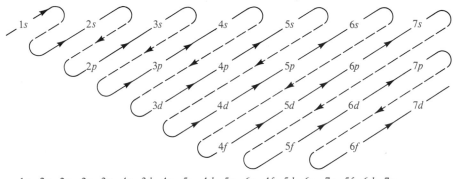

1s 2s 2p 3s 3p 4s 3d 4p 5s 4d 5p 6s 4f 5d 6p 7s 5f 6d 7p

The arrow shows a second way of remembering the order in which sublevels fill.

Electron energy levels of the atom; composition of the first four levels (Section 5.3)

Energy level	Maximum number of electrons in level	Sublevel		
		Type	# orbitals	# electrons
1	2	s	1	2
2	8	s	1	2
		p	3	6
3	18	s	1	2
		p	3	6
		d	5	10
4	32	s	1	2
		p	3	6
		d	5	10
		f	7	14

Orbital shapes (Figures 5.5, 5.7)

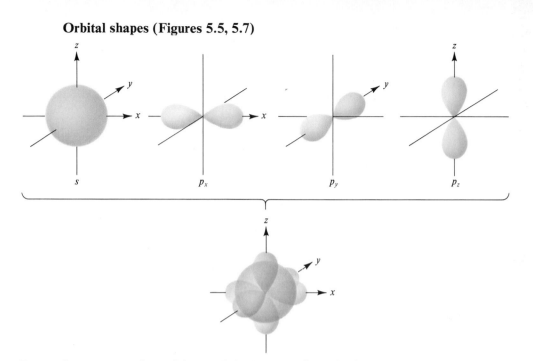

Perspective representations of the s and the three p orbitals of a single energy level. The clouds show the space within which the electron is most apt to be. The lower sketch shows how these orbitals overlap in the energy level.

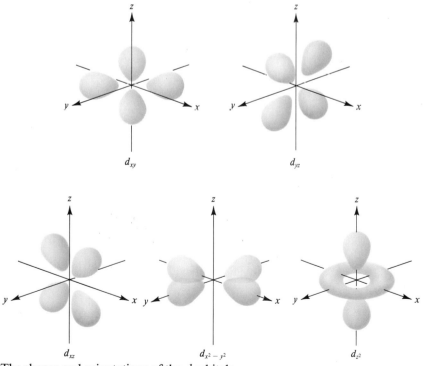

The shapes and orientations of the d orbitals.

Energy

Terms

Enthalpy, H (Section 8.5): The enthalpy change (ΔH) is the change in energy of the system measured at constant pressure. The change is exothermic if ΔH is negative. The change is endothermic if ΔH is positive.

Free energy, G (Section 13.1): Free energy is energy available to do useful work. A process is spontaneous if ΔG is negative. A process is not spontaneous if ΔG is positive.

Entropy, S (Section 13.1): Entropy measures disorder. The entropy of the universe is increasing.

The entropy, free energy, and enthalpy changes that accompany a process are related by the equation $\Delta G = \Delta H - T\Delta S$.

Physical constants related to energy (Sections 10.3, 10.4)

$$\text{Specific heat} = \frac{\text{energy}}{\text{(mass)(temperature change)}}$$

Molar heat of fusion: energy required to change one mole of substance from the solid to the liquid state at the normal melting point. Units: J/mol

Molar heat of vaporization: energy required to change one mole of substance from the liquid to the gaseous state at the normal boiling point and one atm pressure. Units: J/mol

Exponents (Section 2.2B)

Designations

$$3.7 \quad \times \quad 10^{-8}$$
$$\text{coefficient} \quad \text{exponent}$$

Operations

Adding and subtracting: All numbers must have the same exponent.

$$5.4 \times 10^4 + 3.2 \times 10^3 = 0.54 \times 10^3 + 3.2 \times 10^3 = 3.7 \times 10^3$$

Multiplying: The coefficients are multiplied, the exponents added.

$$(2.6 \times 10^{-2})(4.1 \times 10^5) = 10.66 \times 10^3 = 1.1 \times 10^4$$

Dividing: The coefficients are divided, the exponents subtracted.

$$(5.7 \times 10^{-3})/(8.3 \times 10^4) = 0.687 \times 10^{-7} = 6.9 \times 10^{-8}$$

To raise to a power: The coefficient is raised to the power; the exponent is multiplied by the power.

$$(6.3 \times 10^7)^3 = (6.3)^3 \times 10^{7 \times 3} = 250 \times 10^{21} = 2.5 \times 10^{23}$$

To take a root: The root of the coefficient is taken; the exponent is divided by the root. Often the power is changed so that the exponent is an even multiple of the root being taken.

$$\sqrt[3]{1.6 \times 10^{-7}} = \sqrt[3]{160 \times 10^{-9}} = 5.4 \times 10^{-3}$$

Gas Laws (Sections 9.5, 9.6)

Avogadro's Law

Equal volumes of gases at the same temperature and pressure contain the same number of molecules. One mole of any gas at STP occupies 22.4 L.

Boyle's Law

$$P_1 V_1 = P_2 V_2$$

Charles' Law

$$\frac{V_1}{T_1} = \frac{V_2}{T_2}$$

Combined Gas Law

$$\frac{P_1 V_1}{T_1} = \frac{P_2 V_2}{T_2}$$

Ideal Gas Law equation

$PV = nRT$, where R = universal gas constant
$$= 0.0821 \text{ L-atm/mol-K}$$

Dalton's Law of Partial Pressure

$$P_{\text{Total}} = P_1 + P_2 + P_3 + \cdots$$

Graphs, *see* Mathematical Operations

Kinetic Molecular Theory (Section 9.3)

Postulates

1. The volume of a gas sample is enormous compared with the real volume of the gas molecules. The gas molecules can be considered to have negligible volume.

2. There are no attractive forces operating between the molecules of a gas.

3. The molecules of a gas are in constant, rapid, random, straight-line motion.

4. The molecules collide constantly with each other and with the walls of the container. These collisions are perfectly elastic.

5. The average kinetic energy of the molecules in a sample is proportional to the Kelvin temperature of the sample.

Lewis (Electron-Dot) Structures (Section 7.2)

To draw the Lewis structure of a molecule or ion:

1. Arrange the atoms of the molecule or ion symmetrically around the central atom(s).

2. Sum the total number of spaces in the valence shells of the atom(s) in the structure.

3. Sum the number of valence electrons available for the molecule.

4. The bonding electrons = spaces (step 2) − valence electrons (step 3). Place two of these electrons wherever a bond is needed in the molecule (ion). Use any electrons left over to form double or triple bonds wherever needed.

5. Unshared electrons = valence electrons (step 3) − bonding electrons (step 4). Use these unshared electrons to complete the octets of electrons around each atom.

Mathematical Operations

Algebraic equations

To solve an equation, the unknown must be isolated on one side of the equality sign. To rearrange an equation, the same operation must be performed on both sides of the equality sign.

Addition or subtraction: In the equation $3x + 2 = 2x - 9$, to remove the 2 from the left, subtract 2 from both sides.

$$3x + 2 - 2 = 2x - 9 - 2$$
$$3x = 2x - 11$$

To remove the unknown from the right, subtract $2x$ from both sides.

$$3x - 2x = 2x - 11 - 2x$$
$$x = -11$$

Multiplication: In the equation $26x = 78$, divide both sides by 26.

$$26x/26 = 78/26$$
$$x = 3$$

Division: In the equation $x/19 = 6$, multiply both sides by 19.

$$19(x/19) = 6 \times 19$$
$$x = 114$$

Graphs

Direct proportionality: If y is directly proportional to x, a graph of y with respect to x will be a straight line with the equation

$$y = mx + b$$

in which $b = y$ when $x = 0$
$\qquad\qquad = $ the y-intercept
$\qquad m = $ the slope of the line
$\qquad\qquad = $ rate at which y changes with respect to x

Indirect proportionality: If y is inversely proportional to x, the product xy is a constant. The graph will be a curve.

Percent

Percent expresses the ratio of a part to the whole in terms of parts per hundred.

$$\text{Percent} = \frac{\text{part}}{\text{whole}} \times 100\%$$

Rounding off

When a number is rounded off to the correct number of significant figures, all excess digits following the last to be kept are dropped. The digit to be kept is increased by one if the first digit dropped is 5 or greater; it is left unchanged if the first digit dropped is less than 5 (Section 2.2C2).

Significant figures (digits) (Section 2.2C)

1. The significant figures in a measurement are all those that were measured plus one that is estimated.

2. A number that is part of a definition is exact, or infinitely significant. For example, the numbers in the definitions 3 ft = 1 yd and 100 cm = 1 m are infinitely significant; their use in a calculation does not affect the number of significant figures in the results of a calculation.

3. A zero is significant if it represents a measurement; it is not significant if it shows only the location of the decimal point.

4. A zero that disappears when the measurement is expressed exponentially is not significant.

5. In calculations
   a. Addition or subtraction: The answer has as many decimal places as are common to all measurements used in the calculation.
   b. Multiplication and division: The answer has as many significant figures as that measurement in the calculation that contains the fewest significant figures.

Molarity (Section 11.3)

Molarity (M) = moles of solute/volume (L) of solution

Molecular Shapes, *see* Compounds: Bonding and Geometry

Moles (Section 4.4)

One mole = 6.02×10^{23} things
= the formula weight in grams
= 22.4 L of a gas at STP
= the amount of solute in 1.0 L of a 1.0 M solution

Organic Chemistry, Structural Characteristics of Classes of Compounds

Group name	Structural formula	Example
alkane	R—R	ethane, CH_3—CH_3
alkene	RC=CR	ethylene (ethene), CH_2=CH_2
alkyne	RC≡CR	acetylene (ethyne), HC≡CH
alcohol	R—OH	ethyl alcohol (ethanol), C_2H_5OH
aldehyde	R—CHO	acetaldehyde (ethanal), CH_3CHO
amide	$RCONH_2$	acetamide, CH_3CONH_2
amine	RNH	methyl amine, CH_3NH_2
carboxylic acid	RCOOH	acetic acid, CH_3COOH
ester	RCOOR′	ethyl acetate, $CH_3COOC_2H_5$
ether	R—O—R′	diethyl ether, $C_2H_5OC_2H_5$
ketone	RCOR′	methyl ethyl ketone, $CH_3CO\ C_2H_5$

Note: In these formulas, R and R′ may be a hydrogen atom or an alkyl group or in some cases an aryl group.

Oxidation Numbers (Section 6.2A)

The oxidation number of

an uncombined element   is   0

a monatomic ion   is   the charge on the ion

hydrogen ion, combined   is $+1$

oxygen, combined   is -2

In compounds, the sum of the oxidation numbers is 0.
In ions, the sum of the oxidation numbers equals the charge on the ion.

Oxidation–Reduction

Definitions (Section 14.1)

Substance oxidized	Substance reduced
loses electrons	gains electrons
increases its oxidation number	decreases its oxidation number
is the reducing agent	is the oxidizing agent

Balancing oxidation–reduction equations

   A. Using oxidation numbers (Section 14.3A)

   1. Write a skeletal equation for the reaction including only the ions, molecules, and atoms participating in the reaction.

   2. If the reaction occurs in acid, add hydrogen ion and water as either reactant or product. If the reaction occurs in base, add hydroxide ion and water as either reactant or product.

   3. Assign an oxidation number to each element each time it occurs in the equation.

   4. For each element that changes oxidation number, show its electron change.

   5. Using multipliers, equate the loss and gain of electrons.

   6. Multiply the coefficient of the substances oxidized or reduced by the multipliers used to balance electrons.

   Balance the oxygen by changing the number of water molecules in the equation. Balance the hydrogen ions to match the number of water molecules.

   7. Check that the charges are balanced.

   B. Using half-reactions (Section 14.3B)

   1. Write the skeletal equation for the reaction using only the ions, molecules, and atoms that participate in the reaction.

   2. Isolate the product and reactant form of each substance (except H^+) that participates in the reaction.

   3. Balance these half-reactions by mass using hydrogen ions and water if the reaction takes place in acidic solution; use hydroxide ions and water if the reaction takes place in basic solution.

   4. Balance the charge in these half-reactions by adding electrons to the more positive side of the equation.

   5. Using multipliers, equate the loss and gain of electrons in these half-reactions.

   6. Add the multiplied half-reactions. Delete the electrons from both sides.

   7. Check that the overall equation is balanced by mass and by charge.

Percent, *see* **Mathematical Operations**

Periodic Table (Sections 5.5, 5.7)

Trends in properties (Figure 5.16)

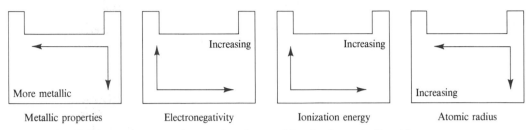

Trends of various atomic properties as related to position in the periodic table.

Relationship to electron configuration (Figure 5.11)

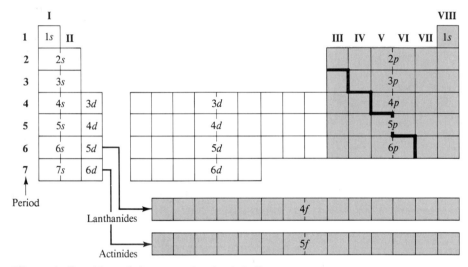

The periodic table and the energy level subshells.

For *s* and *p* blocks,

number of valence electrons = number of column

number of period = number of valence shell

Problem Solving by Unit Analysis (Section 2.3)

1. Determine what quantity is wanted and in what units.
2. Determine what quantity is given and in what units.
3. Determine what conversion factor (or factors) can be used to convert from units given to units wanted.
4. Arrange the given quantity and its units and the chosen conversion factors into an equation in which the unwanted units will cancel and only the wanted units will remain.
5. Do the arithmetic and express the answer to the correct number of significant figures.

SI System

Units (Table 2.1; Section 9.4)

Dimension	Unit	Common units	Conversion factors
length	meter, m	centimeter, cm	1 m = 39.37 in.
mass	kilogram, kg	gram, g	1 kg = 2.204 lb
volume	cubic meter, m^3	liter, L	$1 L = 10^{-3}m^3$ = 0.001 m^3
		milliliter, mL	= 1000 mL = 1000 cm^3
			= 1.057 qt
energy	joule, J	calorie, cal	4.184 J = 1 cal
pressure	pascal, Pa	atmosphere, atm	1.01×10^5 Pa = 1 atm
		torr	1 atm = 760 torr
		mm Hg	= 760 mm Hg
temperature	Kelvin, K	degrees Celsius	K = °C + 273.15
			°C = 5/9(°F − 32)

Prefixes (Table 2.2 is more complete; these are the common ones.)

kilo- 10^3

centi- 10^{-2}

milli- 10^{-3}

Solubility (Table 8.3)

Table 8.3 has a complete set of rules. The following single rule will take care of most situations: All common salts of the following ions are soluble:

Cations: sodium, potassium, ammonium

Anions: nitrate, acetate

Specific Heat, *see* Energy Terms

Stoichiometric Problem Solving (Section 8.4)

1. Write a balanced chemical equation for the reaction involved.

2. Determine what substance is asked for and in what units.

3. Determine what substance is given and in what units.

4. Determine the conversion factors needed to convert
  a. from given amount to moles
  b. from moles of given substance to moles of wanted substance
  c. from moles of wanted substance to units wanted

5. Combine the given substance and the conversion factors chosen into an equation so that the unwanted units cancel and only the wanted substance in the correct units remains.

6. Do the arithmetic, report the answer to the correct number of significant figures, and check that the answer is reasonable.

Water, Physical Properties

Molecular weight: 18.0 g/mol
Melting point: 0°C
Boiling point: 100°C
Density: 1.0 g/mL
Specific heat: solid, 2.09 J/g°C; liquid, 4.184 J/g°C; gas, 1.84 J/g°C
Molar heat of fusion: 6.02 kJ/mol
Molar heat of vaporization: 40.7 kJ/mol

Answer Section

Answers to In-Chapter Problems

Chapter 2

2.1. a. 1 picoHertz $= 10^{-12}$ Hertz **b.** 1 microHertz $= 10^{-6}$ Hertz
c. 1 megaHertz $= 10^6$ Hertz
2.2. a. 5.976×10^{27} g **b.** 7.382×10^9 km
2.3. a. 1.67×10^{-24} g **b.** 1.54×10^{-8} m
2.4. a. 1.10×10^5 **b.** 9.04×10^4 **c.** 6.10×10^{-7} **d.** 7.01×10^{-2}
2.5. 54 mi/hr
2.6. 2.1×10^{-7} %
2.7. 15.5 mL
2.8. 1.47 mL/g
2.9. 41 m
2.10. 325 mg
2.11. 73.6 L
2.12. 2.64 g/cm^3
2.13. 1.46 cm^3
2.14. a. 2.0×10^1°C **b.** 113°F
2.15. a. 378 K **b.** -43°C **c.** 3.10×10^2 K
2.16. Solid
2.17. 7.0 kJ
2.18. 1.02 J/g°C
2.19. 101°C

Chapter 3

3.1. a. strontium, Sr, metal, solid **b.** phosphorus, P, nonmetal
c. chromium, Cr, metal, solid
3.2. a. nitrogen, N_2, nonmetal **b.** strontium, Sr, metal
c. chromium, Cr, metal **d.** helium, He, nonmetal
3.3. C_2H_6O
3.4. A formula unit of calcium nitrate contains 1 calcium atom, 2 nitrogen atoms, and 6 oxygen atoms.
3.5. $2 \, Na + Cl_2 \rightarrow 2 \, NaCl$
3.6. $C_3H_8 + 5 \, O_2 \rightarrow 3 \, CO_2 + 4 \, H_2O$
3.7. $N_2O_5 + H_2O \rightarrow 2 \, HNO_3$

Chapter 4

4.1. 15 protons, 15 electrons, 16 neutrons

4.2. $^{238}_{92}U$: 92 p, 92 e⁻, 146 n
$^{235}_{92}U$: 92 p, 92 e⁻, 143 n

4.3. $^{60}_{27}Co$: 27 p and 33 n in nucleus, 27 e⁻ outside the nucleus

4.4. 63.56 amu

4.5. 285 amu

4.6. neon-20

4.7. 19.5 g C

4.8. 0.204 g U

4.9. 4.24×10^{25} atoms Cu

4.10. $^{226}_{88}Ra \rightarrow {}^{222}_{86}Rn + {}^{4}_{2}He$; product is radon-222

4.11. 8.6×10^{-10} g strontium-90

Chapter 5

5.1. Energy absorbing   Energy emitting

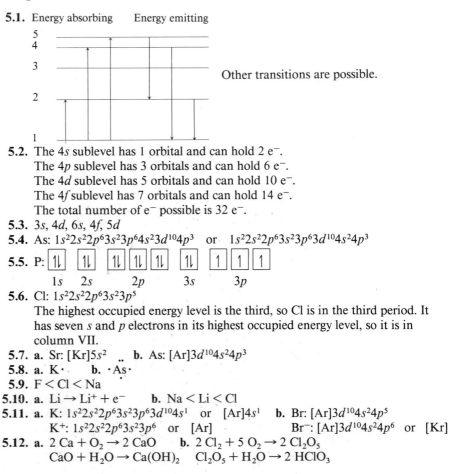

Other transitions are possible.

5.2. The $4s$ sublevel has 1 orbital and can hold 2 e⁻.
The $4p$ sublevel has 3 orbitals and can hold 6 e⁻.
The $4d$ sublevel has 5 orbitals and can hold 10 e⁻.
The $4f$ sublevel has 7 orbitals and can hold 14 e⁻.
The total number of e⁻ possible is 32 e⁻.

5.3. $3s$, $4d$, $6s$, $4f$, $5d$

5.4. As: $1s^22s^22p^63s^23p^64s^23d^{10}4p^3$ or $1s^22s^22p^63s^23p^63d^{10}4s^24p^3$

5.5. P: [1↓] [1↓] [1↓][1↓][1↓] [1↓] [1][1][1]
　　　 $1s$　 $2s$　　 $2p$　　 $3s$　　 $3p$

5.6. Cl: $1s^22s^22p^63s^23p^5$
The highest occupied energy level is the third, so Cl is in the third period. It has seven s and p electrons in its highest occupied energy level, so it is in column VII.

5.7. a. Sr: $[Kr]5s^2$ **b.** As: $[Ar]3d^{10}4s^24p^3$

5.8. a. K·   **b.** ·As·

5.9. F < Cl < Na

5.10. a. Li → Li⁺ + e⁻   **b.** Na < Li < Cl

5.11. a. K: $1s^22s^22p^63s^23p^63d^{10}4s^1$ or $[Ar]4s^1$ **b.** Br: $[Ar]3d^{10}4s^24p^5$
K⁺: $1s^22s^22p^63s^23p^6$ or $[Ar]$　　　　Br⁻: $[Ar]3d^{10}4s^24p^6$ or $[Kr]$

5.12. a. $2 Ca + O_2 \rightarrow 2 CaO$ **b.** $2 Cl_2 + 5 O_2 \rightarrow 2 Cl_2O_5$
$CaO + H_2O \rightarrow Ca(OH)_2$ $Cl_2O_5 + H_2O \rightarrow 2 HClO_3$

Chapter 6

6.1. a. $CaCl_2$ **b.** $Ba(NO_3)_2$ **c.** $Mg_3(PO_4)_2$

6.2. a. oxygen $= -2$, iron $= +3$
    b. sodium $= +1$, manganese $= +7$, oxygen $= -2$
    c. nitrogen $= +4$, oxygen $= -2$

6.3. a. copper(II) oxide   **b.** iron(II) sulfide
    c. strontium chloride

6.4. a. nickel(II) sulfide (nickelous sulfide)
       nickel(III) sulfide (nickelic sulfide)
    b. platinum(II) chloride (platinous chloride)
       platinum(IV) chloride (platinic chloride)

6.5. a. dichlorine monoxide (often dichlorine oxide)
    b. dinitrogen tetroxide   **c.** arsenic trichloride

6.6. a. $(NH_4)_2SO_3$ **b.** $Fe_3(PO_4)_2$ **c.** $CaCO_3$ **d.** HIO_3

6.7. a. silver(I) sulfide   **b.** nickel(II) carbonate
    c. potassium hydrogen sulfite (potassium bisulfite)   **d.** sodium chlorate

6.8. a. 110.3 amu   **b.** 94.97 amu   **c.** 148.3 amu

6.9. 38.5 g CO_2

6.10. 5.64×10^{20} atoms H

6.11. 0.318 mol H_2SO_4

6.12. 28.88% Mg, 14.25% C, 56.92% O

6.13. 28.19% N

6.14. 92.6% Hg

6.15. 85.6% C, 14.4% H

6.16. N_2O

6.17. $FeCl_2$

6.18. $C_5H_{10}O_5$

6.19. $C_4H_4O_4$

6.20. $C_6H_6O_2$

Chapter 7

7.1. Polar covalent, $^{\delta+}C$; $^{\delta-}N$

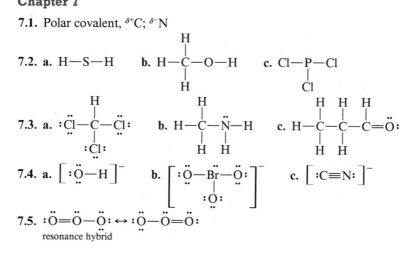

7.2. a. H—S—H   **b.** structure   **c.** Cl—P—Cl

7.3. a., **b.**, **c.** Lewis structures

7.4. a. $\left[:\ddot{O}-H \right]^-$ **b.** structure   **c.** $\left[:C\equiv N: \right]^-$

7.5. $:\ddot{O}=\ddot{O}-\ddot{O}: \leftrightarrow :\ddot{O}-\ddot{O}=\ddot{O}:$
    resonance hybrid

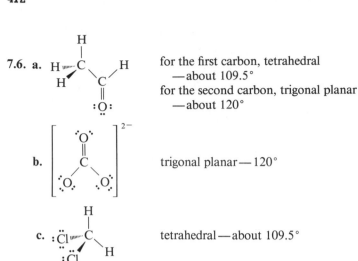

7.6. a. for the first carbon, tetrahedral
—about 109.5°
for the second carbon, trigonal planar
—about 120°

b. trigonal planar— 120°

c. tetrahedral—about 109.5°

d. tetrahedral— 109.5°

7.7. a. H ⇸ B̈r: polar

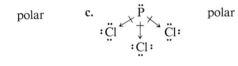

b. polar   **c.**   polar

7.8. a. Yes (HBr would be like HCl, which is ionic in aqueous solution.)
     b. Yes (ionic)
     c. No (molecular compound other than an acid)

Chapter 8

8.1. When two moles of solid mercury(II) oxide are decomposed to form two moles of liquid mercury and one mole of gaseous oxygen, 90.7 kJ of heat are absorbed by the reaction.

8.2. a. $2C(s) + O_2(g) \rightarrow 2\, CO(g)$ **b.** $Li_2O(s) + H_2O(l) \rightarrow 2\, LiOH(aq)$

8.3. a. $2\, Ag_2O(s) \xrightarrow{\Delta} 4\, Ag(s) + O_2(g)$ **b.** $2\, AsH_3(g) \xrightarrow{\Delta} 2\, As(s) + 3\, H_2(g)$

8.4. a. $Zn(s) + H_2SO_4(aq) \rightarrow H_2(g) + ZnSO_4(aq)$
     b. $Hg(l) + AgNO_3(aq) \rightarrow Ag(s) + HgNO_3(aq)$

8.5. a. $Al(OH)_3(aq) + 3\, HC_2H_3O_2(aq) \rightarrow Al(C_2H_3O_2)_3(aq) + 3\, H_2O(l)$
     b. $2\, KOH(aq) + H_2SO_4(aq) \rightarrow K_2SO_4(aq) + 2\, H_2O(l)$
     c. $NH_4OH(aq) + H_3PO_4(aq) \rightarrow NH_4H_2PO_4(aq) + H_2O(l)$

8.6. a. $Ba(C_2H_3O_2)_2$: soluble   **b.** Ag_2S: insoluble
     c. $(NH_4)_3PO_4$: soluble   **d.** $CaCO_3$: insoluble

8.7. a. $CrCl_3(aq) + 3\ NaOH(aq) \longrightarrow Cr(OH)_3(\downarrow) + 3\ NaCl(aq)$
chromium(III) hydroxide   sodium chloride

    b. $H_2SO_4(aq) + BaCl_2(aq) \rightarrow BaSO_4(\downarrow) + 2\ HCl(aq)$
barium sulfate   hydrochloric acid

8.8. a. Not redox   **b.** Yes, Zn oxidized from 0 to +2 and H reduced from +1 to 0

8.9. $Cu(s) + Hg(NO_3)_2(aq) \rightarrow Hg(l) + Cu(NO_3)_2(aq)$
Cu is oxidized from 0 to +2 and is the reducing agent. Hg is reduced from +2 to 0 and is the oxidizing agent.

8.10. a. $2\ CH_3OH(l) + 3\ O_2(g) \rightarrow 2\ CO_2(g) + 4\ H_2O(l)$
    b. $C_7H_{16}(l) + 11\ O_2(g) \rightarrow 7\ CO_2(g) + 8\ H_2O(l)$
    c. $C_4H_8O_2(l) + 5\ O_2(g) \rightarrow 4\ CO_2(g) + 4\ H_2O(l)$

8.11. 13.8 g O_2

8.12. 0.941 g Cl_2

8.13. 1.25×10^{23} molecules O_2

8.14. 43% yield NO

8.15. 81% pure

8.16. 83.9 g Cu

8.17. a. Exothermic; $2\ NH_3(g) \rightarrow N_2(g) + 3\ H_2(g)$ $\Delta H = +46.0$ kJ
    b. Exothermic; $4\ HCl(g) + O_2(g) \rightarrow 2\ H_2O(g) + 2\ Cl_2(g)$ $\Delta H = +120$ kJ
    c. Exothermic; $2\ CO_2(g) + 3\ H_2O(l) \rightarrow C_2H_5OH(l) + 3\ O_2(g)$
$$\Delta H = +1.37 \times 10^3\ \text{kJ}$$

8.18. -724 kJ

8.19. -916 kJ

Chapter 9

9.1. 1.67×10^5 Pa, 4.92×10^4 Pa

9.2. 126 mL

9.3. 1.38 L

9.4. 4.8 L

9.5. 11.9 torr

9.6. 0.179 g/L He

9.7. 2.96×10^{-12} L

9.8. 1.16 g O_2

9.9. 2.71 g/L CO_2

9.10. 114 g/mol

9.11. 0.195 atm O_2(148 torr)   0.004 atm CO_2(3 torr)
0.062 atm H_2O(47 torr)   0.741 atm N_2(563 torr)

9.12. 0.0320 mol CO_2

9.13. 15 L O_2

Chapter 10

10.1. a. Cl_2—dispersion forces   **b.** He—dispersion forces
    CH_3OH—hydrogen bonds   HCl—dipole–dipole interactions
    NaCl—ionic bonds   LiCl—ionic bonds

10.2. 2.2 J/g°C

10.3. 53.7°C

10.4. 22.3 kJ/mol C_4H_{10}
10.5. 2.1×10^2 kJ
10.6. 5.0 kJ
10.7. 26.3 kJ
10.8. 13.4°C

Chapter 11

11.1. 3 g NaCl
11.2. Take 400 mL of ethyl alcohol and dilute to 1.0 L with water.
11.3. 0.0562 mol glucose
11.4. 11.0 g NaCl
11.5. 12 mL of 6.0 M HCl
11.6. 256 mL of 0.195 M HCl
11.7. 0.78 mL of 15 M acetic acid
11.8. 1.90 g Mg
11.9. 0.220 M HCl
11.10. 0.031 L CO_2
11.11. 0.0958 M acetic acid
11.12. **a.** $Co^{2+}(aq) + S^{2-}(aq) \rightarrow CoS(s)$
　　　　　 cobalt(II)　　sulfide
　　　b. $Fe^{2+}(aq) + 2\,OH^-(aq) \rightarrow Fe(OH)_2(s)$
　　　　　 iron(II)　　　hydroxide
　　　c. $Sr^{2+}(aq) + SO_4^{2-}(aq) \rightarrow SrSO_4(s)$
　　　　　 strontium　　sulfate
　　In each case, the spectator ions could be nitrate (NO_3^-) and sodium (Na^+).
11.13. **a.** $Al^{3+}(aq) + 3\,OH^-(aq) \rightarrow Al(OH)_3(s)$　　**b.** $H^+(aq) + OH^-(aq) \rightarrow H_2O(l)$
　　　c. $Zn(s) + Ni^{2+}(aq) \rightarrow Zn^{2+}(aq) + Ni(s)$

Chapter 12

12.1. **a.**　　$NH_4^+ \;+\; H_2O \rightleftarrows NH_3 \;+\; H_3O^+$
　　　　　　 ammonium ion　　　　ammonia　hydronium ion
　　　　$H_2CO_3 \;+\; H_2O \;\rightleftarrows\; HCO_3^- \;+\; H_3O^+$
　　　　 carbonic acid　　　　 hydrogen carbonate
　　　　$HC_2H_3O_2 + H_2O \rightleftarrows C_2H_3O_2^- + H_3O^+$
　　　　 acetic acid　　　　　acetate ion
　　　b.　$NH_3 + H_2O \;\rightleftarrows\; NH_4^+ \;+\; OH^-$
　　　　　 ammonia　　　　 ammonium ion　hydroxide ion
　　　　$H_2O + H_2O \;\rightleftarrows\; H_3O^+ \;+\; OH^-$
　　　　 water　　　　　 hydronium ion
　　　　$HPO_4^{2-} \;+\; H_2O \;\rightleftarrows\; H_2PO_4^- \;+\; OH^-$
　　　　 hydrogen phosphate　　　 dihydrogen phosphate

12.2.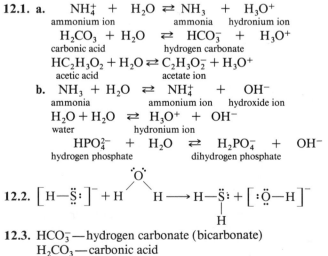

12.3. HCO_3^- —hydrogen carbonate (bicarbonate)
　　　H_2CO_3 —carbonic acid
　　　HI —hydroiodic acid
　　　$HClO_3$ —chloric acid

12.4. a. $H^+(aq) + HCOO^-(aq) \rightarrow HCOOH(aq)$
  b. $H^+(aq) + HCO_3^-(aq) \rightarrow H_2CO_3(aq)$
  c. $2\,H^+(aq) + Fe(s) \rightarrow Fe^{2+} + H_2(g)$
12.5. a. $HBrO_3 \rightarrow H^+ + BrO_3^-$ **b.** $HNO_2 \rightleftarrows H^+ + NO_2^-$
  c. $H_2SO_3 \rightleftarrows HSO_3^- + H^+$
  $HSO_3 \rightleftarrows SO_3^{2-} + H^+$
12.6. a. $HC_6H_8O_7 \rightleftarrows H^+ + C_6H_8O_7^-$ **b.** $K_a = \dfrac{[H^+][C_6H_8O_7^-]}{[HC_6H_8O_7]}$
12.7. $0.1\,M\,H^+$
12.8. $4.2 \times 10^{-3}\,M\,H^+$
12.9. $1.8 \times 10^{-4}\,M\,H^+$
12.10. a. $pH = 3.30$ **b.** $[H^+] = 6.9 \times 10^{-4}\,M$
12.11. $[H^+] = 1.0 \times 10^{-5}\,M;\ pH = 5.00$
12.12. a. $HCO_2^- + H_2O \rightleftarrows HCO_2H + OH^-$ (basic)
  b. $NH_4^+ + H_2O \rightleftarrows NH_3 + H_3O^+$ (acid)
  c. none (neutral)

Chapter 13

13.1.

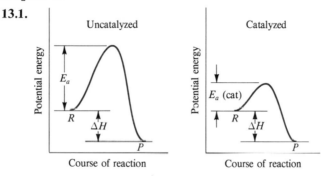

R = reactants P = products E_a = activation energy
ΔH = the change in enthalpy of reaction
13.2. $K_{eq} = \dfrac{[NO_2]^2}{[NO]^2[O_2]}$
13.3. $K_{eq} = [H_2O]$
13.4. a. $K_{eq} = \dfrac{[H_2O][CO]}{[H_2][CO_2]}$
  b. Increased $[CO_2]$ will shift the equilibrium to the right, increasing the amount of products. Decreased $[H_2]$ will shift the equilibrium to the left, decreasing the amount of products.
13.5. a. Shift toward more products   **b.** No effect
13.6. Increased temperature will shift the equilibrium toward more reactants and less products.
13.7. a. $K_{eq} = \dfrac{[H_2]^2[O_2]}{[H_2O]^2}$ **b.** Toward reactants   **c.** Toward products
  d. Toward products   **e.** No effect   **f.** Only (d)—increased temperature
13.8. 1.08×10^{-10}
13.9. 2.8×10^{-16} g/L
13.10. $5 \times 10^{-7}\,M\,Ca^{2+}$

13.11. $2 \times 10^{-11} \, M \, \mathrm{Mg^{2+}}$
13.12. a. 4.05   **b.** 4.00   **c.** 4.10
13.13. From 4.74 to 12.0

Chapter 14

14.1. Mn in $\mathrm{MnO_2}$ is reduced and is the oxidizing agent. $\mathrm{Cl^-}$ is oxidized and is the reducing agent.

14.2. $2 \, e^- + \mathrm{Br_2} \rightarrow 2 \, \mathrm{Br^-}$
$$\underline{\quad\quad 2 \, \mathrm{I^-} \rightarrow \mathrm{I_2} + 2 \, e^- \quad\quad}$$
$$\mathrm{Br_2} + 2 \, \mathrm{I^-} \rightarrow 2 \, \mathrm{Br^-} + \mathrm{I_2}$$

14.3. $\mathrm{PbO_2} + 4 \, \mathrm{H^+} + 2 \, \mathrm{I^-} \rightarrow \mathrm{Pb^{2+}} + \mathrm{I_2} + 2 \, \mathrm{H_2O}$

14.4. $6 \, \mathrm{I^-} + \mathrm{Cr_2O_7^{2-}} + 14 \, \mathrm{H^+} \rightarrow 2 \, \mathrm{Cr^{3+}} + 3 \, \mathrm{I_2} + 7 \, \mathrm{H_2O}$

14.5. $2 \, \mathrm{Br^-} + 4 \, \mathrm{H^+} + \mathrm{SO_4^{2-}} \rightarrow \mathrm{Br_2} + \mathrm{SO_2} + 2 \, \mathrm{H_2O}$

14.6. $14 \, \mathrm{H^+} + \mathrm{Cr_2O_7^{2-}} + 6 \, \mathrm{Fe^{2+}} \rightarrow 2 \, \mathrm{Cr^{3+}} + 6 \, \mathrm{Fe^{3+}} + 7 \, \mathrm{H_2O}$

14.7. $10 \, \mathrm{Cl^-} + 16 \, \mathrm{H^+} + 2 \, \mathrm{MnO_4^-} \rightarrow 5 \, \mathrm{Cl_2} + 2 \, \mathrm{Mn^{2+}} + 8 \, \mathrm{H_2O}$

14.8. Yes, $E^0 + 0.26 \, \mathrm{V}$

14.9. a. Yes, spontaneous   **b.** $\mathrm{Hg} + 2 \, \mathrm{H^+} \rightarrow$ no reaction   **c.** Yes, spontaneous

Answers to Multiple-Choice Questions

Chapter 1

MC1. d   MC2. d   MC3. c   MC4. e   MC5. d   MC6. d
MC7. b   MC8. d   MC9. d   MC10. e

Chapter 2

MC1. b   MC2. b   MC3. e   MC4. a   MC5. c   MC6. d
MC7. e   MC8. b   MC9. b   MC10. a

Chapter 3

MC1. c   MC2. c   MC3. d   MC4. a   MC5. b   MC6. b
MC7. d   MC8. c   MC9. a   MC10. e

Chapter 4

MC1. a   MC2. c   MC3. d   MC4. a   MC5. a   MC6. d
MC7. d   MC8. c   MC9. b   MC10. e

Chapter 5

MC1. c   MC2. a   MC3. b   MC4. d   MC5. d   MC6. a
MC7. b   MC8. b   MC9. b   MC10. a

Chapter 6

MC1. d   MC2. d   MC3. d   MC4. e   MC5. a   MC6. b
MC7. c   MC8. b   MC9. c   MC10. c

Chapter 7

MC1. a   MC2. c   MC3. e   MC4. d   MC5. d   MC6. d
MC7. c   MC8. b   MC9. e   MC10. e

Chapter 8

MC1. e   MC2. d   MC3. c   MC4. a   MC5. c   MC6. b
MC7. a   MC8. b   MC9. c   MC10. b

Chapter 9

MC1. a   MC2. c   MC3. e   MC4. d   MC5. c   MC6. b
MC7. a   MC8. d   MC9. e   MC10. d

Chapter 10

MC1. c   MC2. d   MC3. e   MC4. a   MC5. b   MC6. d
MC7. b   MC8. c   MC9. b   MC10. d

Chapter 11

MC1. d   MC2. b   MC3. d   MC4. e   MC5. b   MC6. d
MC7. d   MC8. a   MC9. d   MC10. a

Chapter 12

MC1. a   MC2. c   MC3. c   MC4. b   MC5. c   MC6. b
MC7. c   MC8. c   MC9. d   MC10. d

Chapter 13

MC1. a   MC2. a   MC3. a   MC4. a   MC5. e   MC6. c
MC7. c   MC8. b   MC9. d   MC10. c

Chapter 14

MC1. e   MC2. b   MC3. c   MC4. b   MC5. a   MC6. b
MC7. a   MC8. b   MC9. a   MC10. a

Chapter 15

MC1. c   MC2. d   MC3. b   MC4. a   MC5. c   MC6. b
MC7. a   MC8. b   MC9. b   MC10. b

Answers to Selected End-of-Chapter Problems

Chapter 1

1.3. Vinegar and baking soda yield bubbles, indicating the formation of a new substance and thus a chemical change. Baking soda dissolves in the water but does not bubble. This change is physical, not chemical.

1.7. Bubbles form when the soft drink is uncapped. This is a physical change of carbon dioxide coming out of solution. Increased temperature will increase the rate at which the carbon dioxide escapes and thus will increase the bubbling.

Chapter 2

2.21. a. 15 in.   **b.** 0.320 lb   **c.** 11.9 m   **d.** 1.5 L   **e.** 103 mi
f. 257 kg   **g.** 32.0 cm   **h.** 0.265 qt   **i.** 0.339 m^2 **j.** 8.63 ft

2.25. a. 1.89 L   **b.** 2.5 min/km   **c.** 1.0 kg   **d.** 2.3 m   **e.** 0.943 L
f. 5.9×10^6 metric tons

2.28. a. 19501   **b.** 0.34   **c.** 28.61   **d.** 0.02115   **e.** 73.5   **f.** 46,000

2.29. a. 14,725   **b.** 0.020   **c.** 0.182   **d.** 618   **e.** 13

2.30. 60 mi/hr; yes

2.36. a. 33 g   **b.** 6.7 mL   **c.** 62 g   **d.** 7.0×10^2 g

2.40. 0.831 cm^3

2.42. 37.0°C or 310.2 K

2.48. a. 6.5×10^2 J   **b.** 76 cal

2.50. 3.77 L

2.55. 1.01 g/mL; liquid

2.57. a. 2×10^{-3} g/L or 2×10^{-3} mg/mL   **b.** 10 mg ascorbic acid
c. Approximately 4.5 times more than in blood

Chapter 3

3.8. Mercury, hydrogen, and neon are elements. Lime, table salt, and water are compounds.

3.14.

natrium	Na	sodium
aurum	Au	gold
stibium	Sb	antimony
plumbum	Pb	lead
ferrum	Fe	iron
argentum	Ag	silver
wolfram	W	tungsten
stannum	Sn	tin
hydrargyrum	Hg	mercury
kalium	K	potassium

3.20.

sodium bromide	NaBr
methane	CH_4
aspirin	$C_9H_8O_4$
ammonia	NH_3
urea	N_2CH_4O

3.25. a. $2\,Ca + O_2 \rightarrow 2\,CaO$ **b.** $Si + O_2 \rightarrow SiO_2$ **c.** $N_2 + O_2 \rightarrow 2\,NO$
d. $2\,Mg + O_2 \rightarrow 2\,MgO$ **e.** $4\,P + 5\,O_2 \rightarrow P_4O_{10}$

3.26. a. $2\,HI \rightarrow H_2 + I_2$ **b.** $2\,Ag_2O \rightarrow 4\,Ag + O_2$
c. $4\,PCl_3 \rightarrow P_4 + 6\,Cl_2$ or $2\,PCl_3 \rightarrow 2\,P + 3\,Cl_2$
3.31. Iron is heavier. The difference is 1.4×10^2 g.
3.33. $-11.4°C \approx 11.4°F$

Chapter 4

4.13.

Element	Atomic no.	Mass no.	# e^-	# p	# n
sodium	11	23	11	11	12
sulfur	16	34	16	16	18
barium	56	137	56	56	81
calcium	20	40	20	20	20
oxygen	8	16	8	8	8

4.18. a. nickel-58   28 p/28 e^-/30 n   metal
b. cobalt-59   27 p/27 e^-/32 n   metal
c. iron-58   26 p/26 e^-/32 n   metal
4.22. 69.7 amu

4.26.

Mass	Moles	Atoms
4.6 g Na	0.20 mol Na	1.2×10^{23}
5.0 g Ba	0.036 mol Ba	2.2×10^{22}
6.7 g Ca	0.17 mol Ca	1.0×10^{23}
16 g K	0.42 mol K	2.5×10^{23}
9.3 g Li	1.3 mol Li	8.1×10^{23}

4.29. a. 22.4 L neon   **b.** 22.4 L krypton   **c.** 22.4 L radon   **d.** 22.4 L argon
4.41. 7×10^{18} atoms mercury

Chapter 5

5.13.

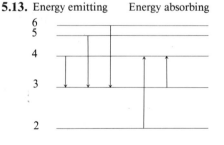

5.15. a. Mg: $1s^2 2s^2 2p^6 3s^2$ **b.** P: $1s^2 2s^2 2p^6 3s^2 3p^3$ **c.** Ar: $1s^2 2s^2 2p^6 3s^2 3p^6$
d. O: $1s^2 2s^2 2p^4$ **e.** Cl: $1s^2 2s^2 2p^6 3s^2 3p^5$ **f.** Na: $1s^2 2s^2 2p^6 3s^1$

5.17. a. Si:

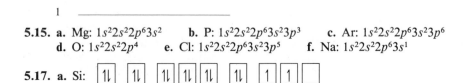

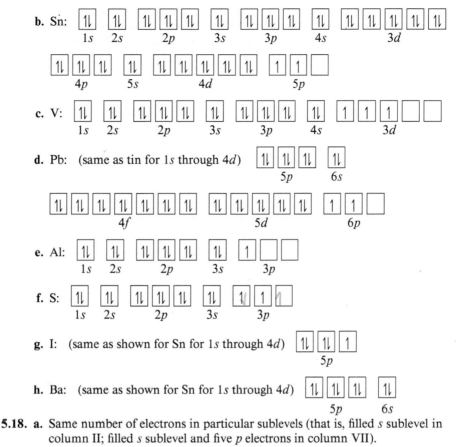

b. Sn: 1s 2s 2p 3s 3p 4s 3d / 4p 5s 4d 5p

c. V: 1s 2s 2p 3s 3p 4s 3d

d. Pb: (same as tin for 1s through 4d) 5p 6s / 4f 5d 6p

e. Al: 1s 2s 2p 3s 3p

f. S: 1s 2s 2p 3s 3p

g. I: (same as shown for Sn for 1s through 4d) 5p

h. Ba: (same as shown for Sn for 1s through 4d) 5p 6s

5.18. a. Same number of electrons in particular sublevels (that is, filled s sublevel in column II; filled s sublevel and five p electrons in column VII).

b. Their valence electrons are all in the same energy level.

c. For all except helium, they have filled s and p sublevels. Helium has a filled $1s$.

5.26. a. ·S̈:   **b.** :C̈l: **c.** Mg: **d.** ·T̈e: **e.** ·Ċ: **f.** Ḃ:

g. Li·   **h.** Ba:

5.30. a. Metallic character increases to the left and down. Nonmetallic character increases to the right and up. Ionization energy increases to the right and up. Atomic radius increases to the left and down.

b. As there are more electrons in a specific energy level (as you move to the right in a period of the periodic table), the elements become less metallic (more nonmetallic); they have higher ionization energies, and their atoms are smaller. As electrons are filling higher energy levels (as you move down a column on the periodic table), the elements become more metallic (less nonmetallic); they have lower ionization energies, and their atoms are larger.

5.32. Electronic configuration: $[Rn] 5f^{14}6d^{10}7s^27p^5$

Element would be in column VII, a halogen, more metallic than astatine. Its atoms would be larger than astatine atoms. It would have a lower ionization energy. Its Lewis structure would be :Ẍ:.

5.33. a. Ce > Cs   **b.** As > Bi   **c.** Si > Al   **d.** Br > I

5.37. Exceptions are (b) Ga^+, (c) Te^{2+}, (d) Bi^{3+}, and (e) Pb^{2+}.

5.40. a. I^- iodide   **b.** O^{2-} oxide   **c.** Br^- bromide   **d.** S^{2-} sulfide
      e. F^- fluoride

5.43. $2\,Be + O_2 \rightarrow 2\,BeO$ $BeO + H_2O \rightarrow Be(OH)_2$
        $2\,Mg + O_2 \rightarrow 2\,MgO$ $MgO + H_2O \rightarrow Mg(OH)_2$
        $2\,Ca + O_2 \rightarrow 2\,CaO$ $CaO + H_2O \rightarrow Ca(OH)_2$
        $2\,Sr + O_2 \rightarrow 2\,SrO$ $SrO + H_2O \rightarrow Sr(OH)_2$
        $2\,Ba + O_2 \rightarrow 2\,BaO$ $BaO + H_2O \rightarrow Ba(OH)_2$
        $2\,Ra + O_2 \rightarrow 2\,RaO$ $RaO + H_2O \rightarrow Ra(OH)_2$

Chapter 6

6.22. a. P_4O_{10} **b.** CaC_2O_4 **c.** BrC_6H_5 (or C_6H_5Br)

6.24. a. $+1$ **b.** -3 **c.** $+3$ **d.** $+5$ **e.** 0 **f.** $+4$

6.26. a. dinitrogen tetroxide   **b.** carbon tetrachloride
      c. dichlorine monoxide   **d.** carbon monoxide   **e.** sulfur trioxide
      f. diphosphorus trioxide   **g.** silicon tetrabromide
      h. dichlorine heptoxide

6.27. a. $AgNO_3$ **b.** NH_4NO_3 **c.** $MgSO_3$ **d.** $CaBr_2$ **e.** $NaClO_4$
      f. Cu_2O **g.** HI   **h.** $PbCl_4$

6.33. a. 119.38   **b.** 132.13   **c.** 40.30   **d.** 152.24   **e.** 168.15

6.36. a. 0.0457 mol NaI   **b.** 6.823 mol SO_2 **c.** 9.61×10^{-3} mol NO_2
      d. 0.248 mol Li_2O **e.** 0.180 mol $C_6H_{12}O_6$ **f.** 8.44×10^{-3} mol O_2
      g. 0.0949 mol $C_{12}H_{22}O_{11}$

6.41. C_2H_5O; $C_4H_{10}O_2$

6.45. C_6H_{12} — molecular formula

6.53. 3.64×10^4 J

Chapter 7

7.9. Ionic: Na—Cl, Mg—Cl
      Polar covalent: Al—Cl, Si—Cl, P—Cl, S—Cl
      Nonpolar covalent: Cl—Cl

7.12. Single bond — 2 shared electrons
      Double bond — 4 shared electrons
      Triple bond — 6 shared electrons

7.15.

7.17.

7.20. a. Tetrahedral   **b.** Bent   **c.** Pyramidal   **d.** Tetrahedral

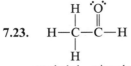

7.23.

tetrahedral   trigonal planar

7.26. $SiCl_4$ has a symmetrical distribution of the polar bonds.

7.27. Electrolytes—LiF and CaI_2; nonelectrolytes—CO and N_2O

Chapter 8

8.26. a. $2\ HCl + Mg(OH)_2 \rightarrow 2\ H_2O\ \ +\ \ MgCl_2$
magnesium chloride

b. $Cu(OH)_2 + H_2SO_4 \rightarrow 2\ H_2O\ \ +\ \ CuSO_4$
copper(II) sulfate

c. $H_3PO_4 + Al(OH)_3 \rightarrow 3\ H_2O\ \ +\ \ AlPO_4$
aluminum phosphate

d. $H_2SO_4 + 2\ KOH \rightarrow 2\ H_2O\ \ +\ \ K_2SO_4$
potassium phosphate

e. $H_2CO_3 + 2\ NaOH \rightarrow 2\ H_2O\ \ +\ \ Na_2CO_3$
sodium carbonate

8.28. a. $Ba(NO_3)_2 + K_2SO_4 \rightarrow BaSO_4(\downarrow) + 2\ KNO_3$
barium sulfate

b. $ZnCl_2 + H_2S \rightarrow ZnS(\downarrow)\ +\ 2\ HCl$
zinc sulfide

c. $Pb(NO_3)_2 + K_2CrO_4 \rightarrow PbCrO_4(\downarrow)\ +\ 2\ KNO_3$
lead(II) chromate

d. $Ni(NO_3)_2 + 2\ NaOH \rightarrow Ni(OH)_2(\downarrow)\ +\ 2\ NaNO_3$
nickel(II) hydroxide

c. $AgNO_3 + HBr \rightarrow AgBr(\downarrow)\ +\ HNO_3$
silver bromide

8.30. a. Cr: reduction   **b.** Mn: reduction   **c.** S: reduction
d. K: oxidation   **e.** Ag: reduction   **f.** N: oxidation

8.32. a. $Ca(OH)_2 + 2\ HCl \rightarrow CaCl_2 + 2\ H_2O$ not redox
b. $SO_3 + BaO \rightarrow BaSO_4$ not redox
c. $2\ AgNO_3 + Fe \rightarrow Fe(NO_3)_2 + 2\ Ag$ redox
d. $Na_2SO_4 + BaCl_2 \rightarrow BaSO_4 + 2\ NaCl$ not redox
e. $2\ NaI + Cl_2 \rightarrow 2\ NaCl + I_2$ redox

8.34. a. Cl in Cl_2 is the oxidizing agent; Cl in $NaClO_2$ is the reducing agent.
b. 4.10 kg ClO_2

8.41. a. $2\ C_8H_{18} + 25\ O_2 \rightarrow 16\ CO_2 + 18\ H_2O$ **b.** 2.0 kg H_2O

8.44. 77%

8.49. 2.59 g MgO

8.53. -52 kJ (released)

Chapter 9

9.26. 0.0900 g/L at STP; 0.0763 g/L at 25°C and 0.925 atm

9.31. a. 1.42 L   **b.** 0.0399 mol

9.33. 117 g/mol

9.36. 272 L O_2

9.41. 84.0%

Chapter 10

10.9. C_3H_8—dispersion forces
Ga—metallic bonds
KNO_3—ionic bonds
SO_2—dipole–dipole interactions
NO—dipole–dipole interactions
Cl_2—dispersion forces
La_2O_3—network covalent

10.13. $0.27 \text{ J/g}°\text{C}$

10.18. 17.3 kJ

10.29. $25°\text{C}$

Chapter 11

11.17. a and b most soluble in water; c, d, and e most soluble in gasoline

11.20. **a.** Dilute 42 mL of 18 M H_2SO_4 to 5.0 L with water.

b. Dissolve 2.2 g KOH in water and dilute with water to 400 mL.

c. Dilute 4 mL of 6 M HCl to 100 mL with water.

d. Dilute 2.6×10^2 mL 95% alcohol to 500 mL.

e. Dissolve 13.5 g glucose in 450 mL water.

11.28. $0.28\ M$ vitamin C

11.34. 6.6×10^2 mL of 12 M HNO_3

11.37. $0.138\ M$ HCl

11.45. $1.08\ M$ H_2SO_4

11.46. **a.** $Cr^{3+}(aq) + 3\ OH^-(aq) \rightarrow Cr(OH)_3(s)$

b. $Pb^{2+}(aq) + 2\ I^-(aq) \rightarrow PbI_2(s)$

c. $Zn^{2+}(aq) + S^{2-}(aq) \rightarrow ZnS(s)$

d. $Ca^{2+}(aq) + CO_3^{2-}(aq) \rightarrow CaCO_3(s)$

11.47. **a.** $H^+(aq) + OH^-(aq) \rightarrow H_2O(l)$

b. $Pb^{2+}(aq) + 2\ Cl^-(aq) \rightarrow PbCl_2(s)$

c. $2\ Ag^+(aq) + SO_4^{2-}(aq) \rightarrow Ag_2SO_4(s)$

d. $H^+(aq) + OH^-(aq) \rightarrow H_2O(l)$

e. $Zn(s) + 2\ H^+(aq) \rightarrow Zn^{2+}(aq) + H_2(g)$

f. $Cl_2(aq) + 2\ I^-(aq) \rightarrow 2\ Cl^-(aq) + I_2(s)$

11.49. 10.0 M glucose; increased concentration of nonvolatile solute decreases the vapor pressure of a solution at a given temperature, thus increasing the temperature necessary to bring the vapor pressure of the liquid up to atmospheric pressure, thus increasing the boiling point.

11.51. **c.** $(NH_4)_3PO_4$; 0.20 M $(NH_4)_3PO_4$ gives 0.80 M ions,
0.20 M Na_2SO_4 gives 0.60 M ions,
0.20 M sucrose gives 0.20 M molecules.
The greater the concentration of solute particles in solution, the lower the freezing point of a solution.

Chapter 12

12.15. **a.** hydrogen phosphate ion $\rightarrow PO_4^{3-}$, phosphate ion

b. nitrous acid $\rightarrow NO_2^-$, nitrite ion

c. hydrogen cyanide $\rightarrow CN^-$, cyanide ion

d. hydrogen carbonate ion $\rightarrow CO_3^{2-}$, carbonate ion

e. ammonium ion $\rightarrow NH_3$, ammonia

12.19. a. $KBrO_4$—potassium perbromate

    b. $HBrO(aq)$—hypobromous acid   **c.** $LiBr$—lithium bromide

    d. $Ca(BrO_3)_2$—calcium bromate   **e.** $Fe(BrO)_2$—iron(II) hypobromite

    f. $Al(BrO_3)_3$—aluminum bromate   **g.** $HBr(aq)$—hydrobromic acid

12.21. a. $H_2SO_4 + 2\ LiOH \rightarrow 2\ H_2O + Li_2SO_4$

                          water   lithium sulfate

    $H^+ + OH^- \rightarrow H_2O$

    b. $2\ HNO_2 + Ca(OH)_2 \rightarrow 2\ H_2O + Ca(NO_2)_2$

                                 calcium nitrite

    $HNO_2 + OH^- \rightarrow H_2O + NO_2^-$

                           nitrite ion

    c. $HC_2H_3O_2 + KOH \rightarrow H_2O + KC_2H_3O_2$

                              potassium acetate

    $HC_2H_3O_2 + OH^- \rightarrow H_2O + C_2H_3O_2^-$

                             acetate ion

    d. $2\ HCl + Zn \rightarrow H_2 + ZnCl_2$

                        zinc chloride

    $2\ H^+ + Zn \rightarrow H_2 + Zn^{2+}$

    e. $H_2SO_4 + Mg \rightarrow H_2 + MgSO_4$

                         magnesium sulfate

    $2\ H^+ + Mg \rightarrow H_2 + Mg^{2+}$

    f. $Ba(OH)_2 + 2\ HNO_3 \rightarrow Ba(NO_3)_2 + 2\ H_2O$

                          barium nitrate

    $OH^- + H^+ \rightarrow H_2O$

12.23. a. $HClO + H_2O \rightleftarrows ClO^- + H_3O^+$

    hypochlorous acid   hypochlorite ion

    b. $HNO_2 + H_2O \rightleftarrows NO_2^- + H_3O^+$

    nitrous acid   nitrite ion

    c. $H_2SO_3 + H_2O \rightleftarrows HSO_3^- + H_3O^+$

    sulfurous acid   hydrogen sulfite

           $HSO_3^- + H_2O \rightleftarrows SO_3^{2-} + H_3O^+$

    hydrogen sulfite   sulfite ion

    d. $HIO_2 + H_2O \rightleftarrows IO_2^- + H_3O^+$

    iodous acid   iodite ion

Chapter 13

13.18.

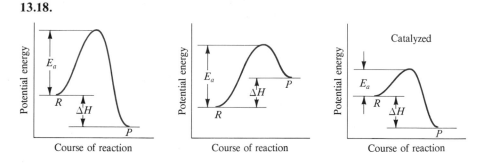

13.19. See answer to 13.18.

13.20. Increased concentration of reactants will lead to increased collisions between reactants, which will increase the forward rate of reaction. Increased pressure of gases brings the gas particles closer together (assuming constant temperature) causing more collisions and thus a greater rate of reaction.

13.30. **a.** $K_{eq} = \dfrac{[NH_3]^2}{[N_2][H_2]^3}$ **b.** Increase NH_3 **c.** Increase NH_3
   d. No change   **e.** Decrease NH_3

13.33. **a.** Increase Cl_2 **b.** Increase Cl_2 **c.** Increase Cl_2
   d. Decrease Cl_2 **e.** No change

13.35. $5 \times 10^{-12} \ M \ Ag^+$

Chapter 14

14.10. **a.** $CO_2 + Ca(OH)_2 \rightarrow CaCO_3 + H_2O$ neutralization

   b. $Zn + H_2SO_4 \rightarrow ZnSO_4 + H_2$
      redox: Zn—oxidized and reducing agent
         H^+ in H_2SO_4—reduced and oxidizing agent

   c. $CaO + 2 HCl \rightarrow CaCl_2 + H_2O$ neutralization

   d. $Pb + Cu(NO_3)_2 \rightarrow Cu + Pb(NO_3)_2$
      redox: Pb—oxidized and reducing agent
         Cu^{2+} in $Cu(NO_3)_2$—reduced and oxidizing agent

   e. $Br_2 + 2 KI \rightarrow I_2 + 2 KBr$
      redox: Br in Br_2—reduced and oxidizing agent
         I^- in KI—oxidized and reducing agent

14.15. **a.** $3 \ e^- + 4 \ H^+ + MnO_4^- \rightarrow MnO_2 + 2 \ H_2O$

   b. $5 \ e^- + 8 \ H^+ + MnO_4^- \rightarrow Mn^{2+} + 4 \ H_2O$

   c. $4 \ e^- + 4 \ H_2O + MnO_4^- \rightarrow Mn^{3+} + 8 \ OH^-$

14.18. **a.** $3 \ Pb + 8 \ HNO_3 \rightarrow 3 \ Pb(NO_3)_2 + 2 \ NO + 4 \ H_2O$

   b. $2 \ Fe^{3+} + 2 \ Br^- \rightarrow 2 \ Fe^{2+} + Br_2$

   c. $2 \ H_2O + 3 \ SO_2 + 2 \ HNO_3 \rightarrow 3 \ H_2SO_4 + 2 \ NO$

14.22. $14 \ MnO_4^- + Mo_{24}O_{37} + 42 \ H^+ \rightarrow 21 \ H_2O + 24 \ MoO_3 + 14 \ Mn^{2+}$

Chapter 15

15.2. $CH_3CH_2CH_2CH_2CH_2CH_3$ $CH_3CHCH_2CH_2CH_3$
                                                         |
                                                        CH_3

$CH_3CH \ CHCH_3$ $CH_3CH_2CHCH_2CH_3$ $CH_3CCH_2CH_3$
      | | | |
     CH_3CH_3 CH_3 CH_3

with additional CH_3 above the last structure.

15.3. $2 \ C_2H_6 + 7 \ O_2 \rightarrow 4 \ CO_2 + 6 \ H_2O$
   $C_3H_8 + 5 \ O_2 \rightarrow 3 \ CO_2 + 4 \ H_2O$
   $2 \ C_4H_{10} + 13 \ O_2 \rightarrow 8 \ CO_2 + 10 \ H_2O$

15.7. $CH_3CH_2CH_2CH_2OH$ CH_3CHCH_2OH $CH_3CH_2CHCH_3$
                                                       | |
                                                      CH_3 OH

$\qquad$ OH
$\qquad$ |
CH_3C-CH_3
$\qquad$ |
$\qquad$ CH_3

15.10. $CH_3CH_2CH_2OH$ $CH_3OCH_2CH_3$
      propyl alcohol   methyl ethyl ether
   The molecular formula of both is C_3H_8O.

15.11. $CH_3CH_2CH_2OH + CH_3COOH \rightarrow CH_3COOCH_2CH_2CH_3 + H_2O$
      propyl alcohol   acetic acid   propyl acetate   water

Index

Absolute temperature, *see* Kelvin temperature scale
Absorption spectrum, 108
Accuracy, 21
Acid dissociation constants, 357, 358, 359
Acids, 141, 348–349, 455
  amino, 449
  anions, 352
  Arrhenius definitions, 348
  binary, 156
  Brønsted-Lowry definitions, 348
  carboxylic, 446
  conjugate, 349
  diprotic, 355
  equilibrium in solution, 357
  ionization, 354
  and nonmetals, 140
  nomenclature, 351
  oxyacids, 184
  polyprotic, 355
  properties, 141
  reactions, 353
  strong and weak, 354
  ternary, 157
  water as an acid, 368
Acid salts, 221
Actinides, 124
Activation energy, 380, 384
Activity series, 428, 429
Alchemists, 8
Alcohols, 444
Aldehyde, 446
Aliphatic, 439
Alkali metals, 130
Alkaline earth metals, 130
Alkanes, 439
Alkenes, 441
Alkyl group, 440
Alkynes, 442
Alpha particles, 79, 87, 88

Amides, 448
Amines, 448
Amino acids, 449
Ammonia, 196
Ammonium ion as weak acid, 359
Anions, 139, 153, 157
Aromatic hydrocarbons, 442
Arrhenius acids, 348
Arrows, in equations, 64
Atmosphere, 262
Atomic mass, 74
Atomic mass unit, 75
Atomic nucleus, 80
Atomic number, 54, 76
Atomic radius, 133
Atomic spectrum, 106
Atomic theory, 51–52
Atomic weight, 82, 85
Atoms:
  composition, 76, 456
  definition, 52
  dimensions, 81
  electron configuration, 109–121
  nucleus, 80
  size, 133
  structure, 80
  subatomic particles, 74–75, 456
Attractive forces:
  in gases, 278
  in liquids, 287
  in solids, 289
Avogadro's hypothesis, 270
Avogadro's number, 84

Baking soda, 2
Balance, 20
Bar (unit of pressure), 262
Barometer, 261
Bases, 141, 348, 455
  Arrhenius definitions, 348

Bases *(contd)*
  Brønsted-Lowry definitions, 348
  conjugate, 348
  from metals, 140
  properties, 141
Bent molecules, 194
Beta particles, 88
Binary compounds, 153–157
  acids, 156
  containing nonmetals, 155
  covalent, 178
  metal and nonmetal, 153
Boiling point, 6, 41
  elevation, 337
  normal, 254, 293
  and physical state, 41
  of solutions, 336
  and vapor pressure, 295
Bond angles, 192–199, 457
Bonding electrons, 178
Bonds, 178–183
  coordinate covalent, 183
  covalent, 178–179
  double, 183
  energy, 287
  hydrogen, 288
  ionic, 179, 290
  metallic, 291
  nonpolar, 179
  and octet rule, 178
  polar covalent, 179
  single, 182
  triple, 183
Box diagrams, 120
Boyle's Law, 264
Branched chain, 437
Breeder reactor, 98
Brønsted-Lowry acids and bases, 348
Buffers:
  calculations involving, 402
  capacity, 405
  characteristics, 401

Buffers *(contd)*
  concentration, 405
  and pH, 405
Buret, 24, 328

Calorie, 16
Calx, 6
Carbonate ion resonance, 191
Carbon isotopes, 77
Carbonyl compounds, 446
Carboxylic acids, 446
Catalyst, 216, 384
Cathode ray tube, 74
Cations, 137–138
Celsius temperature scale, 15, 37–40
Centigrade temperature scale *(see*
  Celsius temperature scale)
Chain reaction, 97
Changes:
  chemical, 212–214
  endothermic, 242
  exothermic, 242
  physical versus chemical, 212–213
  of state, 301
Charges:
  on bonding atoms, 181
  on ions, 136
Charles' Law, 266
Chemical bonds *(see* Bonds)
Chemical changes, 212–214
  combination, 214
  decomposition, 216
  displacement, 218
  double displacement, 220
  versus physical changes, 212–213
Chemical equations *(see* Equations)
Chemical properties, 5, 212
Chemical reactions, 63
Coefficients in an equation, 63
Coinage metals, 126
Colligative properties, 335
Collision theory of reactions, 380
Combination reactions, 214
Combined Gas Law, 268
Combustion reactions, 66, 229
Common-ion effect, 363
Compounds, 51
  binary, 153–157
  empirical formula, 167
  formulas, 62, 150
  ionic, 150, 205, 330
  molecular, 150
  percent composition, 164
Compressibility, 255
Concentration changes:
  effect on equilibrium, 393
  effect on reaction rate, 383
Concentration of solutions, 317–319,
  323

Condensed states, 293
Configuration, electron, 118–124
Conjugate acid-base pairs, 349
Conservation of Energy, Law of, 7
Conservation of Mass, Law of, 6
Conservation of Mass/Energy, Law
  of, 8
Constant composition, 2–3
Continuous light, 107
Contributing structures, 191
Conversion factors, 16, 30, 31
Core notation, 127
Covalent bonds:
  geometry, 192–199
  Lewis structures, 183–192
  polarity, 199–202
Covalent compounds, 206
Crystal structure, 299
Curie, Marie, 53

Dalton, John, 52, 275
d block, 124
Decay, radioactive, 86–91
Decomposition reactions, 216
d electrons, 112, 115
Density, 34, 457
  of different states, 255
  of gases, 271
  of various substances, 35
Diatomic elements, 60
Dimensional analysis *(see* Unit
  analysis)
Dipole, 179
  dipole-dipole interaction, 279
Disorder, 378
Dispersion (London) forces, 279
Displacement reactions, 218, 331
Dissolution, 314
d orbitals, 115
Double bond, 183
Double displacement reactions, 220
Dynamic equilibrium:
  liquid/vapor, 293
  in reversible reactions, 387
  in saturated solutions, 314
  solid/liquid, 300

$E°$, 424
Earth's crust, 57
Effective collisions, 380
Einstein, Albert, 8
Elastic collisions, 260
Electrical forces, 79
Electrochemical cell, 430
Electrodes, 74
Electrolytes, 204–205, 370–371
Electromagnetic radiation, 106

Electron affinity, 136
Electron configuration, 118, 458
  box diagrams, 120
  of elements, 118–119
  and the periodic table, 123–124
Electron dot structures, 128 *(see also*
  Lewis structures)
Electronegativity, 179–181
  and hydrogen bonds, 288
Electron-pair repulsion model, 192
Electron pairs, shared and unshared,
  185
Electrons, 74
  in covalent bonds, 179
  energy changes, 109
  energy levels, 109, 458
  in ionic bonds, 179
  lone pair, 183
  orbitals, 112–115, 458–459
  properties, 75
  quantization of energy, 109
  in redox reactions, 226, 416
  valence electrons, 128
Electron shells, 111
Electron spin, 112
Electrostatic forces, 79
Elements, 51
  atomic weights, 82
  biologically important, 59
  diatomic, 60
  distribution, 57–58
  families of, 123, 130
  historical classification, 129
  inner transition, 126
  names, 53
  and the periodic table, 123
  representative, 124
  spectra, 117
  symbols, 53–54
  transition, 125
Emission spectrum, 107
Empirical formulas, 167
Endothermic reactions, 242, 381
Endpoint, 327
Energy, 7–8, 460
  of activation, 381
  and change of state, 301
  and change of temperature, 387
  conservation of, 8
  kinetic, 7
  levels, 109
  measurement, 37
  nuclear, 96
  potential, 7
  quantization, 109
  radiant, 8, 106
  in reactions, 241
  requirements for reaction, 378
  stoichiometry of changes, 244
  units, 16

Energy levels, 109
    excited states, 117
    ground state, 117
    order of filling, 116
    principal, 111
English system, 16
Enthalpy of reaction, 213, 241, 378
Entropy, 378
Enzymes, 385
Equations:
    balancing, 65
    half-reactions (redox), 416, 464
    mass relationships in, 232
    net ionic, 331
    nuclear, 89
    oxidation–reduction, 417
    parts, 64
    quantitative relations, 232
    writing, 63–65, 212–213
Equilibrium:
    catalysts and, 396
    characteristics, 388
    chemical, 388
    concentration changes and, 393
    constant, 389
    LeChatelier's Principle, 392
    and pressure changes, 394
    and temperature changes, 395
    in water, 368
    in weak acids, 357
Equilibrium constants, 357, 389, 457
    acid dissociation constant, K_a, 357
    ionization constant, 357
    ion product of water, K_w, 368
    solubility product, K_{sp}, 398
    and temperature changes, 391
Esters, 447
Estimated figures, 25
Ethers, 445
Evaporation, 293
Excess reagent, 239
Excited state, 117
Exothermic reactions, 242, 381
Exponential notation, 22, 460
Extranuclear space, 81

Fahrenheit temperature scale, 16,
    37–40
Fatty acids, 447
f block, 124
f electrons, 112
Figures, significant, 25, 463
Fission, nuclear, 96
f orbitals, 115
Forces:
    electrical, 79
    intermolecular, 202–203, 279,
       287–289
    magnetic, 79

Formulas, 62
    acids, 157, 158
    anions, 139
    cations, 137
    empirical, 167
    ionic, 150
    molecular, 169
Formula unit, 59
Formula weight, 160
Free energy, 241, 378, 379
Freezing point (*see* Melting point)
Freezing point depression, 337
Frequency of light, 106
Functional group, 437, 464
Fusion:
    molar heat of, 300
    nuclear, 98

Gamma rays, 88
Gases, 5, 461
    Boyle's Law, 264
    Charles' Law, 266
    Combined Gas Law, 268
    Dalton's Law of Partial Pressures,
       275
    density, 271
    ideal, 260
    ideal gas constant, 272
    ideal gas equation, 272
    kinetic theory of, 259
    molar volume, 271
    noble, 44
    partial pressures, 275
    pressure, 261
    properties, 254
    real, 272, 278
    solubility, 316
    standard pressure, 262
    standard temperature, 260
    stoichiometry, 277
Geometry, molecular, 192, 457
Glucose, 213
Gram, 16
Graphs, 463
Gravity, 79
Ground state, 117

Half-life, 91
Half-reactions, 416, 421
Halogens, 130, 443
Heat:
    of fusion, 300
    measurement, 37
    of reaction, 241
    specific, 42, 295
    of vaporization, 296
Helium, 58
Homogeneous matter, 312

Homologous series, 440
Hydrocarbons, 437
    aliphatic, 439
    aromatic, 442
    normal, 440
    saturated, 439
    unsaturated, 441
Hydrogen, 58, 178
Hydrogen bonds, 288–289
Hydrogen ion concentration,
    361–364, 370–372
Hydrolysis of ions, 371
Hydronium ions, 349
Hydroxides, 141
Hyperbaric, 317
Hypothesis, 9

Ideal gas, 260
Ideal gas equation, 272
Immiscible, 312
Indicator, 141, 327
Inner transition elements, 126
Intermolecular forces:
    in gases, 278
    in liquids, 287
    between polar molecules, 315
    in solids, 289
    in solutions, 315
International Union of Pure and
    Applied Chemistry (IUPAC), 52
    nomenclature of compounds, 151
    nomenclature of ions, 138
Ionic bonds, 290
Ionic compounds (*see also* Com-
    pounds):
    formulas, 150
    insoluble, 222
    names, 150
    properties, 205
    reactions, 380
    in solutions, 203
Ionic equilibrium:
    of acids, 354
    of sparingly soluble salts, 398
Ionization constant, 357
Ionization energy, 134–136
Ionizing radiation, 87
Ions:
    anions, 139
    cations, 137
    definition, 87
    formation, 136
    hydrogen, 384
    hydronium, 348
    monatomic, 137
    naming, 138
    and octet rule, 137
    polyatomic, 139
Isomer, 439

Isotopes, 77, 82
IUPAC (*see* International Union of Pure and Applied Chemistry)

▇▇▇▇▇▇

Joule, 16, 37

▇▇▇▇▇▇

K_a, 357–359
K_{eq}, 357
K_w, 369
Kelvin temperature scale, 16, 37–40
Ketones, 446
Kilogram, 18
Kinetic energy, 7
  average, 260
  distribution, 256
  and temperature, 258, 287
Kinetic molecular theory, 259, 286, 461

▇▇▇▇▇▇

Lanthanides, 124
Lavoisier, 7
Laws:
  Boyle's, 264
  Charles', 266
  Combined Gas, 268
  of Conservation of Energy, 7
  of Conservation of Mass, 6
  of Conservation of Mass/Energy, 8
  Dalton's (of partial pressures), 275
  scientific, 10
LeChatelier's Principle, 392
  and catalysts, 396
  and concentration changes, 393
  and temperature changes, 395
Length, 16
Lewis, G. N., 128
Lewis structures:
  of atoms, 128
  of ions, 189
  of molecules, 183–186
  rules for drawing, 186, 462
Light, 106–108
  kinds of, 107
  and reaction rate, 386
  wavelength, 106
Limiting reactant, 239
Linear molecules, 192
Liquids, 5
  intermolecular forces in, 287
  kinetic molecular theory, 286
  properties, 292
  vapor pressure, 292
Liter, 18
Litmus, 141
Logarithms, 366
London forces, 279

Magnetic forces, 79
Manometer, 263
Mass, 19
  conservation of, 6
  relationships in equations, 382
  relative, 75
  units, 19
  and weight, 19
Mass number, 76
Mathematical operations, 462
Matter, 2
  and energy, 8
  kinetic molecular theory of, 259
  pure versus mixture, 2
  states, 254
Measurement:
  conversion factors, 30
  length, 16, 18
  mass and weight, 19–20
  recording, 21–30
  SI system, 16–21
  temperature, 37
  units, 16–20
  volume, 16, 18
Melting point, 6, 41, 254, 300
Mendeleev, Dmitri, 131
Metallic bonding, 291
Metallic oxides, 140
Metals, 55
  alkali, 130
  alkaline earth, 130
  coinage, 125
  inner transition, 126
  occurrence, 60
  properties, 134
  transition, 125
Metathesis, 220
Meter, 16, 20
Methyl orange, 141
Metric units, 16, 18
Meyer, Lothar, 131
Miscible, 312
Mixtures, 3, 51
  of gases, 275
Molar heat of fusion, 300
Molar heat of vaporization, 296
Molarity, 319
Molar volume of a gas, 271, 274
Mole, 84, 464
  and atomic weight, 85
  Avogadro's number, 84
  of compounds, 162
Molecular formulas, 169
Molecular geometry:
  bent, 194
  bond angles, 192, 457
  electron-pair repulsion model, 192
  linear, 192

Molecular geometry (*contd*)
  pyramidal, 186
  tetrahedral, 195
  trigonal planar, 193
Molecular orientation, 380
Molecular weight, 160
Molecules, 59, 150
  covalent, 206
  polarity, 200
  shape (*see* Molecular geometry)
  symmetry, 184
Monatomic gases, 60
Monomer, 450
Moseley, H. G. J., 132

▇▇▇▇▇▇

Names (*see* Nomenclature)
Net ionic equations, 331
Network covalent compounds, 290
Neutralization, 220
Neutron, 75, 96
Nitrogen, 448
Noble gases, 60, 126, 131
Nomenclature:
  acids, 351
  binary compounds, 153–157
  ionic compounds, 112
  ions, 138
  IUPAC, 131, 151
  ous/ic method, 138, 154
  oxidation numbers in, 152
  Stock System, 153
  ternary acids, 157
  trivial (common), 151
Nonelectrolyte, 205
Nonmetallic oxide, 140
Nonmetals, 55, 135
Nonpolar bond, 179
Nonvolatile solute, 335
Normal hydrocarbon, 440
Nuclear decay, 87–89
Nuclear energy, 95
Nuclear equations, 60
Nuclear fission, 96
Nuclear forces, 81
Nuclear fusion, 98
Nuclear power plant, 96
Nuclear reactions, 89, 96–98
Nucleus, atomic, 80

▇▇▇▇▇▇

Octet rule, 137, 185
Orbitals:
  description, 112
  order of filling, 116
  overlap, 116
  relative energies, 115
  shapes, 113–115
Organic compounds, 437

Osmosis, 337–339
Osmotic pressure, 337–339
Oxidation, 227, 414
Oxidation numbers, 152, 414
  changes in oxidation–reduction, 227
  rules for assigning, 152, 464
  use in balancing redox equations, 418, 465
Oxidation–reduction reactions, 226, 413
  balancing redox equations, 417–418, 421–422, 465
  in electrochemical cells, 430
  half-reactions, 416, 421, 465
  oxidation numbers and, 184, 226
  recognition of, 226
  reduction potentials, 425
Oxidizing agent, 229, 414
Oxyacids, 184
Oxygen, 444

Partial charges, 181
Partial pressure, 275
Parts per billion, 319
Parts per million, 319
Pascal, 262
Pauling, Linus, 180, 190
p block, 124
p electron, 112
Percent composition, 164
Percent concentration, 317
Percent ionic character, 181
Percent purity, 237
Percent yield, 236
Periodic law, 132–133
Periodic properties, 133
Periodic table of the elements, 54, 121–122, 466
  blocks, 124
  columns, 121
  and electron configuration, 123
  history, 131
  periods, 121
  trends, 133–137
Petroleum, 438
pH, 365
  of buffer solutions, 401
  calculation, 366
  interpretation, 367
Phase, 312
Phenolphthalein, 141
Phenols, 445
Philosopher's stone, 8
Photochemical reaction, 386
Physical change, 212
Physical properties, 5, 212
Physical states, 5, 41, 63, 212, 254

pK_a, 367, 368
pK_w, 369
Planar structures, 194
pOH, 365
Polarity:
  of bonds, 179
  and electronegativity, 179, 199
  and intermolecular attraction, 202
  of molecules, 199
Polar solvents, 315
Polyatomic ions, 139
Polyfunctional compounds, 448
Polymer, 450
Polyprotic acids, 221, 355, 358
p orbitals, 112
Potential energy, 7, 287
Power, nuclear, 96
Precipitation reactions, 222–225, 381
Precision, 21
Prefixes, 17, 467
Pressure, 261
  and gas solubility, 316
  and gas volume, 264
  partial, 275
  standard, 262
  vapor, 292
Principal energy levels, 111
Problem solving by unit analysis, 31, 466
Products of a reaction, 63
Properties, 5, 212
Proton, 75
  in Brønsted-Lowry system, 348
  properties, 75
Pseudobinary compounds, 157
Pure substance, 2
Pyramidal molecule, 196

Quanta, 109
Quantized energy, 109

Radiant energy, 8, 106 (*see also* Light)
Radioactivity, 86, 89
  applications, 93
  biological effects, 95
  characteristics, 86
  emissions, 87, 92
  equations, 89
  half-life, 91
  uses, 95
Radioisotopes, 87
Rare earths, 124
Rate of reaction, 382, 383
Rays, gamma, 88
Reactants, 63
Reaction rate, 382
  change of, 383–385

Reaction rate *(contd)*
  effect of catalyst, 384
  effect of concentration changes, 383
  effect of temperature changes, 383
  at equilibrium, 387
  factors affecting, 385–386
  and surface area, 385
Reactions:
  acid–base, 220
  chain, 97
  collision theory of, 381
  combination, 214
  conditions for, 64
  course, 381
  decomposition, 216
  displacement, 218
  double displacement, 220
  endothermic, 242, 381
  energy of, 379
  enthalpy of, 242
  exothermic, 242, 381
  free energy of, 379
  half-reactions, 416
  neutralization, 220
  nuclear, 89
  oxidation–reduction, 417
  precipitation, 222
  rate, 382
  requirements, 378
  reversible, 387
  spontaneous, 379
Real gases, 272, 278
Redox reactions (*see* Oxidation–reduction reactions)
Reducing agents, 229, 414
Reduction, 227, 414
Reduction potentials, 424–427
Relative mass, 75
Representative elements, 124
Resonance, 190, 442
Reversible reactions, 387
Rounding off, 27
Rutherford's experiment, 79

Salts, 159
  hydrolysis, 371
  nomenclature, 159
  solubility, 223, 398
  solubility product constant, 399
  of weak electrolytes, 370–371
Saturated hydrocarbon, 439
Saturated solution, 313
s block, 124
Scientific Law, 10
Scientific method, 8
Scientific notation, 22
Scientific theory, 10
Seaborg, Glenn, 53

s electrons, 112
Semipermeable membrane, 337
Significant figures, 25, 28, 29, 463
Single bond, 182
SI prefixes, 17
SI system (*see* Measurement;
    Système International)
Solids, 5
    bonding, 289
    crystal structure, 299
    properties, 41, 254, 299
Solubility, 222–223, 312
    determining, 313
    factors affecting, 314
    of gases, 315
    of ionic compounds, 222–223
    and pressure, 316
    rules, 223
    and temperature, 316
Solute, 312
Solutions, 312
    boiling point, 337
    colligative properties, 335
    concentration, 314
    freezing point, 337
    of gases, 315
    ionic reactions in, 330
    molarity, 319, 457
    percent by mass, 317
    physical properties, 335
    saturated, 313
    stoichiometry, 323
    unsaturated, 314
Solvent, 312, 315
s orbitals, 112
Specific gravity, 36
Specific heat, 42, 295
Spectator ions, 331
Spectrum:
    absorption, 108
    electromagnetic, 106
    and electrons, 117
    emission, 107
Spin, electron, 112
Spontaneous reactions, 379, 426
Standard pressure, 262
Standard temperature, 260
States of matter, 5
Stock nomenclature, 153
Stoichiometry, 232
    calculations, 233
    of energy changes, 244
    of gases, 277
    limiting reagent problems, 239
    pattern for solving problems, 233,
        467

Stoichiometry *(contd)*
    percent yield, 236
    of solutions, 323
    theoretical yield, 236
STP, 262
Straight chain, 437
Structural formula, 439
Structure, atomic, 74
Subatomic particles, 74
Sublevels, 112
Sublimation, 299
Subscript, 62
Substance, pure, 2
Substitution reaction, 152
Supernatant liquid, 331
Surface area, 385
Surface tension, 298
Symbols:
    of elements, 53
    in equations, 64
    nuclear, 89
Symmetry, 184
Systematic names, 151
Système International (SI), 16–21,
    467 (*see also* Measurement)

Temperature, 37
    absolute (*see* Kelvin temperature
        scale)
    Celsius scale, 15, 37–40
    Fahrenheit scale, 38
    Kelvin scale, 38
    standard, 260
Temperature changes:
    effect on reaction rate, 383
    effect of vapor pressure, 293
    and gas volume, 266
Temporary dipole, 279
Ternary acids, 157, 159
Ternary compounds, 157
Tetrahedral structures, 195
Theoretical yield, 236
Theory, 9
    atomic, 51
    kinetic molecular, 259
    scientific, 10
Thermometer, 37
Titration, 326, 327
Torr, 262
Tracers, radioactive, 94
Transition metals, 125, 138
Trigonal planar, 193
Triple bond, 183

Uncertainty, 23
Unit analysis, 31, 168
Units of measurement, 16–20, 467
    (*see also* Measurement)
Unsaturated hydrocarbons (*see*
    Alkenes; Alkynes)
Unsaturated solutions, 314

Valence electrons, 128
Valence-shell electron-pair repulsion
    model (VSEPR), 192 (*see also*
    Molecular geometry)
Van der Waal's forces, 279
Vaporization, molar heat of, 296
Vapor pressure, 293
    at equilibrium, 293
    of solutions, 335
Viscosity, 298
Voltaic cell, 430
Volume:
    molar, 271, 274
    in SI system, 16–20
Volumetric glassware, 24, 328
VSEPR model, 192

Water:
    geometry, 196
    hydrogen bonding, 289
    K_w, 369
    properties, 305, 468
    uniqueness of, 304
    as weak electrolyte, 368
Wave frequency, 106
Wavelength, 106
Wave properties, 106
Weak acids, 354
    in buffers, 401
    equilibrium in solutions of, 357
    H^+ in solutions of, 361
Weak electrolytes, 205
Weight, 20
    atomic, 82, 85
    formula, 160
    and mass, 20

Yield, 236

Zero as significant figure, 25